MW01101041

Milestones in Drug Therapy
MDT

Valproate

Edited by W. Löscher

Birkhäuser Verlag
Basel · Boston · Berlin

Editor

Prof. Dr. Wolfgang Löscher
Department of Pharmacology,
Toxicology, and Pharmacy
School of Veterinary Medicine
Bünteweg 17, Bldg. 225
D-30559 Hannover

Library of Congress Cataloging-in-Publication Data
Valproate / edited by Wolfgang Löscher.
p. cm. – (Milestones in drug therapy)
Includes bibliographical references and index.
ISBN 3-7643-5836-X (alk. paper).
ISBN 0-8176-5836-X (alk. paper)
1. Valproic acid. I. Löscher. Wolfgang. 1949– . II. Series.
RM666.V33V35 1999 615′.784–DC21 99-18269 CIP

Deutsche Bibliothek - CIP Einheitsaufnahme
Valproate / ed. by Wolfgang Löscher. - Basel ; Boston ; Berlin :
Birkhäuser, 1999
(Milestones in drug therapy)
ISBN 3-7643-5836-X (Basel …)
ISBN 0-8176-5836-X (Boston)

The publisher and editor can give no guarantee for the information on drug dosage and administration contained in this publication. The respective user must check its accuracy by consulting other sources of reference in each individual case.
The use of registered names, trademarks etc. in this publication, even if not identified as such, does not imply that they are exempt from the relevant protective laws and regulations or free for general use.

Printed on acid-free paper produced from chlorine-free pulp. TCF ∞
Printed in Germany
ISBN 3-7643-5836-X
ISBN 0-8176-5836-X

9 8 7 6 5 4 3 2 1

Contents

List of contributors

Frank S. Abbott, Faculty of Pharmaceutical Sciences, The University of British Columbia, 2146 East Mall, Vancouver, B.C., Canada V6T 1Z3; e-mail: fabbott@unixg.ubc.ca

M. Reza Anari, Faculty of Pharmaceutical Sciences, The University of British Columbia, 2146 East Mall, Vancouver, B.C., Canada V6T 1Z3; e-mail: ranari@unixg.ubc.ca

Fritz E. Dreifuss†, Department of Neurology, Box 394, University of Virginia Health Sciences Center, Charlottesville, VA 22908, USA

Olivier Dulac, Hôpital Saint Vincent de Paul, INSERM U29, Université René Descartes, 82 Avenue Denfert-Rochereau, F-75674 Paris Cedex, France; e-mail: o.dulac@svp.ap-hap-paris.fr

Nathan B. Fountain, Department of Neurology, Box 394, University of Virginia Health Sciences Center, Charlottesville, VA 22908, USA; e-mail: nbf2p@virginia.edu

Wolfgang Löscher, Department of Pharmacology, Toxicology and Pharmacy, School of Veterinary Medicine, Bünteweg 17, D-30559 Hannover, Germany; e-mail: wloscher@pharma.tiho-hannover.de

Heinz Nau, Department of Food Toxicology, Center of Food Science, School of Veterinary Medicine, Bischofsholer Damm 15, D-30173 Hannover, Germany; e-mail: hnau@lebensmittel.tiho-hannover.de

Robert M. Post, Biological Psychiatry Branch, NIMH, Bldg. 10, Room 3N212, 10 Center Drive MSC 1272, Bethesda MD 20892–1272, USA; e-mail: robert.post@nih.gov

Matthias Radatz, Department of Food Toxicology, Center of Food Science, School of Veterinary Medicine, Bischofsholer Damm 15, D-30173 Hannover, Germany; e-mail: maradatz@lebensmittel.tiho-hannover.de

N. Sarisjulis, Hospital de Niños Sor Maria Ludovica, Calle 14e/65 y 66, CP 1900 La Plata, Argentina

Dieter Schmidt, Emeritus Professor of Neurology, Epilepsy Research Group, Goethestrasse 5, D-14163 Berlin, Germany; e-mail: dbschmidt@t-online.de

Danny D. Shen, Department of Pharmaceutics, University of Washington, Box 357610, H272 Health Sciences Building, Seattle, WA 98195, USA; e-mail: ds@u.washington.edu

Stephen D. Silberstein, Thomas Jefferson University School of Medicine, Jefferson Headache Center, 111 South Eleventh Street, Gibbon Building, Suite 8130, Philadelphia, PA 19107, USA; e-mail: stephen.silberstein@mail.tju.edu

James L. Willmore, Saint Louis University School of Medicine, 1402 South Grand Blvd., Suite M226, St. Louis, MO 63104, USA; e-mail: Willmore@slu.edu

Dedicated to the memory of Fritz E. Dreifuss (1926–1997)

Fritz E. Dreifuss

Preface

Since the fortuitous discovery of its anticonvulsant activity in 1962, valproate has established itself worldwide as a major antiepileptic drug against several types of epileptic seizures. Clinical experience with valproate has continued to grow in recent years, including use of valproate for diseases other than epilepsy, for example in bipolar disorders and migraine. In this volume on valproate emphasis is placed on the scientific background leading to the discovery of valproate, its subsequent pharmacological and toxicological characterization, and its clinical development into one of the most widely and successfully used antiepileptic drugs, a real milestone in drug therapy. The current state of knowledge of valproate will be reviewed by experts in the field, including new hypotheses about its mechanisms of action, its metabolism into pharmacologically active metabolites, its unique distribution characteristics, its unwanted hepatotoxic and teratogenic adverse effects and its various clinical uses. Furthermore, the wide variety of available pharmaceutical formulations of valproate, including novel controlled-release formulations, will be outlined. The monograph is aimed at a broad readership, particularly neurologists, psychiatrists and basic scientists working in the field of epilepsy research. Because the monograph also deals with structure-activity relationships of valproate as well as of its metabolites and analogs, the book should also serve for researchers working in medicinal chemistry, particularly in the pharmaceutical industry.

Based on the multidisciplinary effort provided by this volume, the cast of authors is broad and international. The editor is indebted to all authors for their contributions and to the staff of Birkhäuser, particularly Katrin Serries, for their assistance in the preparation of this monograph. The many discussions with Dr. Dieter Schmidt on the outline and contents of the book are gratefully acknowledged.

During the preparation of this volume, one of its authors, Dr. Fritz Dreifuss, died of lung cancer at his home in Charlottesville, Virginia, on 18 October 1997, at the age of 71. Fritz Dreifuss was Professor of Epileptology and Neurology at the Department of Neurology, University of Virginia at Charlottesville, which he joined in 1959, heading the Child Neurology Program of the State of Virginia. By the mid-1960s Fritz Dreifuss was working on collaborative projects in epilepsy with the National Institutes of Health (NIH). In the late 1960s, after the arrival of J. Kiffin Penry at the NIH, he intensified his interest in seizure disorders, eventually becoming one of the world's most influential epileptologists. His many invaluable contributions in epileptology have been recently appreciated by R. J. Porter [1]. With respect to valproate, Fritz Dreifuss coauthored one of the earliest controlled trials on the use of this medication and, more important, he elucidated the patients at greatest risk for developing hepatotoxicity during valproate treatment and the unspecific warning symptoms that allow recognition of incipient hepatic failure [2]. As a consequence, by taking respective precautions, the risk of fatal hepatotoxicity, which is the major safety concern of valproate, decreased in recent years. Because of this and his many other contributions to the clinical management of patients with epilepsy, this volume on valproate is dedicated to Fritz Dreifuss.

W. Löscher

References

1 Porter RJ (1998) In memoriam Fritz E. Dreifuss, 1926–1997. *Epilepsia* 39:556–559
2 Dreifuss FE, Santilli N, Langer DH, Sweeny KP, Moline KA, Meander KB (1987) Valproic acid hepatic fatalities: a retrospective review. *Neurology* 37:397–400

Foreword

This excellent new volume edited by Wolfgang Löscher addresses one of the most important therapies for the treatment of epilepsy, the use of valproate. The history of the development of the drug in the United States, which follows briefly below, summarizes some of the many hurdles that a new medication must overcome.

In the late 1950s, at the end of the discovery and development of the last of the cyclic ureides such as ethotoin and ethosuximide, a considerable period of dormancy occurred in the United States in the marketing of new antiepileptic drugs. Two decades later, however, two major drugs would emerge and would revolutionize the therapy of seizure control. Both of these drugs had their roots in Europe and were, only after a prolonged period, approved for use in the United States. The first of these, carbamazepine, would be approved in the United States for trigeminal neuralgia long before it was acceptable for the treatment of epilepsy. And in some ways, the arrival of carbamazepine was relatively slow in the United States, where phenytoin, which largely shares its spectrum of activity, was and is a dominant player.

The second of these new drugs to arrive was valproate, which was first synthesized in the 19th century. Its antiseizure action was uncovered in 1962, and trials were conducted in France as early as 1964. It became available in several countries in Europe by the late 1960s. In many ways, valproate was a much more significant advance than carbamazepine. It has proven to have a remarkable spectrum of activity – from absence to myoclonias to generalized tonic-clonic seizures and even partial seizures. Many patients are relieved of their seizures only because of the availability of this medication, as other drugs do not control the epilepsy.

But it was 1978 before the drug was approved in the United States, and the course to approval was tortuous and controversial. The first problem was finding a U.S. company that would accept the drug from the European manufacturer, develop it and submit it to the U.S. Food and Drug Administration (FDA) for approval. In this regard, J. Kiffin Penry, Chief of the Epilepsy Branch of the National Institutes of Health, became die champion. When Abbott Laboratories finally agreed to take on the drug in 1974, it was clear that the clinical trials conducted abroad would not withstand the scrutiny of the FDA and that more work would need to be done.

As the mid- to late-1970s arrived, the reports from abroad about the remarkable effectiveness of valproate became deafening in the U.S. epilepsy community, especially among parents of children with intractable seizures. An unpleasant cycle began in which the lay community demanded availability of the drug, the pharmaceutical firm was desperately trying to finish its work to file its application and the FDA was caught in the middle-tagged as source of the delay. A number of patients went to Mexico to obtain the drug; most were successful (I was once called personally by a customs agent for my opinion on a specific case; my best efforts failed, and the drug was confiscated). A prominent industrialist took his 6-year-old daughter to England for valproate; when her myoclonic seizures responded with remarkable efficiency, he wrote to the President of the United States and made every effort to publicize his frustration. NBC News ran a series of commentaries on television, complete with patient testimonials. By October 1977, the FDA's Advisory Committee on Neurologic Drugs voted unanimously to recommend FDA approval in spite of the public comment by FDA officials who thought that the application was premature and that further study was needed. In December, approval was formally denied, spurring a request for congressional hearings.

In the middle of this controversy, a curious parallel process began to emerge. In 1975, this was deemed a drug with "virtually no serious side effect" [1]; it was being taken by more than 100,000 persons worldwide. By late 1977, however, the drug had been charged with its first serious safety concern. Hepatotoxicity had been seen in five cases in England, and four had bad a fatal outcome. These data began to be available just at the time of the FDA deliberations and made the approval decision even more difficult. It would be several years before the outstanding collaboration between Abbott Laboratories and Dr. Fritz Dreifuss would elucidate the patients at greatest risk: those who were very young and who were taking multiple medications.

U.S. approval occurred on 28 February 1978. In testimony to the acrimony of the previous year, the Assistant Secretary for Health called the decision "an excellent example of the FDA scientific regulatory process working efficiently and objectively in the face of strong emotions and partisan pressures" [2]. The FDA identified the pivotal study by Dr. B.J. Wilder, submitted just 3 weeks earlier, as the key factor in the approval. At that time the drug was approved for the treatment of absence (petit mal) and mixed seizure which include absence. But in the United States, where the physician has license to use medications at his or her discretion, the drug was now, in fact, available for treatment of any patient deemed to benefit.

The problems for the manufacturer did not end with approval. In addition to the continuing reports of hepatotoxicity, a formulation problem came clearly to the fore. In Europe, the marketed formulation, sodium valproate, was exceedingly hygroscopic, and each tablet was individually wrapped in foil to assure its integrity. Abbott had carefully studied this problem and determined that the free acid of valproate in a corn oil capsule would be much better; this form, marketed as Depakene in the United States, was the first available form. Most patients were very grateful to have it. But this formulation had a considerable incidence of gastrointestinal side effects (for example, I had one Italian patient who came to the United States and tried to switch to the free acid and had terrible problems). Clearly, the formulation needed to be improved, and Abbott proved equal to the task. The Depakote formulation combined a molecule of sodium valproate with a molecule of the free acid and added enteric coating. It has remained the highly successful U.S. formulation. Another successful development has been sustained (or controlled) release formulations of valproate which have been marketed in many countries and, more recently, also in the United States; for many patients this permits once per day dosing.

This volume is an excellent summary of both the basic and the clinical aspects of valproate. Wolfgang Löscher has assembled an outstanding faculty who provide a comprehensive overview of this remarkable medication. The book is of great value to students of drug development, to physicians who treat patients with epilepsy and to the many physicians who use the drug for other disorders – from migraine to psychiatric disorders.

References

1 Noronha MJ and Bevan PLT (1976) A literature review of unwanted effects of treatment with Epilim. In: J Legg (ed): *Clinical and pharmacological aspects of sodium valporate (Epilim) in the treatment of epilepsy.* MCS Consultants, England
2 Epilepsy Foundation of America (1978) *National Spokesman* 11(3)

Roger J. Porter
9 October 1998

Milestones in Drug Therapy
Valproate
ed. by W. Löscher

The discovery of valproate

Wolfgang Löscher

Department of Pharmacology, Toxicology and Pharmacy, School of Veterinary Medicine, Bünteweg 17, D-30559 Hannover, Germany

Like many important discoveries in the history of pharmacology, valproate was not developed by any "rational strategy," but its anticonvulsant activity was serendipitously discovered by Pierre Eymard in France in 1962. Valproic acid (VPA; valproate; di-*n*-propylacetic acid, DPA; 2-propylpentanoic acid, or 2-propylvaleric acid) was first synthesized in 1882, by Burton [1], but there was no known clinical use until its anticonvulsant activity was fortuitously discovered by Eymard. Because valproic acid is a liquid, it was used as a lipophilic vehicle to dissolve water-insoluble compounds during preclinical drug testing. As part of his thesis in 1962, Eymard had synthesized a number of khelline derivatives in the laboratory of G. Carraz at the School of Medicine and Pharmacy in Grenoble, France [2]. Two colleagues, H. Meunier and Y. Meunier, working for a small company, Berthier Laboratories, in Grenoble, had used valproate for a long time as a vehicle for dissolving of a bismuth salt. So the three scientists Eymard, Meunier and Meunier had the idea to use this vehicle also for dissolving some of the khelline derivatives synthesized by Eymard. In order to evaluate the pharmacological activities of the khelline derivatives, Carraz proposed to test the most active derivative in the pentylenetetrazole (PTZ) seizure test. By doing this, the researchers found that the vehicle, valproate, alone exerted an anticonvulsant effect. Similarly, Meunier, by dissolving a coumarin derivative in valproate also found an anticonvulsant effect of the vehicle valproate alone in the PTZ test in rabbits. This unexpected finding was first presented at a meeting of the *French Society of Therapeutics and Pharmacodynamics* on December 19, 1962, and published by Meunier, Carraz, Meunier, Eymard and Aimard in the French journal *Thérapie* in 1963 [2]. The French patent (CAM 244) was obtained by Berthier Laboratories in 1969.

In the first paper on valproate by Meunier et al. [2], valproate was given in the form of its sodium salt in doses of 165, 200, 250, and 420 mg/kg by the intraperitoneal (i.p.) route in groups of 4 rabbits per dose, 40 min before intravenous injection of PTZ, 30 mg/kg. While the two higher doses protected all animals against seizures, two of four rabbits were protected at 200 mg/kg, and no rabbit was protected at 165 mg/kg, but the duration of seizures was reduced by the latter dose. The authors also described the duration of anticonvulsant activity of valproate in rabbits by injecting valproate at a dose of 200 mg/kg i.p. at different times before intravenous administration of PTZ. An anticonvulsant effect was seen up to 3–4 h following valproate, the most marked effect was seen after 15 min. In subsequent experiments, valproate was given by the rectal route in rabbits and PTZ was injected after 105 min. A dose of 400 mg/kg valproate protected all rabbits against seizures. Finally, the anticonvulsant activity of valproate after i.p. or rectal administration was demonstrated in the mouse [2].

Following this first description of valproate's anticonvulsant activity, the group of Carraz published a series of subsequent preclinical studies on various pharmacodynamic effects of valproate, including its anticonvulsant activity against maximal electroshock seizures and seizures induced by strychnine and picrotoxin in mice and audiogenic seizures in rats, its interactions with other drugs, its adverse effects, several biochemical effects, and its acute and chronic toxicity [3–6]. The promising data of these preclinical studies led to clinical evaluation of valproate in patients with epilepsy. It was again the group of Carraz which first demonstrated the anticonvulsant effect of valproate in epileptic patients. The results of the earliest clinical trials of the sodium salt of valproate were reported in 1964 by Carraz et al. [7]. Valproate was first marketed for treatment of epilepsy as "Depakine" in France in 1967 (i.e. only 5 years after the discovery of its anticonvulsant activity in rodents) and was commercialized subsequently in more than 100 other countries (in the United States in 1978; see the Foreword to this volume) for use in epilepsy under trade names such as Ergenyl, Depakene, Depakane, Depakote, Depakon, Convulex, Epilim, Orfiril, Labazene, Leptilan, Eurekene, and Leptilan. Since then, valproate has established itself worldwide as a major antiepileptic drug against several types of epileptic seizures. Clinical experience with valproate has continued to grow in recent years, including use of valproate for diseases other than epilepsy, e.g. in bipolar disorders and migraine. In the United States, valproate has been approved for mania and migraine by the Food and Drug Administration. Furthermore, treatment with valproate bas been improved by development of new pharmaceutical formulations, such as controlled-release or sustained-release formulations, resulting in improved tolerability and reduction of the number of dally intakes. Based on the enormous success and clinical importance of this "milestone in drug therapy," more than 5000 publications have been devoted to its study.

References

1 Burton BS (1882) On the propyl derivatives and decomposition products of ethylacetoacetate. *Am Chem J* 3: 385–395

2 Meunier H, Carraz G, Meunier Y, Eymard P, Aimard M (1963) Propriétés pharmacodynamiques de l'acide n-dipropylacétique. 1er Mémoire: Propriétés antiépileptiques. *Thérapie* 18: 435–438

3 Carraz G, Darbon M, Lebreton S, Bériel H (1964) Propriétés pharmacodynamiques de l'acide n-dipropylacétique et de ses dérivés. *Thérapy* 19: 468–476

4 Lebreton S, Carraz G, Meunier H, Bériel H (1964) Propriétés pharmacodynamiques de l'acide N-dipropylacétique. Deuxième mémoire sur les propriétés antiépileptiques. *Thérapie* 19: 451–476
5 Lebreton S, Carraz G, Bériel H, Meunier H (1964) Propriétés pharmacodynamiques dc l'acide N-dipropylacétique. Troisième mémoire. *Thérapie* 19: 457–467
6 Carraz G, Bériel H, Luu-Duc, Lebreton S (1965) Approches dans la pharmacosynamie biochimique de la structure N-dipropylacétique. *Thérapie* 20: 419–426
7 Carraz G, Fau R, Chateau R, Bonnin J (1964) Communication à propos des premiers essais cliniques sur l'activité anti-épileptique de l'acide n-dipropylacétiques (sel de Na). *Ann Med Psychol (Paris)* 122: 577–585

The pharmacology of valproate

Milestones in Drug Therapy
Valproate
ed. by W. Löscher

Pharmacological effects and mechanisms of action

Wolfgang Löscher

Department of Pharmacology, Toxicology and Pharmacy, School of Veterinary Medicine, Bünteweg 17, D-30559 Hannover, Germany

Introduction

Valproate (valproic acid; usually used as its sodium salt), also referred to as di-*n*-propylacetic acid, is a simple eight-carbon branched-chain fatty acid with unique anticonvulsant properties. Valproic acid was first synthesized in 1882 by Burton [1], but there was no known clinical use until its anticonvulsant activity was fortuitously discovered by Pierre Eymard in 1962 in the laboratory of G. Carraz, which was published by Meunier et al. [2]. The first clinical trials of the sodium salt of valproate were reported in 1964 by Carraz et al. [3]. It was marketed in France in 1967, and was released subsequently in more than 100 other countries (in the United States in 1978) for the treatment of epilepsy. Since then, valproate has established itself worldwide as a major antiepileptic drug against several types of epileptic seizures. Clinical experience with valproate has continued to grow in recent years, including use of valproate for diseases other than epilepsy, e.g. in bipolar disorders and migraine. In the following, the preclinical pharmacological profile of valproate and the putative mechanisms involved will be reviewed.

Anticonvulsant effects of valproate in animal models

As noted above and described in more detail in another chapter of this volume, the anticonvulsant properties of valproate were serendipitously discovered in France in 1962 [2]. By using valproate as a lipophilic vehicle for dissolving water-insoluble khelline derivatives, a significant anticonvulsant effect against pentylenetetrazol(PTZ)-induced seizures was observed in the vehicle controls. Subsequent clinical trials substantiated the anticonvulsant activity of valproate in epileptic patients, and nowadays valproate is one of the major drugs for treatment of different types of epileptic seizures.

Experimentally, valproate exerts anticonvulsant effects in almost all animal models of seizure states examined in this respect (Tabs. 1–3), including models of different types of generalized seizures as well as focal

Table 1. Anticonvulsant potency of valproate in different animal models of generalized clonic or tonic seizures with chemical seizure induction

Model			Efficacy of valproate			Ref. (Examples)
Convulsant (mg/kg)	Seizure type	Spezies	Time[a] (h)	Route of application	ED_{50}[b] (mg/kg)	
PTZ[c] (85–100 s.c.)	Clonic	Mouse	0.25	i.p.	120–150	[4, 5, 6]
			0.5	p.o.	220–420	[4, 5, 7]
PTZ (70 s.c.)	Clonic	Rat	0.5	i.p.	74–260	[4, 8]
			0.5	p.o.	180	[8]
Picrotoxin (3.2 s.c.)	Clonic	Mouse	0.25	i.p.	390	[8]
Bicuculline (2.7 s.c.)	Clonic	Mouse	0.25	i.p.	360	[8]
3-MP[d] (66 s.c.)	Clonic	Mouse	0.5	i.p.	290	[9]
Allylglycine (400 i.v.)	Clonic	Mouse	0.5	i.p.	200	[10]
Isoniazide (200 s.c.)	Clonic	Mouse	0[e]	p.o.	280	[11]
DMCM[f] (15 i.p.	Clonic/ Tonic	Mouse	0.5	i.p.	60	[12]
Strychnine (1.2 s.c.)	Tonic	Mouse	0.25	i.p.	290	[8]
NMDLA[g] (340 s.c.)	Clonic	Mouse	0.5	i.p.	340	[13]

[a] Time from application of valproate to injection of convulsant.
[b] Dose which protects 50% of animals from seizures.
[c] Pentylenetetrazol.
[d] 3-Mercaptopropionic acid.
[e] Valproate and isoniazide were injected simultaneously.
[f] Methyl-6,7-dimethoxy-4-ethyl-β-carbolin-3-carboxylate (an inverse agonist at central benzodiazepine receptors).
[g] N-Methyl-D, L-aspartate (an agonist at the NMDA subtype of glutamate receptors).

seizures. The anticonvulsant potency of valproate strongly depends on the species, the route of administration, the type of seizure induction, and the time interval between drug administration and seizure induction. Because of the rapid penetration into the brain but the short half-life of valproate in most species [20], the most marked effects are obtained shortly, i.e. 2–15 min, after parenteral (e.g. i.p.) injection. Depending on the preparation, onset of action after oral administration may be somewhat retarded. In most laboratory animal species, the duration of anticonvulsant action of valproate is only short so that high doses of valproate are needed to suppress long-lasting or repeatedly occurring seizures in animal models. In general, the anticonvulsant potency of valproate increases in parallel with the size of the animal. In rodents, the highest anticonvulsant potencies are obtained in genetically seizure susceptible species, such as gerbils and rats with spon-

Table 2. Anticonvulsant potency of valproate in animal models of generalized tonic-clonic or focal seizures with electrical seizure induction

Model				Efficacy of valproate			Ref. (Examples)
Name	Stimulus	Seizure type	Species	Time[a] (h)	Route of administration	ED_{50}[b] (mg/kg)	
MES[c]	50 mA	Tonic	Mouse	0.25	i.p.	235–270	[4, 5, 6]
				0.5	p.o.	315	[5]
MES	150 mA	Tonic	Rat	0.5	i.p.	140–170	[4, 8]
				1.0	p.o.	320–490	[4, 8]
MES		Tonic	Rabbit,	0.5	i.p.	235	[4]
MES		Tonic	Cat	0.5	i.p.	67	[4]
Amygdala-Kindling	500 μA	Gen. clonus[d]	Rat	0.25	i.p.	190	[14]
		Complex-focal (clinical)	Rat	0.25	i.p.	220	[14]
		Focal (EEG)	Rat	0.25	i.p.	300	[14]

[a] Time between administration of valproate and electrical stimulation.
[b] Dose which protects 50% of animals from seizures.
[c] Maximal electroshock seizure.
[d] Secondarily generalized clonus following focal onset of seizures.

Table 3. Anticonvulsant potency of valproate in genetic animal models of epilepsy

Model			Efficacy of valproate		Ref. (Exampls)
Species	Seizure type	Induction	Route of administration	ED_{50}[a] (mg/kg)	
Epileptic gerbils	Myoclonic	Air blast	p.o.	210	[15]
	Clonic/Tonic	Air blast	p.o.	280	[15]
	Clonic/Tonic	Air blast	i.p.	73	[16]
Epileptic Rats	Petit mal (spike/waves)	Spontaneous seizures	i.p.	81	[16]
Epileptic Rats	Clonic/Tonic	Audiogenic	i.p.	115–150	[17]
DBA/2 Mice	Clonic	Audiogenic	i.p.	55–300	[18]
Photosensitive Baboons	Myoclonic	Photic	i.v.	200	[18]
Epileptic Dogs	Tonic/Clonic	Spontaneous seizures	Ineffective because of too short action		[19]

[a] Dose which protects 50% of animals from seizures.

taneously occurring spike-wave discharges (Tab. 2), and against DMCM-induced seizures in mice (Tab. 1).

Three animal models that are commonly used in characterization of anticonvulsant drugs are the maximal electroshock seizure (MES) test, the s.c. PTZ seizure test, and kindling. The MES test, in which tonic hindlimb seizures are induced by bilateral corneal or transauricular electrical stimulation, is thought to be predictive of anticonvulsant drug efficacy against generalized tonic-clonic seizures, while the PTZ test, in which generalized myoclonic and clonic seizures are induced by systemic (usually s.c.) administration of convulsant doses of PTZ, is thought to represent a valid model for generalized absence and/or myoclonic seizures in humans [21]. The kindling model with electrical stimulation *via* chronically implanted electrodes in amygdala or hippocampus is probably the best suited model for focal seizures, particularly complex focal seizures as occurring in temporal lobe epilepsy [21], so that by use of these three models the major types of epileptic seizures are covered. As shown in Tables 1 and 2, valproate is effective in these three models, which reflects the wide spectrum of anticonvulsant activity against different types of seizures and epilepsy. In Table 4, the activity of valproate in these models is compared with respective activities of other “old” or “first generation” drugs, i.e. drugs developed and introduced before 1970, and “new” drugs or “second generation” drugs, i.e. drugs developed and introduced after 1970. Furthermore, the clinical spectrum of anticonvulsant activities is shown in this table. It can be seen that only few drugs compete with valproate in terms of its wide spectrum of anticonvulsant activity both preclinically and clinically, thus illustrating the unique profile of this milestone in drug therapy.

In addition to animal models of generalized or focal seizures, valproate also has been evaluated in models of status epilepticus. As shown by Hönack and Löscher [23] in a mouse model of generalized convulsive (grand mal) status epilepticus, i.v. injection of valproate was as rapid as benzodiazepines to suppress generalized tonic-clonic seizures, which was related to the instantaneous entry of valproate into the brain after this route of administration. We therefore proposed that an intravenous formulation of valproate might be a useful alternative to other antiepileptic drugs such as phenytoin as a non-sedative anticonvulsant for diazepam-resistant grand-mal status [23]. In view of the different mechanisms presumably involved in anticonvulsant activity of valproate against different seizure types (see below), the situation may be different for other types of *status epilepticus*, because not all cellular effects of valproate occur rapidly after administration. This is substantiated by accumulating clinical experience with parenteral formulations of valproate in treatment of different types (e.g. convulsive vs. nonconvulsive) of status epilepticus. In a monkey model for *status epilepticus*, in which a status of focal and secondarily generalized tonic-clonic seizures was induced by focal brain injection of

Table 4. Anticonvulsant effect of old (first generation) and new (second generation) antiepileptic drugs against different types of seizures in animal models and in human epilepsy

Drug	Anticonvulsant activity in experimental models			Clinical efficacy			
	MES test (Mice or rats, tonic seizures)	s.c. PTZ test (Mice or rats, clonic seizures)	Amygdala-kindling (Rats, focal seizures)	Partial seizures	Generalized seizures		
					Tonic-clonic	Absence	Myoclonic
First generation drugs							
Valproate	+	+	+	+	+	+	+
Carbamazepine	+	NE	+	+	+	NE	NE
Phenytoin	+	NE	+	+	+	NE	NE
Phenobarbital	+	+	+	+	+	NE	+
Primidone	+	+	+	+	+	NE	+
Benzodiazepines	+	+	+	+	+	+	+
Ethosuximide	NE	+	NE	NE	NE	+	+/–
Second generation drugs							
Lamotrigine	+	NE	+	+	+	+	+
Topiramate	+	NE	+	+	+	+/–	+
Oxcarbazepine	+	+/–	?	+	?	NE	NE
Felbamate	+	+	+	+	+	+/–	+
Vigabatrin	NE	+	+	+	?	NE	NE
Tiagabine	NE	+	+	+	?	?	NE
Gabapentin	+/–	+/–	+	+	?	NE	NE
NMDA antagonists	+	+/–	NE	NE	?	?	?

Effect is indicated by + = effective; +/– = inconsistent data; NE = not effective; ? = no data available (or found). MES, maximal electroshock seizure; PTZ, pentylenetetrazole. Adapted from Löscher [22].

aluminia gel combined with systemic administration of 4-deoxypyridoxine, i.v. administration of valproate delayed seizures but did not prevent their occurrence [23a]. In rats with cortical cobalt lesions injected with homocysteine thiolactone to induce a status of secondarily generalized tonic-clonic seizures, i.p. injection of valproate blocked seizures, although only at relatively high doses [23b].

In addition to the *acute* short-lasting anticonvulsant effects of valproate in diverse animal models after single dose administration (Tabs. 1–4), several studies have examined the anticonvulsant efficacy of valproate during *chronic* administration. During the first days of treatment of amygdala-kindled rats, a marked increase in anticonvulsant activity was observed which was not related to alterations in brain or plasma drug or metabolite levels [24, 25]. Similarly, when anticonvulsant activity was measured by means of timed intravenous infusion of PTZ, prolonged treatment of mice with valproate resulted in marked increases in anticonvulsant activity on the second day of treatment and thereafter compared to the acute effect of valproate, although plasma levels measured at each seizure threshold determination did not differ significantly [26]. This "late effect" of valproate developed irrespective of the administration protocol (once per day, three times per day, continuous infusion) used for treatment with valproate in the animals. Such an increase in anticonvulsant activity during chronic treatment was also observed in epileptic patients and should be considered when *acute* anticonvulsant doses or concentrations of valproate in animal models are compared to effective doses or concentrations in epileptic patients during chronic treatment (Tab. 5). In other words, doses or plasma levels being ineffective after acute administration can become effective during chronic administration. The possible mechanisms involved in "early" (i.e. occurring immediately after first administration of an effective dose) and "late" (i.e. developing during chronic administration) anticonvulsant effects of valproate will be discussed later in this review. In this respect, it is important to note that early and late effects of valproate have been also observed in *in vitro* preparations [27].

All effects of valproate in animal models of seizures or epilepsy summarized so far were acute or chronic *anticonvulsant* effects. In fact, all currently available drugs are anticonvulsant (antiseizure) rather than antiepileptic. Accurately, the latter term should only be used for drugs which prevent or treat epilepsy and not solely its symptoms. Traditionally, pharmacological strategies for treatment of epilepsy have aimed at seizure initiation and propagation rather than the processes leading to epilepsy. As a result, none of the currently available antiepileptic drugs clinically evaluated in this respect (e.g. phenytoin and carbamazepine) is capable of preventing epilepsy, e.g. after brain injury [27a]. Furthermore, there is increasing evidence that – despite early onset of treatment and suppression of seizures – antiepileptic drugs do not affect the progression or underlying natural history of epilepsy [27b]. Thus, one important goal for the future will be to develop anti-epileptogenic drugs, i.e. drugs which prevent or treat epilepsy and not only its symptoms and thus can be referred to as truly antiepileptic [22]. With respect to valproate, data from the kindling model indicate that this drug may exert antiepileptogenic effects [27c], and clinical trials are currently being conducted with this drug [27a].

The "active concentrations" of valproate in the brain or plasma strongly depend on the model examined. When a valproate-sensitive model, such as the threshold for clonic seizures determined by i.v. infusion of PTZ in mice, is used, the drug concentrations in brain tissue after administration of effective doses are near the range of effective concentrations determined in brain biopsies of epileptic patients (Tab. 5). However, it should be noted that because of the marked differences in pharmacokinetics of valproate between rodents and humans (rodents eliminate valproate about 10 times more rapidly than humans), the doses which have to be administered to reach these brain concentrations in mice or rats are much higher than respective doses in humans (Tab. 5). Such determinations of effective brain concentrations are important for interpretation of *in vitro* data on valproate, since the neurochemical or neurophysiological effects of valproate found *in vitro* are only of interest if they occur in concentrations which are reached *in vivo* at anticonvulsant (nontoxic) doses.

In addition to evaluating valproate's anticonvulsant activity in models of pharmacosensitive types of seizures or epilepsy, this drug also has been evaluated in a model of pharmacoresistant epilepsy [29]. In view of the fact that 20 to 30% of patients with epilepsy develop chronic or intractable epilepsy, i.e. the seizures persist despite accurate diagnosis and carefully monitored treatment with anticonvulsant drugs, it is important to develop and use animal models of epilepsy allowing selection of pharmacoresistant and pharmacosensitive subgroups of animals [30]. Such models would be a valuable tool to study mechanisms of intractability and to develop more effective treatment strategies. We developed such a model by selecting subgroups of rats from large groups of amygdala kindled rats, i.e. an important

Table 5. Concentrations of valproate in plasma and brain after administration of anticonvulsant doses

Mouse				Humans			
TIC_{50} [a]				During chronic oral treatment (average doses 15–20 mg/kg)			
Plasma		Brain		Plasma		Brain	
µg/ml	µMol	µg/g	µMol	µg/ml	µMol	µg/g	µMol
120–150	830–1040	25–40	170–280	40–100	280–690	6–27 [b]	42–190

[a] Concentration which increases the threshold for clonic PTC seizures by 50%; determined 5 min after doses (TID_{50}) of 80–100 mg/kg i.p. (Löscher and Fiedler, unpublished data).
[b] [28].

model of temporal lobe epilepsy, which is often refractory to treatment by antiepileptic drugs [30]. These subgroups of kindled rats were selected by the ability of phenytoin to reproducibly increase the focal seizure threshold (afterdischarge threshold, ADT). In phenytoin responders, phenytoin always increased ADT after repeated acute (single dose) administration at weekly intervals, while no significant ADT increases were seen in phenytoin nonresponders, albeit plasma concentrations of phenytoin did not differ between the two subgroups [31]. Valproate significantly increased ADT in phenytoin responders but not in nonresponders, indicating that the resistance to phenytoin extended to other anticonvulsant drugs such as valproate [29]. Similarly, phenobarbital and the novel antiepileptic vigabatrin were anticonvulsant in phenytoin responders but not nonresponders [29]. We currently investigate the reasons for these differences in anticonvulsant activity of major antiepileptic drugs such as valproate in these subgroups of kindled rats [30].

Anticonvulsant effects of valproate in *in vitro* models

A number of *in vitro* preparations such as brain slices or neuronal cultures are used in anticonvulsant drug evaluations, particularly when mechanisms of action are characterized [cf., 32–34]. The effects of valproate in such *in vitro* preparations have been reviewed by Chapman et al. [35], Jurna [32], Cotariu et al. [36], Rogawski and Porter [37], Evans [38], Clark and Wilson [39], Fromm [40], and Fariello et al. [41]. In general, valproate is effective in several *in vitro* models used to predict anticonvulsant activity, such as sustained repetitive neuronal firing in cultured neurons, penicillin-induced spikes in slice preparations, epileptiform bursts in the zero magnesium hippocampal slice model, epileptiform activity induced by high potassium in hippocampal slices, and seizure-like activity in the low calcium hippocampal slice model [39]. Some slice preparations have been proposed to be suited as models for intractable epilepsy [33, 34, 42–44]. Valproate is effective in most of these models. For instance, short recurrent discharges induced by lowering Mg^{2+} in area CA1 and CA3 of the hippocampus were sensitive to valproate but not to any of the other drugs of primary choice in clinical practice [45]. The only exception from the anticonvulsant activity of valproate in such models are late recurrent discharges observed in the low Mg^{2+} model in the entorhinal cortex, which are resistant to all clinically antiepileptic drugs including valproate [46].

Other pharmacodynamic effects of valproate

As shown in Table 6, in addition to its anticonvulsant activity, valproate exerts several other pharmacodynamic effects in animal models, including

Table 6. Other Pharmacodynamic effects of valproate

Effect	Species	Ref. (Examples)
Anxiolytic	Rat	[47]
Antiaggressive	Mouse	[48, 49]
Anticonflict	Rat	[50]
Sedative	Mouse, Rat, Dog etc.	[20]
Antidystonic	Hamster (dystonic mutant)	[51]
"Quasi morphine abstinence behaviour" ("wet dog shakes," piloerection etc.)	Rat	[52, 53]
Antidopaminergic (?) (Inhibition of hyperactivity induced by bilateral microinjection of dopamine in *N. accumbens*)	Rat	[54]
Antinociceptive	Mouse	[5]
	Rat	[55]
Immune stimulating	Rabbit, Mouse	[56, 57]
Antihypertensive	Rat	[58]

anxiolytic, antiaggressive, anticonflict, antidystonic, antinociceptive, sedative/hypnotic, immunestimulating and antihypertensive actions. Several of these preclinical actions are in line with valproate's therapeutic potential in indications other than epilepsy [58a]. Thus, the antinociceptive effect is in accordance with preliminary clinical findings suggesting that valproate may be of value in the management of severe refractory pain, including trigeminal neuralgia, lancinating pain, and neuropathic pain associated with advanced cancer [58a]. The action of valproate in a behavioural model of dopaminergic hyperactivity [54] is in agreement with the antimanic effects in patients with bipolar and schizoaffective disorders [58a, 59]. Furthermore, the recently described antidystonic effect of valproate in a hamster mutant is consistent with the antidystonic effect observed in humans [60]. Other effects, e.g. the sedative action or induction of "quasi morphine abstinence behaviour," which closely resembles the "serotonin syndrome" in rodents, are obviously adverse effects and can be attenuated or prevented by dose reduction. A detailed review of valproate's adverse and toxic effects will be given in another chapter of this book.

Mechanisms of action of valproate

Before going into details of the putative mechanisms involved in the pharmacodynamic actions of valproate, some mechanistic concepts of anticonvulsant drug action are shortly reviewed, including comparison of valproate with other antiepileptic drugs.

With respect to mechanisms of action of old (first generation) antiepileptic drugs, Macdonald has proposed that these drugs can be divided mechanistically into at least three classes based on ability to block sustained high-frequency repetitive firing of action potentials by blockade of voltage-dependent Na^+ channels, to enhance GABAergic inhibition, and to block slow, pacemaker-driven, repetitive firing by blocking T-Ca^{2+} current [61–63]. Macdonald [61] further suggested that the ability of an anticonvulsant drug to block generalized tonic-clonic seizures and some forms of partial seizures may correlate with the ability of the drug to block sustained repetitive firing, while drugs with a broader spectrum of anticonvulsant activity such as valproate block both sustained repetitive firing and enhance GABAergic inhibition (Tab. 7). Anti-absence drugs such as ethosuximide may act *via* their effect on thalamic T-Ca^{2+} current (Tab. 7). However, by this concept it is difficult to explain how valproate, benzodiazepines or lamotrigine act on absence seizures. Furthermore, the concept ignores several additional cellular mechanisms of anticonvulsant drugs which could explain the differences in pharmacology between drugs in the same class [37, 64–66]. For instance, while carbamazepine and phenytoin are generally thought to act by their effect on Na^+ channels [63], an epileptic patient being resistant to one of these drugs may respond favorably to alternative treatment with the other of the two drugs, clearly indicating that these drugs act by more than one mechanism [67]. Similarly, subgroups of amygdala-kindled rats resistant to phenytoin ("phenytoin nonresponders") respond to the anticonvulsant activity of carbamazepine [29, 31], again demonstrating that the concept of Macdonald oversimplifies the complex effects of these and other anticonvulsant drugs both experimentally and clinically [22].

It has been proposed that new anticonvulsant drugs act by new mechanisms [63]. In this respect, inhibition of glutamatergic excitation, particularly that mediated by the N-methyl-D-aspartate (NMDA) and non-NMDA types of glutamate receptors, has been suggested to play a significant role [63]. Indeed, as shown in Table 7, for most of the new anticonvulsant drugs there is evidence for an anti-glutamatergic action. However, as also shown in this table, there is increasing evidence that several of the old drugs may affect glutamatergic excitation, too. Thus, apparently there is no real difference in mechanisms between the old and new drugs. Except the GABA-mimetics vigabatrin and tiagabine, all of the new drugs appear to act by several cellular mechanisms, and so do most of the old drugs. Indeed, this combination of several mechanisms may be the explanation for the clinical efficacy of these drugs in human epilepsy, because epilepsy is certainly a disease with multiple etiologies [64].

Apart from the cellular mechanisms illustrated in Table 7, both old and new drugs exert additional effects on other neurotransmitters or ion channels which may be important for their pharmacodynamic effects [37, 64–66]. In view of the fact that most of the old and new drugs possess more than one mechanism, the individual combination of the "old" mechanisms might be the real difference between old and new drugs.

Table 7. Some cellular mechanisms of action of old (first generation) and new (second generation) antiepileptic drugs

Anticonvulsant drug	Blockade of voltage-dependent Na^+ channels	Potentiation of GABAergic mechanisms	Blockade of thalamic T-type Ca^{2+} channels	Blockade of glutamatergic mechanisms
First generation drugs				
Valproate	+/–	+	NE	+/–
Phenytoin	++	+/–	NE	+/–
Carbamazepine	++	+/–	?	+/–
Phenobarbital	+	+	NE	+
Benzodiazepines	+	++	NE	NE
Ethosuximide	NE	NE	++	NE
Second generation drugs				
Lamotrigine	++	NE	NE	+/–
Oxcarbazepine	++	NE	?	?
Topiramate	+	+	?	+
Felbamate	+	+	?	++
Vigabatrin	?	++	?	?
Tiagabine	?	++	?	?
Gabapentin	+/–	+	NE	+/–

Effect is indicated by +, ++ = effective; +/– = inconsistent data; NE = no effective in therapeutically relevant concentrations; ? = no data available (or found). Adapted from Löscher [22].

With respect to valproate, it is already clear from Table 7 and the respective discussion above that this drug clearly acts by more than one mechanism, which may be the explanation for its wide spectrum of preclinical and clinical activities. In the following, neurochemical and neurophysiological effects of valproate will be reviewed in more detail. This review is not meant to be an exhaustive survey on this topic; detailed summaries on various aspects of valproate's actions are already available [20, 35, 36, 68–78]. This chapter extends the previous reviews and summarizes information on valproate's actions that appear to be of particular importance for its diverse therapeutic effects.

Neurochemical effects of valproate on the GABA system

It seems generally accepted that impairment of GABAergic inhibitory neurotransmission can lead to convulsions, whereas potentiation of GABAergic transmission results in anticonvulsant effects [79]. It is thus not astonishing that the initial reports on brain GABA increase by valproate [80, 81] led to the assumption that enhancement of GABAergic neurotransmission is the mechanism of anticonvulsant action of valproate. Since it was first postulated in 1968, this GABA hypothesis has been a matter of repeated and still ongoing dispute in the literature. For instance, it has been claimed that increases of GABA levels in the brain of rodents are only seen after high doses of valproate, whereas lower doses, which still exert anticonvulsant effects, do not change GABA levels [74]. However, more recent regional and subcellular studies on valproate's effects on GABA levels have resolved these apparent discrepancies. As shown in Table 8, doses as low as 125 mg/kg (see anticonvulsant doses in Tabs. 1–3) were shown in mice to increase GABA concentration in isolated brain nerve endings (synaptosomes), which were prepared from the animals shortly after treatment. Regional brain studies in rats showed marked differences in valproate's effects on GABA levels across brain areas, with significant GABA increases in midbrain regions, such as substantia nigra, which are thought to be critically involved in seizure generation and propagation [55, 82]. In the nigra, the GABA increases induced by valproate occurred predominantly in nerve terminals, i.e. in the "neurotransmitter pool" of GABA [55, 83, 93]. The onset of valproate's effects on presynaptic GABA levels in brain regions was very rapid (significant increases were already observed after 5 min), and the time-course of anticonvulsant and antinociceptive activity correlated with that of the nerve terminal alterations in GABA levels [55].

More recently, microdialysis was used to measure extracellular levels of GABA in the hippocampus of rats after administration of valproate (86, 87). As shown in Table 8, high doses of valproate significantly increased extracellular GABA. Since alterations in extracellular GABA as measured by microdialysis *in vivo* are often proposed to indicate changes in GABA release, these data will be discussed in more detail below.

In *in vivo* experiments in dogs, in which valproate was infused continuously to obtain plasma levels in the range known from chronic oral treatment in humans, GABA increases were determined in cerebral cortex and cerebrospinal fluid (CSF) at an infusion rate of 25 mg/kg/h (Tab. 8). Accordingly, significant increases in CSF GABA levels were also found during treatment of epileptic or schizophrenic patients with valproate (Tab. 8). Furthermore, significant GABA increases were determined in plasma of dogs and humans (Tab. 8). In dogs, the plasma GABA increases parallelled those in CSF and brain tissue, thus indicating that plasma GABA determination might be suited as an indirect indicator of alteration in CNS GABA levels in response to valproate [84].

Although there is substantial evidence that valproate increases GABA concentrations at clinically relevant dosages, the mechanism and functional meaning of the increase in brain GABA levels is still a matter of debate. The increase of *presynaptic* GABA levels induced by valproate could be explained by three different mechanisms: (a) an inhibitory effect of valproate on GABA degradation; (b) an enhancement of GABA synthesis; or (c) no *direct* effects of valproate on synthesis or degradation of GABA but an *indirect* effect on presynaptic GABA levels by direct potentiation of

Table 8. Neurochemical effects of valproate as possible mechanism of anticonvulsant action. 1. GABA-System: (a) GABA concentrations

Material	Species	Effect	Effective doses	Ref. (Examples)
Brain	Mouse	Increase (*in vivo*)	400 mg/kg i.p.	[80]
Brain (Different regions)	Rat	Increase (*in vivo*)	200–400 mg/kg i.p.	[82]
Brain (Cerebral cortex)	Dog	Increase (*in vivo*)	60 mg/kg i.v. (bolus) plus infusion with 25 mg/kg/h	[84]
Synaptosomes[a] (Isolated from whole brain)	Mouse	Increase (*in vivo*)	125–250 mg/kg i.p.	[85]
Synaptosomes[a] (Isolated from different brain regions)	Rat	Increase (*in vivo*)	200 mg/kg i.p.	[55]
Extracellular fluid (Microdialysis from hippocampus)	Rat	Increase (*in vivo*)	400 mg/kg i.p.	[86, 87]
CSF (cisternal)	Dog	Increase (*in vivo*)	60 mg/kg i.v. (bolus) plus infusion with 25 mg/kg/h	[84]
CSF (lumbar)	Schizophrenic patients	Increase (*in vivo*)	Chronic oral treatment with 900–2100 mg/day	[88]
CSF (lumbar)	Epileptic patients	Increase (*in vivo*)	Chronic oral treatment with 15–52 mg/kg/d	[89, 90]
Plasma	Dog	Increase (*in vivo*)	60 mg/kg i.v. (bolus) plus infusion with 25 mg/kg/h	[84]
Plasma	Humans (healthy and epileptic)	Increase (*in vivo*)	Chronic oral treatment with 900–3000 mg/d	[91, 92]

[a] Isolated from brain tissue of the animals after treatment with valproate.

postsynaptic GABAergic function leading to feedback inhibition of GABA turnover and thereby to increases in nerve terminal GABA.

Shortly after the initial reports on the GABA-elevating effect of valproate, several groups examined the action of this drug on GABA degradation (Tabs. 9 and 10). GABA is synthesized in GABAergic nerve terminals by decarboxylation of glutamate and is degraded in nerve terminals, glia cells and also postsynaptic neurons (after diffusion) by transamination to succinic semialdehyde (SSA). SSA can either be oxidized to succinate or it can be reduced to gamma-hydroxy butyrate (GHB). The relative importance of these two degradative pathways *in vivo* is unclear, although it appears that the reduction to GHB is generally a minor route of metabolism. The GABA-elevating effect of valproate was originally attributed to inhibition of GABA transaminase (GABA-T), which catalyses the degradation of GABA to SSA [81]. Yet, as shown in Table 9, most studies on *in vitro* in-

Table 9. Neurochemical effects of valproate as possible mechanism of anticonvulsant action. 1. GABA-System: (b) the GABA degrading enzyme GABA aminotransferase (GABA-T)

Material	Species	Effect	Effective doses or concentrations	Ref. (Examples)
Brain homogenate	Rat	Inhibition (*in vitro*)	9–25 mM	[81]
Purified enzyme (From brain homogenates)	Rat	Inhibition (*in vitro*)	K_i = 9.5 mM	[94]
Purified enzyme (From brain homogenates)	Humans	Inhibition (*in vitro*)	K_i = 40 mM	[94]
Partially purified enzyme (from brain homogenates)	Ox	Inhibition (*in vitro*)	K_i = 28 mM	[95]
Brain homogenate	Mouse	Inhibition (*in vitro*)	K_i = 87 mM	[96]
Brain homogenate (Cerebral cortex)	Mouse	Inhibition (*in vitro*)	IC_{50} = 2.6 mM	[97]
Cerebral neurons in culture	Mouse	Inhibition (*in vitro*)	IC_{50} = 0.63 mM	[97]
Astrocytes in culture	Mouse	Inhibition (*in vitro*)	IC_{50} = 1.2 mM	[97]
Brain homogenate (Hippocampus)	Human	No effect (*in vitro*)	0.2–1 mM	[98]
Brain homogenate	Rat	No effect (*ex vivo*)	400 mg/kg i.p.	[81]
Brain homogenate	Mouse	No effect (*ex vivo*)	125–400 mg/kg i.p.	[85, 99]
Synaptosomes (Isolated from whole brain)	Mouse	Inhibition (*ex vivo*)	125–290 mg/kg i.p.	[85]
Synaptosomes (From different brain regions)	Rat	Inhibition (*ex vivo*)	200 mg/kg i.p.	[100]
Blood platelets	Human	No effect (*in vitro*)	0.4–1 mM	[98]

hibition of GABA-T by valproate found inhibitory effects only at very high concentrations which are not reached *in vivo* (see Tab. 5). Indeed, when valproate was administered to rodents and GABA-T was determined in brain homogenates *ex vivo*, no inhibition of the enzyme was found [81, 85, 99]. However, a significant reduction of GABA-T activity was found in synaptosomes prepared from brain tissue after valproate administration in mice (Tab. 9). Similarly, in rats valproate treatment induced a significant inhibition of synaptosomal GABA-T activities in several brain regions, including substantia nigra, hippocampus, hypothalamus, pons, and cerebellum [100]. These data might be explained by assuming that the presynaptic (nerve terminal) GABA-T is different from glial GABA-T (which predominates in whole tissue homogenates) in terms of susceptibility to valproate. Alternatively, the significant reduction in synaptosomal GABA-T observed *in vivo* is not due to a direct inhibition of GABA-T by valproate but is a secondary effect caused by alterations in the subsequent steps of GABA metabolism. The first assumption is substantiated by recent experiments showing that valproate is much more potent to inhibit GABA-T in neurons (IC_{50} 630 μM) than in astrocytes or whole tissue homogenates (Tab. 9). The second assumption has been extensively discussed previously [101–103]. Thus, *in vitro* valproate exerts a more potent inhibitory effect against SSA dehydrogenase (SSADH), the enzyme responsible for degradation of SSA to succinic acid, than against GABA-T (Tab. 10). It has been postulated that accumulation of SSA by inhibition of SSADH either initiates the reverse reaction of GABA-T, thus increasing the levels of GABA *via* conversion of SSA, or that the increased levels of SSA inhibit the degradation of GABA, since SSA has a strong inhibitory effect on the forward reaction of GABA-T [101]. However, studies of Maitre et al. [102] and Simler et al. [103] suggested that it is not possible to raise brain GABA levels by inhibition of SSADH. Thus, the GABA-T reduction observed in synaptosomes but not whole tissue homogenates of rodents after treatment with valproate (Tab. 9) is most certainly due to a higher susceptibility of nerve terminal GABA-T compared to the enzyme outside nerve terminals. Inhibition of nerve terminal GABA-T could explain the increase of presynaptic GABA levels by valproate, although the reduction of synaptosomal GABA-T activity was not marked [85, 100].

In addition to valproate's effects on GABA-T and SSADH, valproate was shown to be a potent inhibitor of NADPH-dependent aldehyde reductase (Tab. 10). Aldehyde reductase is presumably identical to nonspecific SSA reductase (SSAR) [104]. Whereas the specific SSAR is thought to reduce SSA to GHB, the nonspecific SSAR is thought to be responsible for the catabolism of GHB to SSA [104]. In contrast to the potent effect of valproate on nonspecific SSAR, specific SSAR is not affected by the drug [104]. However, Whittle and Turner [105], using rat brain homogenates, demonstrated that valproate inhibited the formation of GHB *in vitro*, which

Table 10. Neurochemical effects of valproate as possible mechanism of anticonvulsant action. 1. GABA-System: (c) enzymes which are involved in the metabolism of succinic acid semialdehyde in the brain

Enzyme	Species	Effect	Effective concentrations	Ref. (Examples)
SSADH[a]	Rat (Brain)	Inhibition (*in vitro*)	$K_i = 0.5$ mM	[101]
SSADH	Rat (Brain)	Inhibition (*in vitro*)	$K_i = 4$ mM	[102]
SSADH	Ox (Brain)	Inhibition (*in vitro*)	$K_i = 4.8$ mM	[95]
Aldehyde reductase[b]	Ox (Brain)	Inhibition (*in vitro*)	$K_i = 38–85$ μM (with SSA as substrate)	[95]

[a] Succinate semialdehyde dehydrogenase.
[b] Identical to nonspecific succinic semialdehyde reductase (SSAR); specific SSAR is not affected by valproate.

would indicate that the specific SSAR is not exclusively responsible for GHB formation but that the nonspecific (valproate sensitive) aldehyde reductase may also contribute to a significant extent to this metabolic pathway. Inhibition of GHB formation by valproate could be of considerable interest since this amino acid has been shown to produce epileptogenic effects in several species [106]. Administration of valproate to rats has been shown to increase (rather than decrease) the brain level of GHB *in vivo* [107], which may be a consequence of valproate's effects on GHB release (see below).

Besides effects of valproate on GABA degradation, an activation of GABA synthesis could be another likely explanation for the GABA-elevating action of this drug (Tab. 11). Godin et al. [81] measured the relative incorporation of ^{14}C into GABA in rat brain following the s.c. injection of [^{14}C]glucose. Thirty minutes after administration of valproate, 400 mg/kg i.p., the incorporation of ^{14}C into the GABA molecule was increased by 30%, which, however, was not significant on account of the small number of animals studied. In similar experiments in mice, Taberner et al. [111] found that valproate, 80 mg/kg i.p., produced a significant increase in the rate of production of GABA by 90%. More recent studies on GABA turnover in various rat brain regions demonstrated that the most marked increase in GABA synthesis by valproate is found in substantia nigra (Tab. 11), which could be explained by the fact that the nigra is one of the regions with the highest rates of GABA synthesis. Indeed, the increase in GABA synthesis by valproate most likely relates to an activation of the GABA synthesizing enzyme glutamic acid decarboxylase (GAD). Increase of GAD activity by valproate has been demonstrated *ex vivo* after administration in mice and rats (Tab. 11). Interestingly, in rats GAD was not activated in all regions, indicating a regional specificity of valproate's effects [99]. Increase of GAD by valproate is rapid in onset, and the time-course of GAD activation matches that of the GABA increase and the anticonvulsant effect [114]. The rapid activation of GAD by valproate may indicate that valproate converts the inactive apoenzyme to the active holoenzyme [99]. However, at high, toxic doses valproate has been shown to inhibit GAD activity and to reduce GABA synthesis [20].

Activation of GAD by valproate has also been reported *in vitro* (Tab. 11). Interestingly, in rats GAD from neonatal animals was more sensitive to activation by valproate than GAD from adult animals [110]. In neonatal rat brain slices, valproate significantly increased the activity of the GABA shunt, which was related to an increase of GAD [115]. However, high (toxic) concentrations of valproate (7 mM) significantly decrease GAD [108].

More recent neurochemical experiments in beef brain preparations have shown that the coenzyme A (CoA) ester of valproate, which is rapidly formed from valproate *in vivo*, is a potent inhibitor of α-ketoglutarate dehydrogenase complex (KDHC) (Tab. 11). Since decreased KDHC activity would reduce substrate flux through the citric acid cycle and may increase flux into GABA synthesis, this finding adds to the accumulating evidence

Table 11. Neurochemical effects of valproate as possible mechanism of anticonvulsant action. 1. GABA-System: (d) GABA synthesis

Parameter	Species	Effect	Effective doses or concentrations	Ref. (Examples)
GAD[a] (Brain homogenate)	Mouse (*in vitro*)	No sign. effect (trend to increase)	0.3–5 mM	[108]
GAD (Purified)		Increase (*in vitro*)	0.25–2.5 mM	[109]
GAD (Brain homogenate)	Rat (adult)	No effect (*in vitro*)	0.01–1 mM	[110]
GAD (Brain homogenate)	Rat (neonatal)	Increase (*in vitro*)	1 mM	[110]
GAD (Brain homogenate)	Mouse	Increase (*ex vivo*)	125–290 mg/kg i.p.	[85]
GAD (Synaptosomes from whole brain[b])	Mouse	Increase (*ex vivo*)	125–290 mg/kg i.p.	[85]
GAD (Brain regions)	Rat	Increase (*ex vivo*)	400 mg/kg i.p.	[99]
GABA-synthesis (Brain; measured by application of ^{14}C-Glucose)	Mouse	Increase (*in vivo*)	80–160 mg/kg i.p.	[111]
GABA-turnover (*S. nigra*) (Measured by inhibition of GABA-T)	Rat	Increase (*in vivo*)	200 mg/kg i.p.	[112]
KDHC[c] (Purified from brain homogenates)	Beef	Inhibition (*in vitro*)	by the CoA-ester of valproate (K_i 2.9 μM); valproate itself is inactive (up to 10 mM)	[113]

[a] Glutamate decarboxylase.
[b] Isolated from brain tissue of the animals after treatment with valproate.
[c] α-Ketoglutarate dehydrogenase complex; inhibition results in reduced substrate flux through the citric cycle and may increase flux into GABA synthesis.

that valproate increases GABA levels predominantly by enhancing the synthesis of this amino acid [113].

An increase of presynaptic GABA levels by valproate would only potentiate GABAergic neurotransmission if the release of GABA into the synaptic cleft would be increased also. Recently, the first direct evidence for enhanced GABA release by valproate came from studies on cortical slices prepared from valproate-treated animals and from studies on neuronal culture (Tab. 12). Thus, in the cortical slice from valproate-treated rats, the potassium induced release of GABA was increased, which was potentiated further by the $GABA_B$ receptor antagonist phaclofen [116]. Similarly, valproate increased the potassium induced release of GABA from cortical neurons in culture in clinically relevant concentrations [117]. Similarly to the biphasic effects of valproate on GABA synthesis (i.e. increase at low doses but decrease at high doses), high concentrations of valproate seem to inhibit GABA release. The uptake of GABA from the synaptic cleft is not affected by valproate (Tab. 12).

Indirect evidence for enhanced GABA release by valproate comes from *in vivo* studies in rats, using microdialysis to measure extracellular GABA levels in the hippocampus [86, 87]. Biggs et al. [86] reported biphasic effects of valproate on extracellular GABA levels which were dependent on the dose used. At 100 mg/kg, valproate transiently reduced GABA concentrations by 50% when compared to basal, 200 mg/kg valproate had virtually no effect, whilst 400 mg/kg valproate raised extracellular GABA levels to 200% of basal. Such biphasic effects of valproate on extracellular GABA levels have also been found by Wolf et al. [119] using local application of valproate into the rat preoptic area *via* push-pull cannulae. Similar to the study of Biggs et al. [86], Rowley et al. [87] reported that valproate, 400 mg/kg, significantly increased extracellular levels of GABA measured by microdialysis in the hippocampus of freely moving rats. Furthermore, valproate prevented decreases in GABA in response to MES in these animals [87]. Using the push-pull technique to measure extracellular GABA in the substantia nigra of rats, Farrant and Webster [120] found no effect of valproate, 200 mg/kg, on the spontaneous release of GABA into the perfusate. However, as recently pointed out by Timmermann and Westerink [121], all these techniques of measuring extracellular GABA do not allow to draw direct conclusions on drug effects on GABA release, because of the marked compartmentation of this neurotransmitter.

An enhancement of GABA release at clinically relevant concentrations of valproate is indirectly indicated by the increase in CSF GABA levels observed in different species, including humans (Tab. 8). In view of the different recent reports demonstrating an increase of GABA turnover and release by valproate (Tabs. 11 and 12), the previous hypothesis that the

Table 12. Neurochemical effects of valproate as possible mechanism of anticonvulsant action. 1. GABA-System: (e) release and re-uptake of GABA

Parameter	Species	Effect	Effective doses or concentrations	Ref. (Examples)
GABA-release (Cortical slice)	Rat	Increase (*ex vivo*)	600 mg/kg i.p.	[116]
GABA-release (Neuronal culture)	Mouse	Increase (*in vitro*)	300 μM (= 40 μg/g)	[117]
GABA concentration in CSF and extracellular fluid	Dog, humans, rats	Increase (*in vivo*)	see Table 8	see Table 8
GABA-reuptake (Brain)	Rat	No effect (*in vitro*)	up to 10 mM	[96]
GABA-reuptake (Brain)	Rat	No effect (*ex vivo*)	in doses from 100–300 mg/kg (acute or chronic)	[118]

increased brain GABA concentrations induced by valproate treatment are only secondary due to feedback inhibition of GABA turnover because of direct postsynaptic effects of the drug has to be rejected.

In contrast to valproate's effects on GABA synthesis and degradation, it does not exert direct effects on the major components of the postsynaptic $GABA_A$ receptor complex. Thus, *in vitro*, valproate did not alter GABA binding, benzodiazepine binding or the binding of the selective chloride ionophore ligand TBPS (Tab. 13). However, *in vivo* valproate has recently been shown to reduce TPBS binding and to increase benzodiazepine binding (Tab. 13), which is most likely secondary due to the increase in GABA levels produced by valproate *in vivo* [124]. The functional meaning of the alteration in benzodiazepine receptor binding is not clear, since benzodiazepine receptor antagonists do not reduce the anticonvulsant potency of valproate [130]. On the other hand, prolonged pretreatment of mice with benzodiazepines reduced the anticonvulsant potency of valproate, thus demonstrating development of cross-tolerance between benzodiazepines and valproate [131]. Furthermore, the anticonflict action of valproate was

Table 13. Neurochemical effects of valproate as possible mechanism of anticonvulsant action. 1. GABA-System: (f) GABA/benzodiazepine-receptor/chloride ionophore-complex ($GABA_A$ receptor complex) and $GABA_B$ receptors

Parameter	Species	Effect	Effective doses or concentrations	Ref. (Examples)
$GABA_A$ receptor complex				
Binding to GABA recognition site (brain)	Rat	None (*in vitro*)	up to 0 mM	[96, 122]
Binding to GABA recognition site (several brain regions)	Rat	None (*ex vivo*)	Chronic treatment	[123]
Binding to benzodiazepine-recognition site (brain)	Rat	None (*in vitro*)	up to 1 mM	[122]
Binding to benzodiazepine-recognition site (brain)	Rat	Increase[a] (*ex vivo*)	at 55 mg/kg	[124, 125]
Binding to chloride ionophore (TBPS-binding[b]) (brain)	Rat	None (*in vitro*)	up to 10 mM	[126, 127]
Binding to chloride ionophore (TBPS-binding) (brain)	Rat	Reduction (*ex vivo*)	200 mg/kg i.p.	[128]
$GABA_B$ receptors				
Binding to $GABA_B$ receptor (Cerebral cortex)	Rat	Increase (*ex vivo*)	Chronic treatment with 100 mg/kg/d i.p.	[129]
Binding to $GABA_B$ receptor (Hippocampus)	Rat	Increase (*ex vivo*)	Chronic treatment	[123]

[a] By "GABA-shift"? (i.e., increase of benzodiazepine receptor sensitivity through increase of GABA concentration).
[b] [^{35}S]t-butylbicyclophosphorothionate.

reversed by benzodiazepine antagonists [132], indicating that enhanced binding of benzodiazepines to the $GABA_A$ receptor complex may be involved in this pharmacodynamic effect of valproate.

Interestingly, two groups independently reported increase of $GABA_B$ receptor binding by chronic treatment of rats with valproate (Tab. 13). In the recent study of Motohashi [123], acute treatment with valproate had no effect on ^{3}H-baclofen binding in frontal cortex, hippocampus and thalamus, while chronic treatment enhanced binding in the hippocampus. ^{3}H-muscimol binding to the $GABA_A$ receptor did not change after valproate in any region [123]. Because similar effects were observed with lithium and carbamazepine, Motohashi [123] concluded that one common mechanism of action of mood stabilizers may be mediated by $GABA_B$ receptors in the hippocampus.

Taken together, the numerous neurochemical reports of valproate's effects on the GABA system indicate that increases in GABA function may be involved in several pharmacodynamic effects of this drug, including the anticonvulsant, anticonflict, and antimanic actions. Furthermore, in view of the role of GABA in antinociception [133], increased GABAergic function by valproate may be involved in its antinociceptive effects.

Neurochemical effects of valproate on amino acids other than GABA

Compared to the numerous studies on valproate's effects on the GABA system, only relatively few neurochemical studies were done on other transmitter amino acids (Tab. 14). As already described above, valproate increases the levels of GHB in the brain. This increase is time- and dose-dependent, and appears to be due to a reduction in synaptic release of GHB [104]. Since GHB produces absence-like epileptic seizures in animals [106], reduction of GHB release could be an important factor in the anti-absence action of valproate [104].

Glutamate concentrations in regional brain homogenates or in extracellular fluid obtained by microdialysis from hippocampus or substantia nigra were not significantly altered by systemic administration of valproate [86, 87, 120]. However, Dixon and Hokin [135] recently reported that valproate stimulates glutamate release in mouse cerebral cortex slices at therapeutic concentrations. This effect was discussed as being involved in the antimanic action of valproate [135]. Nilsson et al. [140] reported that valproate inhibits the transport of glutamate and aspartate in astroglial cells in primary cultures from newborn rat cerebral cortex.

With respect to aspartate, valproate has been shown to reduce the concentration and release of the excitatory amino acid aspartate in rat and mouse brain (Tab. 14). Furthermore, some reports found that concentrations of glycine and taurine increase in brain tissue (Tab. 14). However, there is as yet no evidence that these effects on amino acids other than GABA are relevant to the anticonvulsant effect of valproate.

Table 14. Neurochemical effects of valproate as possible mechanism of anticonvulsant action. 2. Effects on amino acidergic neurotransmitters other than GABA

Parameter	Species	Effect	Effective doses or concentrations	Ref. (Examples)
GHB-release (Hippocampus and striatum)	Rat	Reduction (*in vitro*)	250–500 µM	[104]
Glutamate concentration (Brain regions)	Rat	No effect (*in vivo*)	200 mg/kg	[134]
Glutamate concentration (Extracellular in hippocampus)	Rat	No effect (*in vivo*)	100–400 mg/kg	[86, 87]
Glutamate release (Cerebral cortex slices)	Mouse	Increase (*in vivo*)	Therapeutic concentrations	[135]
Aspartate concentration (Brain)	Rat, Mouse	Reduction (*in vivo*)	200–400 mg/kg i.p.	[35]
Aspartate concentrations (Brain regions)	Rat	Reduction (*in vivo*)	200 mg/kg	[134]
Aspartate concentration (Extracellular in hippocampus)	Rat	No effect (*in vivo*)	100–400 mg/kg	[86]
Aspartate-release (Brain)	Rat	Inhibiton (*in vitro*)	IC_{50} = um 100 µM	[136]
Glycine concentration (Plasma, urine, brain, spinal cord)	Rat	Increase (*in vivo*)	Acute (700 mg/kg i.p.) and chronic application (70–430 mg/kg b.i.d.)	[137]
Glycine concentration (Pons/medulla)	Rat	Decrease (*in vivo*)	200 mg/kg	[134]
Glycine concentration (Plasma, urine, CSF)	Humans	Increase (*in vivo*)	Chronic oral treatment with 14–41 mg/kg/d	[138]
Taurine concentration (Brain)	Rat	Increase (*in vivo*)	Chronic treatment (300 mg/kg/d)	[82]
Taurine concentration (Hippocampus)	Rat	Increase (*in vivo*)	200 mg/kg	[134]
Taurine concentration (Extracellular in hippocampus)	Rat	No effect (*in vivo*)	100–400 mg/kg	[86]

Neurophysiological effects of valproate on amino acidergic neurotransmitter functions

Macdonald and Bergey [141] were the first to describe that valproate potentiates neuronal responses to GABA by a postsynaptic effect (Tab. 15). However, valproate was examined after microiontophoretic application so that the local (extracellular) drug concentration was unknown. Subsequent *in vitro* studies showed that increased postsynaptic GABA responses are only obtained with very high valproate concentrations (Tab. 15). There is

Table 15. Neurophysiological effects of valproate as possible mechanism of anticonvulsant action. 1. Effects on the function of inhibitory and excitatory amino acids

Parameter	Species	Effect	Effective doses or concentrations	Ref. (Examples)
GABA responses (Spinal cord neurons in cell culture)	Mouse	Potentiation (*in vitro*)	Not known (Iontophoretic application)	[141]
GABA responses (Spinal cord neurons in cell culture)	Mouse	No effect (*in vitro*)	up to 1 mM	[142]
GABA responses (Cuneate afferent) fibers)	Rat	Potentiation (*in vitro*)	only at 3–10 mM	[143]
GABA responses (Locus coeruleus neurons)	Rat	Potentiation (*in vitro*)	0.1–1 mM	[144]
GABA-mediated inhibition (Hippocampal pyramidal cells)	Rat	No effect (*in vitro*)	0.5–2 mM	[145]
GABA responses (Medullary reticular formation)	Rat	Potentiation (*in vivo*)	200–400 mg/kg i.v.	[147]
GABA responses (Rat $GABA_A$ receptors expressed in *Xenopus* oocytes)		No effect (*in vitro*)	up to 7 mM	[146]
Glycine responses (Spinal cord neurons in cell culture)	Mouse	Inconsistent effects (*in vitro*)	Not known (Iontophoretic application)	[141]
Glutamate responses (Spinal cord neurons in cell culture)	Mouse	Inconsistent effects (*in vitro*)	Not known (Iontophoretic application)	[141]
Glutamate responses (Neocortex)	Rat	Reduction (*in vitro*)	5–10 mM	[148]
NMDA-evoked transient depolarizations (Neocortex)	Rat	Suppression (*in vitro*)	0.1–1 mM	[148]
NMDA responses (Central neurons in culture)	Mouse	Suppression in 50% of neurons (*in vitro*)	60 µM	[149]
NMDA responses (Rat NMDA receptors expressed in *Xenopus* oocytes)		Reduction (*in vitro*)	1–7 mM	[146]
NMDA responses (Amygdala slices)	Rat	Suppression (*in vitro*)	0.2–10 mM	[150]

only one report which demonstrated a GABA potentiation at therapeutically relevant concentrations of valproate *in vitro* [144]. The authors, using locus coeruleus neurons for their experiments, suggested that the difference between their data and those of other groups may be due to the different brain regions examined in these studies. Indeed, based on neurophysiological data, a regionally specific action of valproate in the brain was also suggested by Baldino and Geller [151].

In *in vivo* experiments, valproate was shown to lead to a potentiation of postsynaptic GABA responses at doses of 200–400 mg/kg (Tab. 15). Since brain concentrations of valproate after these doses are much lower than the concentrations which potentiate GABA responses *in vitro*, the *in vivo* effect of valproate was likely not related to a direct postsynaptic action but more probably was due to the presynaptic effects, i.e. enhanced GABA turnover.

Experiments on central mouse neurons in culture indicated that neuronal responses to glycine or excitatory amino acids, such as glutamate, are not altered by valproate at relevant concentrations (Tab. 15). However, a recent study showed that valproate suppresses glutamate responses and, much more potently, NMDA-evoked transient depolarizations in rat neocortex [148]. The authors suggested that attenuation of NMDA receptor-mediated excitation is an essential mode of action for the anticonvulsant effect of valproate. This view is substantiated by more recent studies, using different preparations to study synaptic responses mediated by the NMDA subtype of glutamate receptors (Tab. 15). In all studies, valproate blocked these responses, indicating that antagonism of NMDA receptor mediated neuronal excitation may be an important mechanism of valproate. In this respect, it is interesting to note that valproate, but not phenobarbital, phenytoin or ethosuximide, blocked seizures induced by NMDLA in rodents (Tab. 1). In contrast to NMDA receptors, valproate had no effect on membrane responses mediated by kainate or quisqualate-sensitive receptors [146].

Neurophysiological effects of valproate on neuronal membranes

As shown in Table 16, valproate exerts several direct effects on neuronal membranes. The spontaneous firing of neurons is usually inhibited only by high doses or concentrations of valproate [35]. However, in substantia nigra pars reticulata (SNR) of rats, a rapid and sustained reduction in the firing rate of GABAergic neurons was found *in vivo* already after i.p. administration of doses as low as 50–100 mg/kg [120, 160, 161]. This inhibitory effect on SNR neurons might be due to the selective increase in GABA turnover induced by valproate in the nigra of rats [112] rather than to direct effects of valproate on GABAergic neurons in SNR.

At much lower concentrations than those depressing normal neuronal cell activity, valproate has been shown to diminish high frequency repetitive firing of action potentials of central neurons in culture (Tab. 16). It has been suggested that this effect might be critically involved in the anticonvulsant action of valproate on generalized tonic-clonic seizures [142]. The effect of valproate on sustained repetitive firing (SRF) was similar to limitation of SRF produced by phenytoin and carbamazepine [142]. The most likely explanation for this effect of valproate would be a use-dependent reduction of inward sodium current [142]. However, in most studies carried out to date, effects on Na^+ channels were inferred indirectly from changes

Table 16. Neurophysiological effects of valproate as possible mechanism of anticonvulsant action. 2. Direct effects on neuronal membranes

Parameter	Species	Effect	Effective doses or concentrations	Ref. (Examples)
Sustained repetitive firing (Spinal cord and cortical Neurons in culture)	Mouse	Diminished (*in vitro*)	6–200 μM	[142]
Sodium currents (Peripheral nerve)	*Xenopus laevis*	Reduction (*in vitro*)	0.25–2.4 mM	[152]
Sodium currents (Cultured hippocampal neurons)	Rat	Reduction (*in vitro*)	1 mM	[153]
NMDA- or quisqualate-stimulated Ca^{2+}-influx (Hippocampal slice)	Rat	Reduction (*in vitro*)	5 mM	[154]
Veratrine-stimulated Ca^{2+}-influx (Coritcal brain slice)	Rat	No effect (*in vitro*)	0.1 μM–10 mM	[136]
Low-threshold (T) calcium current (Thalamic neurons)	Rats, Guinea pigs	No effect (*in vitro*)	up to 1 mM	[155]
Low-threshold (T) calcium current (nodose ganglion neurons)	Rat	Reduction (*in vitro*)	100–1000 μM	[156]
K^+-conductance	Aplysia	Activation	5–30 mM	[157]
Late K^+ outward current (Neurons)	*Helix pomatia*	Increased amplitude	About 1 mM	[158]
Neuronal membranes (Brain)	Mouse	Membrane disorderinga[a] (by partition into cell membranes) (*in vitro*)	in the mMol-range	[159]

[a] In therapeutic concentrations, valproate also interacts with liver and kidney mitochondia by modifying the structural organization of the internal mitochondrial membrane, essentially the membrane protein conformation [97].

in the maximal rate of increase of Na^+-dependent action potentials. In a more recent electrophysiological study using cultured rat hippocampal neurons, valproate indeed strongly delayed the recovery from inactivation of sodium channels, which would be consistent with reduction of sodium conductance [153]. Studies using non-vertebrate preparations also indicated that valproate has a direct inhibitory effect on voltage-sensitive sodium channels (Tab. 16). However, the concept that valproate mediates its main anticonvulsant effect by slowing the recovery from inactivation of voltage-dependent sodium channels has been recently questioned, because, in contrast to cultured neurons, valproate had no effects on the refractory period and, consequently, the bursting behavior of neurons when the rat hippocampal slice was used for studying neurophysiological effects of valproate [162]. The latter authors concluded that at least in the hippocampal slice the drug's principal anticonvulsant effect cannot be explained by its action on voltage-dependent sodium channels.

Furthermore, although the currently available data suggest that valproate inhibits SRF of cultured neurons in a manner superficially similar to that of carbamazepine and phenytoin, it seems that this effect is not achieved through a direct effect on sodium channels [76]. It is possible that SRF could be reduced through an effect on potassium channels involved in action potential repolarization. Enhancement of potassium channel function tends to hyperpolarize neurons, and an activating effect on potassium conductance has been repeatedly discussed as a potential mechanism for the action of valproate [74, 154], although such an effect has only been demonstrated at high drug concentrations (Tab. 16). Recent experiments using various potassium channel subtypes from vertebrate brain expressed in oocytes of *Xenopus laevis* have substantiated that the effects of valproate on potassium currents are too small to be of significance in its mechanism of anticonvulsant action [163].

In regard to calcium channels, the anti-absence drugs ethosuximide and dimethadione (the major active metabolite of trimethadione) have been shown to block use-dependent activation of T-type Ca^{2+} channels in thalamic neurons, which have been implicated in the generation of spike-wave activity associated with absence epilepsy [155]. However, valproate did not affect this T current in thalamic neurons (Tab. 16), although valproate is as effective as ethosuximide to block absence seizures. In contrast to thalamic neurons, valproate was shown to block low threshold T calcium channels in peripheral ganglion neurons (Tab. 16). Effects of valproate on other types of calcium currents are shown in Table 16. Veratrine-stimulated calcium influx in brain slices was not altered by valproate, whereas the drug reduced NMDA- or quisqualate-stimulated calcium influx at 5 mM [136, 154]. In such high, mmolar concentrations of valproate, this lipophilic compound interferes with membrane functions by partition into cell membranes ([159, 164], Tab. 16), which might explain many of the neurochemical or neurophysiological effects of the drug in studies involving this high concentration range.

Neurochemical effects of valproate on non-amino acidergic neurotransmitters

As already mentioned above, valproate induces in rats a behavioral syndrome with "wet dog shakes" and other symptoms reminiscent of the "serotonin syndrome" induced by serotonin precursors or receptor agonists in rodents. Indeed, recent microdialysis studies in rats have demonstrated that valproate enhances the extracellular concentration of serotonin in hippocampus and striatum of rats (Tab. 17). However, in contrast to the increase in anticonvulsant efficacy during prolonged treatment, wet-dog-shake

Table 17. Effects of valproate on non-amino acidergic neurotransmitters and cyclic nucleotides

Parameter	Species	Effect	Effective doses or concentrations	Ref. (Examples)
Serotonin metabolites (Brain)	Rat, Mouse	Increase in 5-HIAA [a] (*in vivo*)	400–600 mg/kg i.p.	[165, 166]
Serotonin and 5-HIAA levels (Striatum)	Mouse	Increase (*in vivo*)	180–360 mg/kg	[167]
Serotonin levels (Brain regions)	Rat	Increase (*in vivo*)	Chronic treatment (200 mg/kg/d)	[168]
Serotonin extracellular levels (Hippocampus, striatum)	Rat	Increase (*in vivo*)	200–400 mg/kg i.p.	[169]
Dopamine metabolites (Brain)	Mouse	Increase in HVA [b] (*in vivo*)	400–600 mg/kg i.p.	[165]
Dopamine and HVA levels (Striatum)	Mouse	Increase (*in vivo*)	180–360 mg/kg	[167]
Dopamine levels (Brain regions)	Rat	Increase (*in vivo*)	Chronic treatment (200 mg/kg/d)	[168]
Dopamine and metabolite extracellular levels (Hippocampus)	Rat	Increase in dopamine, DOPAC [c] but not HVA (*in vivo*)	200–400 mg/kg i.p.	[169]
Dopamine metabolites (CSF)	Schizophrenic patients	Increase in HVA (*in vivo*)	Chronic oral treatment with 900–2100 mg per day	[88]
Dopamine metabolites (CSF)	Schizophrenic patients	Decrease in HVA (*in vivo*)	Chronic oral treatment with 400–600 mg per day	[170]
Cyclic 3′,5′-GMP (Cerebellum)	Mouse	Reduction (*in vivo*)	400 mg/kg i.p.	[171, 172]
Cyclic 3′,5′-GMP (Cortex)	Mouse	Increase (*in vivo*)	400 mg/kg i.p.	[172]
Cyclic 3′,5′-GMP (CSF)	Schizophrenic patients	Increase (*in vivo*)	Chronic oral treatment with 400–600 mg per day	[170]

[a] 5-Hydroxyindoleacetic acid.
[b] Homovanillic acid.
[c] Dihydroxyphenylacetic acid.

behaviour induced by valproate is markedly diminished within some days of treatment, thus indicating that activation of serotonergic transmission is not related to the anticonvulsant action of the drug [24]. Accordingly, Horton et al. [165] showed that pretreatment of mice with p-chloro-phenylalanine, which blocked serotonin synthesis and prevented the increase in serotonin metabolism by valproate, did not diminish the anticonvulsant action of valproate.

Similar to serotonin, microdialysis studies have also demonstrated an increased extracellular level of dopamine in response to valproate (Tab. 17).

Thus, the initial assumption [165] that valproate does not exert effects on serotonin or dopamine levels but only blocks outward transport of their metabolites from the CNS has to be rejected. Similar to serotonin, the alterations in dopamine levels seem not to be associated with valproate's anticonvulsant effect, since pretreatment of mice with α-methyl-p-tyrosine to inhibit dopamine synthesis did not diminish the anticonvulsant action of valproate [165]. However, alterations of dopaminergic functions by valproate might be important for the antipsychotic effects of the drug. Indeed, similar to typical neuroleptics, valproate was shown to increase levels of the dopamine metabolite HVA in CSF of schizophrenic patients (Tab. 17), which could be a reflection of an anti-dopaminergic action, i.e. compensatory elevation of dopamine turnover due to inhibition of postsynaptic dopamine receptor function. As described above, an anti-dopaminergic action of valproate was also suggested by animal experiments (Tab. 6), which could be explained by the GABAmimetic effects of valproate. Indeed, GABAergic agents are known to potentiate the action of neuroleptics and to exert some intrinsic antidopamine activity such as inhibition of stereotypies or hyperactivity induced by dopaminergic drugs, which is due to GABAergic inhibition of dopamine neurons. In this respect, it should be noted that the neurochemical effects of valproate on dopamine metabolism in schizophrenic patients differ as a function of valproate dosage. Thus, in contrast to the HVA increases after high doses of valproate, a clinical study with low doses of valproate in schizophrenic patients reported a decrease of HVA levels in the CSF (Tab. 17).

In addition to dopamine, valproate's effects on serotonin has been suggested to be involved in its antimanic properties [173]. Indeed, studies on effects of subchronic therapy with valproate on central serotonin metabolism in manic patients indicated that valproate increases serotonergic neurotransmission and that this stimulation may play a role in the antimanic effects of valproate [173].

Guanosine 3′, 5′-monophosphate (cGMP) has been implicated as a second messenger in a variety of cellular events [174]. For instance, the levels of cGMP in the cerebellum and cortex are known to increase sharply at the onset of experimentally induced seizures, and it has been proposed that an elevated cerebellar cGMP level is involved in initiating or maintaining seizure activity *via* the regulation of Purkinje cell activity [171, 172]. Valproate was shown to decrease the cerebellar cGMP level during the time of anticonvulsant activity, whereas the cortical cGMP level was elevated [171, 172]. In contrast to cGMP, cAMP levels were not altered by valproate.

In patients with epilepsy or schizophrenia, the level of cGMP in the CSF is not altered [170, 175]. Valproate has been reported to increase cGMP levels in CSF of schizophrenic patients (Tab. 17), although this was not confirmed in another study [88]. Since levels of cGMP in the CNS are altered by several neurotransmitters, including amino acids [174], the effects of valproate on cGMP might be secondary due to the various alterations of neurotransmitter systems described above or *vice versa*.

Neurochemical and neurophysiological effects of active metabolites of valproate

Since valproate is rapidly metabolized to various pharmacologically active metabolites *in vivo* [176], these substances have to be considered when mechanisms of action of valproate are discussed. One of the major active metabolites of valproate in plasma and CNS of different species, including humans, is the trans isomer of 2-en-valproate (E-2-en-valproate) [176]. This compound is the most potent and most extensively studied active metabolite of valproate [177, 178]. For instance, by comparing the effect of trans-2-en-valproate and valproate on the threshold for clonic PTZ seizures in rats, Semmes and Shen [177] reported that trans-2-en-valproate was two to three times more potent than the parent drug. Accordingly, in most neurochemical and neurophysiological experiments with trans-2-en-valproate, the compound exerted more potent effects than valproate (Tab. 18). An exception might be limitation of SRF, since McLean and Macdonald reported that the 2-en-metabolite of valproate did not limit SRF at 12 to 120 µM [142]. However, these authors used a mixture of the cis and trans isomers of 2-en-valproate, although the cis isomer is only a minor metabolite of valproate and is less potent as an anticonvulsant than the trans isomer [178]. More recent experiments with the trans-isomer showed that trans-2-en-valproate is as potent as valproate to limit SRF in mouse central neurons in cell culture [180]. In various rat brain slide preparations of epileptiform activity, trans-2-en-valproate was as potent or more potent than valproate to block this activity [46]. Similarly, in *in vivo* single-unit recordings from the SNR of anesthetized rats, trans-2-en-valproate was more potent to inhibit spontaneous firing of SNR neurons than valproate [161].

An interesting behavioral difference between valproate and trans-2-en-valproate is that the metabolite does not induce wet-dog-shake behaviour in rats [24, 180a]. Neurochemical experiments showed that both trans-2-en-valproate and valproate increase serotonin metabolism in several brain regions of rats, thus indicating that wet-dog-shake behavior in response to valproate is not mediated by alterations in serotonin [180a]. Dopamine metabolism was differentially altered by the two compounds, with marked increases in DOPAC or HVA seen in frontal cortex and brainstem after valproate but not the metabolite, while the latter compound but not valproate significantly increased DOPAC in the SN and HVA in the amygdala. Levels of noradrenaline were not significantly altered in any of the eight brain regions examined [180a].

It has been suggested that trans-2-en-valproate might be responsible for the persistence of neurochemical and pharmacodynamic effects observed after termination of treatment with valproate, since the metabolite was

Table 18. Neurochemical and neurophysiological effects of trans-2-en-valproate, the major active metabolite of valproate

Parameter	Species	Effect	Effective doses or concentrations	Ref. (Examples)
GABA, GAD (Brain tissue, synaptosomes)	Mouse	Increase (*in vivo*)	200 mg/kg i.p.	[179]
GABA-T (Purified from brain)	Rat	Inhibition (*in vitro*)	K_i = 0.5 mM (Valproate 9.5 mM)	[94]
GABA-T (Purified from brain)	Humans	Inhibition (*in vitro*)	K_i = 4.5 mM (Valproate 40 mM)	[94]
KDHC (Purified from brain)	Beef	Inhibition (*in vitro*)	K_i 40 µM (as CoA-ester 6 µM[a])	[113]
$GABA_A$-receptor (Brain)	Rat	Displacement of GABA (*in vitro*)	at 1 mM 56% displacement of GABA (Valproate was not effective at 1 mM)	[176]
GABA synthesis (Brain slices)	Rat (neonatal)	Increase (*in vitro*)		[115]
GABA responses (Cortical neurons)	Rat	Potentiation (*in vivo*)	Not known (Iontophoretic application)	Felix and Löscher (unpublished data)
Serotonin and 5-HIAA levels (Brain regions)	Rat	Increase (*in vivo*)	200 mg/kg i.p.	[180 a]
Dopamine and metabolite levels (Brain regions)	Rat	Increase (*in vivo*)	200 mg/kg i.p.	[180 a]
Sustained repetitive firing of action potentials (Central neurons in culture)	Mouse	Limitation (*in vitro*)	IC_{50} = 1.2 mM (Lower after incubation)	[180]
Spontaneous firing of SNR neurons	Rat	Inhibition (*in vivo*)	50–200 mg/kg i.v.	[161]

[a] The CoA-ester of trans-2-en-valproate is rapidly formed *in vivo*.

longer present in brain tissue than the parent drug [176, 181]. A carry over of antiepileptic activity has been observed also after termination of chronic treatment with valproate in patients. In view of the similarities in the spectrum of anticonvulsant activities of trans-2-en-valproate and valproate, but the lack of serious toxic effects, such as teratogenicity, of trans-2-en-valproate, this compound might be an interesting drug of its own [176–178].

Putative mechanisms involved in the early and late anticonvulsant effects of valproate

As outlined in this chapter, valproate has both extracellular (e.g. ion channels) and intracellular (e.g. GABA synthesis) sites of action. The access to these sites will determine how rapidly valproate acts after systemic action. We were the first to describe that valproate enters and leaves the brain by active, carrier-mediated transport at the blood-brain barrier [182]. Whereas it was initially thought that this transport is mediated by the monocarboxylic acid carrier [182], more recent experiments suggest that a medium-chain fatty acid transporter rather than the short-chain monocarboxylic acid carrier at the blood-brain barrier is involved in the uptake of GABA from blood to brain [183]. This active transport explains why valproate despite its physicochemical properties (highly ionized at physiological pH, highly plasma protein bound) enters the brain so quickly [20]. Assuming that valproate has to enter neurons by passive diffusion, its rapid anticonvulsant effect in some seizure models after parenteral administration of acute doses is most likely explained by an effect on extracellular sites. The late anticonvulsant effects observed both preclinically and clinically (see section 2) are then most likely explained by slow access to intracellular sites of action. This view is corroborated by neurophysiological experiments. Thus, in the buccal ganglia Helix pomatia preparation, extracellular application of valproate decreased frequency of occurrence of PTZ-induced epileptic depolarizations immediately (early effect) and, with a delay, led to a decay in paroxysmal depolarizations (late effect) [27]. This late effect was obtained immediately when valproate was applied intracellularly (E.-J. Speckmann, personal communication), substantiating that the delay in this effect after extracellular application was due to slow penetration of valproate into the neuron. Slow diffusion into and out of neurons could also be involved in carry-over effects observed both preclinically and clinically, because while extracellular levels of valproate will rapidly leave brain or CSF by active outward transport, elimination from neurons may be retarded.

Conclusions

As shown in this review, valproate exerts a variety of effects on neurotransmitter-dependent and -nondependent cellular events. In view of the diverse molecular and cellular events that underlie different seizure types, the combination of several neurochemical and neurophysiological mechanisms in a single drug molecule might explain the broad antiepileptic efficacy of valproate. Furthermore, by acting on diverse regional targets thought to be involved in ictogenesis, valproate may antagonize epileptic activity at several steps of its organization. The fact that valproate exerts not only anticonvulsant but also several other pharmacodynamic and pharmacotherapeutic effects, including antimanic and migraine-prophylactic efficacy, certainly relates to the multiplicity of valproate's effects on neuronal functions [123, 135, 173, 184]. Because of the different pharmacodynamic effects of valproate, it is difficult to ascertain which specific neurochemical or neurophysiological action(s) are related to the anticonvulsant activity of the drug. There is now ample evidence that valproate increases GABA turnover and thereby potentiates GABAergic functions in some specific brain regions thought to be involved in the control of seizure generation and propagation. Furthermore, the effect of valproate on neuronal excitation mediated by the NMDA subtype of glutamate receptors might be important for its anticonvulsant effects. In contrast, the relevance of often cited effect of valproate on SRF in cultured neurons is doubtful, because there is no convincing evidence that valproate blocks voltage-dependent sodium currents and SRF in more conventional preparations [162]. Whereas the GABA potentiation and glutamate/NMDA inhibition could be a likely explanation for the anticonvulsant action on focal and generalized motor seizures, they do not explain the effect of valproate on non-convulsive seizures, such as absences. In this respect, the reduction of GHB release reported for valproate could be of interest. In any event, in contrast to various experimental and clinical papers, valproate should not be used or considered as a specific "GABAmimetic" drug, but the various other effects which are induced concomitantly with alterations of GABAergic functions should be referred to.

References

1 Burton BS (1882) On the propyl derivatives and decomposition products of ethylaceto-acetate. *Am Chem J* 3: 385–395
2 Meunier H, Carraz G, Meunier Y, Eymard P, Aimard M (1963) Propriétés pharmacodynamiques de l'acide n-dipropylacétique. 1er Mémoire: Propriétés antiépileptiques. *Thérapie* 18: 435–438
3 Carraz G, Fau R, Chateau R, Bonnin J (1964) Communication à propos des premiers essais cliniques sur l'activité anti-épileptique de l'acide n-dipropylacétiques (sel de Na). *Ann Med Psychol (Paris)* 122: 577–585
4 Swinyard EA (1964) The pharmacology of dipropylacetic acid sodium with special emphasis on its effects on the central nervous system. University of Utah College of Pharmacy, Salt Lake City, Utah, 1–25
5 Shuto K, Nishigaki T (1970) The pharmacological studies on sodium dipropylacetate anticonvulsant activities and general pharmacological actions (in Japanese). *Pharmacometrics* 4: 937–949
6 Krall RL, Penry JK, White BG, Kupferberg HJ, Swinyard EA (1978) Antiepileptic drug development: II. Anticonvulsant drug screening. *Epilepsia* 19: 409–428
7 Frey H-H, Löscher W (1976) Di-n-propylacetic acid – profile of anticonvulsant activity in mice. *Arzneim-Forsch (Drug Res)* 26: 299–301
8 Kupferberg HJ (1980) Sodium valproate. In: GH Glaser, JK Penry, DM Woodbury (eds) *Antiepileptic drugs: Mechanism of action.* Raven Press, New York, 643–654
9 Löscher W (1980) A comparative study of the pharmacology of inhibitors of GABA-metabolism. *Naunyn-Schmiedeberg's Arch Pharmacol* 315: 119–128
10 Worms P, Lloyd KG (1981) Functional alterations of GABA synapses in relation to seizures. In: PL Morselli, KG Lloyd, W Löscher et al. (eds) *Neurotransmitters, seizures, and epilepsy.* Raven Press, New York, 37–46
11 Löscher W, Frey H-H (1977) Effect of convulsant and anticonvulsant agents on level and metabolism of γ-aminobutyric acid in mouse brain. *Naunyn-Schmiedeberg's Arch Pharmacol* 296: 263–269
12 Petersen EN (1983) DMCM: a potent convulsive benzodiazepine receptor ligand. *Eur J Pharmacol* 94: 117–124
13 Czuczwar SJ, Frey H-H, Löscher W (1985) Antagonism of N-methyl-D,L-aspartic acid-induced convulsions by antiepileptic drugs and other agents. *Eur J Pharmacol* 108: 273–280

14 Löscher W, Jäckel R, Czuczwar SJ (1986) Is amygdala kindling in rats a model for drug-resistant partial epilepsy? *Exp Neurol* 93: 211–226
15 Frey H-H, Löscher W, Reiche R, Schultz D (1983) Anticonvulsant potency of common antiepileptic drugs in the gerbil. *Pharmacology* 27: 330–335
16 Löscher W, Nau H, Marescaux C, Vergnes M (1984) Comparative evaluation of anticonvulsant and toxic potencies of valproic acid and 2-en-valproic acid in different animal models of epilepsy. *Eur J Pharmacol* 99: 211–218
17 Dailey JW, Jobe PC (1985) Anticonvulsant drugs and the genetically epilepsy-prone rat. *Fed Proc* 44: 2640–2644
18 Löscher W, Meldrum BS (1984) Evaluation of anticonvulsant drugs in genetic animal models of epilepsy. *Fed Proc* 43: 276–284
19 Löscher W, Schwartz-Porsche D, Frey H-H, Schmidt D (1985) Evaluation of epileptic dogs as an animal model of human epilepsy. *Arzneim -Forsch (Drug Res)* 35: 82–87
20 Löscher W (1985) Valproic acid. In: H-H Frey, D Janz (eds) *Antiepileptic Drugs*. Springer Verlag, Berlin, 507–536
21 Löscher W, Schmidt D (1988) Which animal models should be used in the search for new antiepileptic drugs? A proposal based on experimental and clinical considerations. *Epilepsy Res* 2: 145–181
22 Löscher W (1998) New visions in the pharmacology of anticonvulsion. *Eur J Pharmacol* 342: 1–13
23 Hönack D, Löscher W (1992) Intravenous valproate: onset and duration of anticonvulsant activity against a series of electroconvulsions in comparison with diazepam and phenytoin. *Epilepsy Res* 13: 215 – 221
23 a Lockard JS, Levy RH, Koch KM et al. (1983) A monkey model for status epilepticus: Carbamazepine and valproate compared to three standard anticonvulsants. In: AV Delgado-Escueta, CG Wasterlain, DM Treiman et al. (eds) *Status epilepticus*. Raven Press, New York, 411–419; b Walton NY, Treiman DM (1992) Valproic acid treatment of experimental status epilepticus. *Epilepsy Res* 12: 199–205
24 Löscher W, Fisher JE, Nau H, Hönack D (1988) Marked increase in anticonvulsant activity but decrease in wet-dog shake behaviour during short-term treatment of amygdala-kindled rats with valproic acid. *Eur J Pharmacol* 150: 221–232
25 Löscher W, Fisher JE, Nau H, Hönack D (1989) Valproic acid in amygdala-kindled rats: alterations in anticonvulsant efficacy, adverse effects and drug and metabolite levels in various brain regions during chronic treatment. *J Pharmacol Exp Ther* 250: 1067– 1078
26 Löscher W, Hönack D (1995) Comparison of anticonvulsant efficacy of valproate during prolonged treatment with one and three daily doses or continuous ("controlled release") administration in a model of generalized seizures in rats. *Epilepsia* 36: 929–937
27 Altrup U, Gerlach G, Reith H, Said MN, Speckmann E-J (1992) Effects of valproate in a model nervous system (buccal ganglia of *Helix pomatia*: I. Antiepileptic actions. *Epilepsia* 743–752
27 a Hernandez TD (1997) Preventing post-traumatic epilepsy after brain injury: weighing the costs and benefits of anticonvulsant prophylaxis. *Trends Pharmacol Sci* 18: 59–62; b Shinnar S, Berg AT (1996) Does antiepileptic drug therapy prevent the development of "chronic" epilepsy? *Epilepsia* 37: 701–708; c Silver JM, Shin C, McNamara JO (1991) Antiepileptogenic effects of conventional anticonvulsants in the kindling model of epilepsy. *Ann Neurol* 29: 356–363
28 Vajda FJ, Donnan GA, Phillips J, Bladin PF (1981) Human brain, plasma and cerebrospinal fluid concentration of sodium valproate after 72 hours of therapy. *Neurology* 31: 486–487
29 Löscher W, Rundfeldt C, Hönack D (1993) Pharmacological characterization of phenytoin-resistant amygdala-kindled rats, a new model of drug-resistant partial epilepsy. *Epilepsy Res* 15: 207–219
30 Löscher W (1997) Animal models of intractable epilepsy. *Prog Neurobiol* 53: 239–258
31 Löscher W, Rundfeldt C (1991) Kindling as a model of drug-resistant partial epilepsy: selection of phenytoin-resistant and nonresistant rats. *J Pharmacol Exp Ther* 258: 483–489
32 Jurna I (1985) Electrophysiological effects of antiepileptic drugs. In: H-H Frey, D Janz (eds) *Antiepileptic drugs*. Springer, Berlin, 611–658
33 Schwartzkroin PA (1986) Hippocampal slices in experimental and human epilepsy. *Adv Neurol* 44: 991–1010

34 Heinemann U, Draguhn A, Ficker E, Stabel J, Zhang CL (1994) Strategies for the development of drugs for pharmacoresistant epilepsies. *Epilepsia* 35 (Suppl. 5): S10–S21
35 Chapman A, Keane PE, Meldrum BS, Simiand J, Vernieres JC (1982) Mechanism of anticonvulsant action of valproate. *Progr Neurobiol* 19: 315–359
36 Cotariu D, Zaidman JL, Evans S (1990) Neurophysiological and biochemical changes evoked by valproic acid in the central nervous system. *Progr Neurobiol* 34: 343–354
37 Rogawski MA, Porter RJ (1990) Antiepileptic drugs: pharmacological mechanisms and clinical efficacy with consideration of promising developmental stage compounds. *Pharmacol Rev* 42: 223–286
38 Evans MS (1992) Overview of actions of antiepileptic drugs on repetitive neuronal firing. In: CL Faingold, GH Fromm (eds): *Drugs for control of epilepsy: Actions on neuronal networks involved in seizure disorders.* CRC Press, Boca Raton, 69–88
39 Clark S, Wilson WA (1992) Brain slice model of epilepsy: Neuronal networks and actions of antiepileptic drugs. In: CL Faingold, GH Fromm (eds): *Drugs for control of epilepsy: Actions on neuronal networks involved in seizure disorders.* CRC Press, Boca Raton, 89–124
40 Fromm GH (1992) Antiepileptic actions of valproate. In: CL Faingold, GH Fromm (eds): *Drugs for control of epilepsy: Actions on neuronal networks involved in seizure disorders.* CRC Press, Boca Raton, 453–462
41 Fariello RG, Varasi M, Smith MC (1995) Valproic acid. Mechanisms of action. In: RH Levy, RH Mattson, BS Meldrum (eds): Antiepileptic drugs. Fourth edition. Raven Press, New York, 581–604
42 Thomson SM (1993) Consequence of epileptic activity in vitro. *Brain Pathol* 3: 413–419
43 Schmitz D, Zhang CL, Chatterjee SS, Heinemann U (1995) Effects of methysticin on three different models of seizure like events studied in rat hippocampal and entorhinal cortex slices. *Naunyn-Schmiedeberg's Arch Pharmacol* 351: 348–355
44 Zhang CL, Dreier JP, Heinemann U (1995) Paroxysmal epileptiform discharges in temporal lobe slices after prolonged exposure to low magnesium are resistant to clinically used anticonvulsants. *Epilepsy Res* 20: 105–111
45 Dreier JP, Heinemann U (1990) Late low magnesium-induced epileptiform activity in rat entorhinal cortex slides becomes insensitive to the anticonvulsant valproic acid. *Neurosci Lett* 119: 68–70
46 Sokolova S, Schmitz D, Zjang CL, Löscher W, Heinemann U (1998) Comparison of effects of valproate and trans-2-en-valproate on different forms of epileptiform activity in rat hippocampal and temporal cortex slices. *Epilepsia* 39: 251–258
47 Lal H, Shearman GT (1980) Effect of valproic acid on anxiety-related behaviours in the rat. *Brain Res Bull* 5 (Suppl. 2): 575–577
48 Simler S, Puglisi-Allegra S, Mandel P (1983) Effects of n-di-propylacetate on aggressive behavior and brain GABA level in isolated mice. *Pharmacol Biochem Behav* 18: 717–720
49 Simler S, Ciesielski L, Klein M, Mandel P (1982) Anticonvulsant and antiaggressive properties of di-n-propylacetate after repeated treatment. *Neuropharmacology* 21: 133–140
50 Vellucci SV, Webster RA (1984) The role of GABA in the anticonflict action of sodium valproate and chlordiazepoxide. *Pharmacol Biochem Behav* 21: 845–851
51 Fredow G, Löscher W (1991) Effects of pharmacological manipulation of GABAergic neurotransmission in a new mutant hamster model of paroxysmal dystonia. *Eur J Pharmacol* 192: 207–219
52 De Boer T, Metselaar HJ, Briunvels J (1977) Suppression of GABA-induced abstinence behaviour in naive rats by morphine and bicuculline. *Life Sci* 20: 933–942
53 Cowan A, Watson T (1978) Lysergic acid diethylamide antagonizes shaking induced in rats by five chemically different compounds. *Psychopharmacology* 57: 43–46
54 Kuruvilla A, Uretsky NJ (1981) Effect of sodium valproate on motor function regulated by the activation of GABA receptors. *Psychopharmacology* 72: 167–172
55 Löscher W, Vetter M (1985) In vivo effects of aminooxyacetic acid and valproic acid on nerve terminal (synaptosomal) GABA levels in discrete brain areas of the rat. Correlation to pharmacological activities. *Biochem Pharmacol* 34: 1747–1756
56 Carraz G, Fiorina S (1967) Activation de la formation d'anti corps par le système réticuloendothelial. *Ann Biol Clin* (Paris) 76: 187

57 De Souza Queiroz ML, Mullen PW (1980) The effects of phenytoin, 5-(parahydroxyphenyl)-5-phenylhydantoin, and valproic acid on humoral immunity in mice. *Int J Immunopharmacol* 2: 224–225
58 Rotiroli D, Palella B, Losi E, Nistico G, Caputi AP (1982) Evidence that a GABAergic mechanism influences the development of DOCA-salthypertension in the rat. *Eur J Pharmacol* 83: 153–154
58 a Balfour JA, Bryson HM (1994) Valproic acid – A review of its pharmacology and therapeutic potential in indications other than epilepsy. *Cns Drugs* 2: 144–173
59 Emrich HM, von Zerssen D, Kissling W, Moeller HJ, Windorfer A (1980) Effect of sodium valproate on mania. The GABA-hypothesis of affective disorders. *Arch Psychiatr Nervenkr* 229: 1–16
60 Jancovic J, Fahn S (1988) Dystonic syndromes. In: J Jancovic, E Tolosa (eds): *Parkinson's disease and movement disorders.* Urban and Schwarzenbek, Baltimore, 283–314
61 Macdonald RL (1989) Antiepileptic drug action. Epilepsia 30 (Suppl. 1): S19–S28
62 Macdonald RL, Kelly KM (1995) Antiepileptic drug mechanisms of action. *Epilepsia* 36: S2–S12
63 Macdonald RL, Meldrum BS (1995) Principles of antiepileptic drug action. In: RH Levy, RH Mattson, BS Meldrum (eds): *Antiepileptic drugs,* Fourth edition. Raven Press, New York, 61–78
64 Löscher W, Schmidt D (1994) Strategies in antiepileptic drug development: is rational drug design superior to random screening and structural variation? *Epilepsy Res* 17: 95–134
65 Schachter SC (1995) Review of the mechanisms of action of antiepileptic drugs. *Cns Drugs* 4: 469–477
66 White HS (1997) Clinical significance of animal seizure models and mechanism of action studies of potential antiepileptic drugs. *Epilepsia* 38: S9–S17
67 Schmidt D, Gram L (1995) Monotherapy versus polytherapy in epilepsy: A reappraisal. *Cns Drugs* 3: 194–208
68 Pinder RM, Brogden RN, Speight TM, Avery GS (1977) Sodium valproate: A review of its pharmacological properties and therapeutic efficacy in epilepsy. *Drugs* 13: 81–123
69 Meldrum B (1980) Mechanism of action of valproate. *Brain Res Bull* 5 (Suppl. 2): 579–584
70 Turner AJ, Whittle SR (1980) Sodium valproate, GABA and epilepsy. *Trends Pharmacol Sci* 1: 257–260
71 Hammond EJ, Wilder BJ, Bruni J (1981) Central actions of valproic acid in man and in experimental models of epilepsy. *Life Sci* 29: 2561–2574
72 Kerwin RW, Taberner PV (1981) The mechanism of action of sodium valproate. *Gen Pharmacol* 12: 71–75
73 Johnston D (1984) Valproic acid: update of its mechanism of action. *Epilepsia* 24 (Suppl. 1): S1–S4
74 Morre M, Keane PE, Vernires JC, Simiand J, Ronucci R (1984) Valproate: Recent findings and perspectives. *Epilepsia* 25 (Suppl. 1): S5–S9
75 Macdonald RL, Mclean MJ (1986) Anticonvulsant drugs: mechanisms of action. In: AV Delgado-Escueta, AAJ Ward, DM Woodbury et al. (eds): *Basic mechanisms of the epilepsies. Molecular and cellular approaches.* Raven Press, New York, 713–736
76 Löscher W (1993) Effects of the antiepileptic drug valproate on metabolism and function of inhibitory and excitatory amino acids in the brain. *Neurochem Res* 18: 485–502
77 Davis R, Peters DH, Mctavish D (1994) Valproic acid – a reappraisal of its pharmacological properties and clinical efficacy in epilepsy. *Drugs* 47: 332–372
78 Löscher W (1991) Anticonvulsant Drug Mechanisms. In: MR Klee, HD Lux, E-J Speckmann (eds): *Physiology, pharmacology and development of epileptogenic phenomena.* Springer-Verlag, Berlin, 193–200
79 Löscher W (1989) GABA and the epilepsies. Experimental and clinical considerations. In: NG Bowery, G Nisticò (eds): *GABA. Basic research and clinical applications.* Pythagora Press, Rome, 260–300
80 Simler S, Randrianarisoa H, Lehman A, Mandel P (1968) Effects du di-n-propylacétate sur les crises audiogènes de la souris. *J Physiol (Paris)* 60: 547
81 Godin Y, Heiner L, Mark J, Mandel P (1969) Effects of di-n-propylacetate, an anticonvulsive compound, on GABA metabolism. *J Neurochem* 16: 869–873

82 Iadarola MJ, Raines A, Gale K (1979) Differential effects of n-dipropylacetate and aminooxyacetic acid on γ-aminobutyric acid levels in discrete areas of rat brain. *J Neurochem* 33: 1119–1123
83 Iadarola MJ, Gale K (1981) Cellular compartments of GABA in brain and their relationship to anticonvulsant activity. *Mol Cell Biochem* 39: 305–330
84 Löscher W (1982) GABA in plasma, CSF and brain of dogs during acute and chronic treatment with y-acetylenic GABA and valproic acid. In: Y Okada, E Roberts (eds): *Problems in GABA research – from Brain to Bacteria.* Exerpta Medica, Amsterdam, 102–109
85 Löscher W (1981) Valproate induced changes in GABA metabolism at the subcellular level. *Biochem Pharmacol* 30: 1364–1366
86 Biggs CS, Pearce BR, Fowler LJ, Whitton PS (1992b) The effect of sodium valproate on extracellular GABA and other amino acids in the rat ventral hippocampus: an in vivo microdialysis study. *Brain Res* 594: 138–142
87 Rowley HL, Marsden CA, Martin KF (1995) Differential effects of phenytoin and sodium valproate on seizure-induced changes in gamma-aminobutyric acid and glutamate release in vivo. *Eur J Pharmacol* 294: 541–546
88 Zimmer R, Teelken AW, Gündürewa M, Rüther E, Cramer H (1980) Effect of sodium valproate on CSF GABA, cAMP, cGMP and homovanillic acid levels in men. *Brain Res Bull* 5 (Suppl. 2): 585–588
89 Löscher W, Siemes H (1984) Valproic acid increases γ-aminobutyric acid in CSF of epileptic children. L*ancet* II: 225
90 Löscher W, Siemes H (1985) Cerebrospinal fluid γ-aminobutyric acid levels in children with different types of epilepsy: effect of anticonvulsant treatment. *Epilepsia* 26: 314–319
91 Löscher W, Schmidt D (1980) Increase of human plasma GABA by sodium valproate. *Epilepsia* 21: 611–615
92 Löscher W, Schmidt D (1981) Plasma GABA levels in neurological patients under treatment with valproic acid. *Life Sci* 28: 2383–2388
93 Iadarola MJ, Gale K (1979) Dissociation between drug-induced increases in nerve terminal and non-nerve terminal pools of GABA in vivo. *Eur J Pharmacol* 59: 125–129
94 Maitre M, Ciesielski L, Cash C, Mandel P (1978) Comparison of the structural characteristics of the 4-aminobutyrate: 2-oxoglutarate transaminases from rat and human brain, and of their affinities for certain inhibitors. *Biochim Biophys Acta* 522: 385–399
95 Whittle SR, Turner AJ (1978) Effects of the anticonvulsant sodium valproate on γ-aminobutyrate and aldehyde metabolism in ox brain. *J Neurochem* 31: 1453–1459
96 Löscher W (1980) Effect of inhibitors of GABA transaminase on the synthesis, binding, uptake, and metabolism of GABA. *J Neurochem* 34: 1603–1608
97 Larsson OM, Gram L, Schousboe I, Schousboe A (1986) Differential effects of gamma-vinyl GABA and valproate on GABA-transaminase from cultured neurons and astrocytes. *Neuropharmacology* 25: 617–625
98 Kumlien E, Sherif F, Ge L, Oreland L (1995) Platelet and brain GABA-transaminase and monoamine oxidase activities in patients with complex partial seizures. *Epilepsy Res* 20: 161–170
99 Phillips NI, Fowler LJ (1982) The effects of sodium valproate on γ-aminobutyrate metabolism and behavior in naive and ethanolamine-O-sulphate pretreated rats and mice. *Biochem Pharmacol* 31: 2257–2261
100 Löscher W (1993) In vivo administration of valproate reduces the nerve terminal (synaptosomal) activity of GABA aminotransferase in discrete brain areas of rats. *Neurosci Lett* 160: 177–180
101 Van der Laan JW, De Boer T, Bruinvels J (1979) Di-n-propylacetate and GABA degradation. Preferential inhibition of succinic semialdehyde dehydrogenase and indirect inhibition of GABA-transaminase. *J Neurochem* 32: 1769–1780
102 Maitre M, Ossola L, Mandel P (1976) In vitro studies into the effect of inhibition of rat brain succinic semialdehyde dehydrogenase on GABA synthesis and degradation. *FEBS Lett* 72: 53–57
103 Simler S, Ciesielski L, Klein M, Gobaille S, Mandel P (1981) Sur le mècanisme d'action d'un anticonvulsivant, le dipropylacétate de sodium. *CR Soc Biol (Paris)* 175: 114–119
104 Vayer P, Cash CD, Maitre M (1988) Is the anticonvulsant mechanism of valproate linked to its interaction with the cerebral γ-hydroxybutyrate system? *Trends Pharmacol Sci* 9: 127–129

105 Whittle SR, Turner SJ (1982) Effects of anticonvulsants on the formation of γ-hydroxybutyrate from γ-aminobutyrate in rat brain. *J Neurochem* 38: 848–851
106 Snead OC III (1988) γ-Hydroxybutyrate model of generalized absence seizures: further characterization and comparison with other absence models. *Epilepsia* 29: 361–377
107 Snead OC III, Bearden LJ, Pegram V (1980) Effect of acute and chronic anticonvulsant administration on endogenous γ-hydroxybutyrate in rat brain. *Neuropharmacology* 19: 47–52
108 Löscher W, Frey H-H (1977) Zum Wirkungsmechanismus von Valproinsäure. *Arzneimittel-Forsch* 27: 1081–1082
109 Taylor CP, Vartanian MG, Andruszkiewicz R, Silverman RB (1992) 3-Alkyl GABA and 3-alkylglutamic acid analogues: two new classes of anticonvulsant agents. *Epilepsy Res* 11: 103–110
110 Wikinski SI, Acosta GB, Rubio MC (1996) Valproic acid differs in its in vitro effect on glutamic acid decarboxylase activity in neonatal and adult rat brain. *Gen Pharmacol* 27: 635–638
111 Taberner PV, Charington CB, Unwin JW (1980) Effects of GAD and GABA-T inhibitors on GABA metabolism in vivo. *Brain Res Bull* 5 (Suppl. 2): 621–625
112 Löscher W (1989) Valproate enhances GABA turnover in the substantia nigra. *Brain Res* 501: 198–203
113 Luder AS, Parks JK, Frerman F, Parker WDJ (1990) Inactivation of beef brain -ketoglutarate dehydrogenase complex by valproic acid and valproic acid metabolites. *J Clin Invest* 86: 1574–1581
114 Nau H, Löscher W (1982) Valproic acid: Brain and plasma levels of the drug and its metabolites, anticonvulsant effects and GABA metabolism in the mouse. *J Pharmacol Exp Ther* 220: 654–659
115 Bolanos JP, Medina JM (1993) Evidence of stimulation of the gamma-aminobutyric acid shunt by valproate and E-delta-2-valproate in neonatal rat brain. *Mol Pharmacol* 43: 487–490
116 Ekwuru MO, Cunningham JR (1990) Phaclofen increases GABA release from valproate treated rats. *Brit J Pharmacol* 99 (Suppl.): 251P
117 Gram L, Larsson OM, Johnsen AH, Schousboe A (1988) Effects of valproate, vigabatrin and aminooxyacetic acid on release of endogenous and exogenous GABA from cultured *neurons. Epilepsy Res* 2: 87–95
118 Ross SM, Craig CR (1981) Studies on γ-aminobutyric acid transport in cobalt experimental epilepsy in the rat. *J Neurochem* 36: 1006–1011
119 Wolf R, Tscherne U, Emrich HM (1988) Suppression of preoptic GABA release caused by push-pull-perfusion with sodium valproate. *Naunyn-Schmiedeberg's Arch Pharmacol* 338: 658–663
120 Farrant M, Webster RA (1989) Neuronal activity, amino acid concentration and amino acid release in the substantia nigra of the rat after sodium valproate. *Brain Res* 504: 49–56
121 Timmerman W, Westerink BHC (1997) Brain microdialysis of GABA and glutamate: What does it signify? *Synapse* 27: 242–261
122 Olsen RW (1981) The GABA postsynaptic membrane receptor-ionophore complex. Site of action of convulsant and anticonvulsant drugs. *Mol Cell Biochem* 39: 261–279
123 Motohashi N (1992) GABA receptor alterations after chronic lithium administration – Comparison with carbamazepine and sodium valproate. *Prog Neuro-Psych Biol Psych* 16: 571–579
124 Miller LG, Greenblatt DJ, Barnhill JG, Summer WR, Shader RI (1988) "GABA shift" in vivo: enhancement of benzodiazepine binding in vivo by modulation of endogenous GABA. *Eur J Pharmacol* 148: 123–130
125 Mimaki T, Yabucchi W, Laird H, Yamamura HI (1984) Effects of seizures and antiepileptic drugs on benzodiazepine receptors in rat brain. *Pediatr Pharmacol* 4: 205–211
126 Squires RF, Casida JE, Richardson M, Saederup E (1983) [^{35}S]-Butylbicyclophosphorothionate binds with high affinity to brain-specific sites coupled to γ-aminobutyric acid-A and ion recognition sites. *Mol Pharmacol* 23: 326–336
127 Pitkänen A, Saano V, Tuomisto L, Riekkinen PJ (1987) Effect of anticonvulsant drugs on [^{35}S]t-butylbicyclophosphorothionate binding in vitro and ex vivo. *Pharmacol Toxicol* 61: 103–106

128 Concas A, Mascia MP, Sanna E, Santoro G, Serra M, Biggio G (1991) "In vivo" administration of valproate decreases t[^{35}S]butylbicyclophosphorothionate binding in the rat brain. *Naunyn-Schmiedeberg's Arch Pharmacol* 343: 296–300
129 Lloyd KG, Thuret F, Pilc A (1985) Upregulation of gamma-aminobutyric (GABA) B binding sites in rat frontal cortex: a common action of repeated administration of different classes of antidepressants and electroshock. *J Pharmacol Exp Ther* 235: 191–199
130 Nutt DJ, Cowen PJ, Little HJ (1982) Unusual interactions of benzodiazepine receptor antagonists. *Nature* 295: 436
131 Gent JP, Bentley M, Feely M, Haigh JRM (1986) Benzodiazepine cross-tolerance in mice extends to sodium valproate. *Eur J Pharmacol* 128: 9–15
132 Liljequist S, Engel JA (1984) Reversal of anticonflict action of valproate by various GABA and benzodiazepine antagonists. *Life Sci* 34: 2525–253
133 DeFeudis FV (1984) Gamma-aminobutyric acid-ergic analgesia: implications for gamma-aminobutyric acid-ergic therapy for drug addiction. *Drug Alcohol Depend* 14: 101–111
134 Löscher W, Hörstermann D (1994) Differential effects of vigabatrin, gamma-acetylenic GABA, aminooxyacetic acid, and valproate on levels of various amino acids in rat brain regions and plasma. *Naunyn-Schmiedeberg's Arch Pharmacol* 349: 270–278
135 Dixon JF, Hokin LE (1997) The antibipolar drug valproate mimics lithium in stimulating glutamate release and inositol 1,4,5- trisphosphate accumulation in brain cortex slices but not accumulation of inositol monophosphates and bisphosphates. *Proc Natl Acad Sci USA* 94: 4757–4760
136 Crowder JM, Bradford HF (1987) Common anticonvulsants inhibit Ca^{2+} uptake and amino acid neurotransmitter release in vitro. *Epilepsia* 28: 378–382
137 Martin-Gallard A, Rodriguez P, Lopet M, Benavides J, Ugarte M (1985) Effects of dipropylacetate on the glycine cleavage enzyme system and glycine levels. *Biochem Pharmacol* 34: 2877–2882
138 Similae S, von Wendt L, Linna SL, Saukkonen AL, Huhtaniemi I (1979) Dipropylacetate and hyperglycemia. *Neuropaediatrie* 1: 158–160
139 Patsalos PN, Lascelles PT (1981) Changes in regional brain levels of amino acid putative neurotransmitters after prolonged treatment with the anticonvulsant drugs diphenylhydantoin, phenobarbitone, sodium valproate, ethosuximide, and sulthiame in the rat. *J Neurochem* 36: 688–695
140 Nilsson M, Hansson E, Ronnback L (1992) Interactions between valproate, glutamate, aspartate, and GABA with respect to uptake in astroglial primary cultures. *Neurochem Res* 17: 327–332
141 Macdonald RL, Bergey GK (1979) Valproic acid augments GABA-mediated postsynaptic inhibition in cultured mammalian neurons. *Brain Res* 170: 558–562
142 McLean MJ, Macdonald RL (1986) Sodium valproate, but not ethosuximide, produces use- and voltage-dependent limitation of high frequency repetitive firing of action potentials of mouse central neurons in cell culture. *J Pharmacol Exp Ther* 237: 1001–1011
143 Harrison NL, Simmonds MA (1982) Sodium valproate enhances responses to GABA receptor activation only at high concentrations. *Brain Res* 250: 201–204
144 Olpe HR, Steinmann MW, Pozza MF, Brugger F, Schmutz M (1988) Valproate enhances GABA-A mediated inhibition of locus coeruleus neurons in vitro. *Naunyn-Schmiedeberg's Arch Pharmacol* 338: 655–657
145 Perreault P, Tancredi V, Avoli M (1989) Failure of the antiepileptic drug valproic acid to modify synaptic and non-synaptic responses of Ca1 hippocampal pyramidal cells "in vitro". *Epilepsy Res* 3: 227–231
146 Muáhoff U, Madeja M, Düsing R, Speckmann E-J (1996) Valproate affects glutamate but not GABA receptors. *Eur J Neurosci* 9 (Suppl.): 205
147 Gent JP, Phillips NI (1980) Sodium di-n-propylacetate (valproate) potentiates responses to GABA and muscimol on single central neurons. *Brain Res* 197: 275–278
148 Zeise ML, Kasparaow S, Zieglgansberger W (1991) Valproate suppresses N-methyl-D-aspartate evoked, transient depolarizations in the rat neocortex in vitro. *Brain Res* 544: 345–348
149 Wamil AW, Mclean MJ (1991) Effect of anticonvulsant medications on responses to NMDA by mouse central neurons in cell culture. *Epilepsia 32* (Suppl. 3): 42
150 Gean PW, Huang CC, Hung CR, Tsai JJ (1994) Valproic acid suppresses the synaptic response mediated by the NMDA receptors in rat amygdalar slices. *Brain Res Bull* 33: 333–336

151 Baldino F, Geller HM (1981) Effect of sodium valproate on hypothalamic neurons in vivo and in vitro. *Brain Res* 219: 231–237
152 VanDongen AMJ, VanErp MG, Voskuyl RA (1986) Valproate reduces excitability by blockade of sodium and potassium conductance. *Epilepsia* 27: 177–182
153 Van den Berg RJ, Kok P, Voskuyl RA (1993) Valproate and sodium currents in cultured hippocampal neurons. *Exp Brain Res* 93: 279–287
154 Franceschetti S, Hannon B, Heinemann U (1986) The action of valproate on spontaneous epileptiform activity in the absence of synaptic transmission and on evoked changes in $[Ca^{2+}]_0$ and $[K^+]_0$ in the hippocampal slice. *Brain Res* 386: 1–11
155 Coulter DA, Huguenard JR, Prince DA (1989) Characterization of ethosuximide reduction of low-threshold calcium current in thalamic neurons. *Ann Neurol* 25: 582–593
156 Kelly KM, Gross RA, Macdonald RL (1990) Valproic acid selectively reduces the low-threshold (T) calcium current in rat nodose neurons. *Neurosci Lett* 116: 1–2
157 Slater GE, Johnston D (1978) Sodium valproate increases potassium conductance in Aplysia neurons. *Epilepsia* 19: 379–384
158 Walden J, Altrup U, Reith H, Speckmann E-J (1993) Effects of valproate on early and late potassium currents of single neurons. *Eur Neuropsychopharmacol* 3: 137–141
159 Perlman BJ, Goldstein DB (1984) Membrane-disordering potency and anticonvulsant action of valproic acid and other short-chain fatty acids. *Mol Pharmacol* 26: 83–89
160 Kerwin RW, Olpe HR, Schmutz M (1980) The effect of sodium-n-dipropylacetate on γ-aminobutyric acid-dependent inhibition in the rat cortex and substantia nigra in relation to its anticonvulsant activity. *Br J Pharmacol* 71: 545–551
161 Rohlfs A, Rundfeldt C, Koch R, Löscher W (1996) A comparison of the effects of valproate and its major active metabolite E-2-en-valproate on single unit activity of substantia nigra pars reticulata neurons in rats. *J Pharmacol Exp Ther* 277: 1305–1314
162 Albus H, Williamson RW (1998) Electrophysiological analysis of the actions of valproate on pyramidal neurons in the rat hippocampal slice. *Epilepsia* 39: 124–139
163 Roderfeld H-J, Altrup U, Düsing R, Lorra C, Madeja M, Mußhoff U, Pongs O, Speckmann E-J, Spener F (1994) Effects of the antiepileptic drug valproate on cloned voltage-dependent potassium channels. *Pflügers Arch* 426 (Suppl.): R32
164 Rumbach L, Mutet C, Cremel G, Marescaux CA, Micheletti G, Warter JM, Waksman A (1986) Effects of sodium valproate on mitochondrial membranes: electron paramagnetic resonance and transmembrane protein movement studies. *Mol Pharmacol* 30: 270–273
165 Horton RW, Anlezark GM, Sawaya MCB, Meldrum BS (1977) Monoamine and GABA metabolism and the anticonvulsant action of di-n-propylacetate and ethanolamine-O-sulphate. *Eur J Pharmacol* 41: 387–397
166 Hwang EC, van Woert MH (1979) Effect of valproic acid on serotonin metabolism. *Neuropharmacology* 18: 391–397
167 Vriend JP, Alexiuk NAM (1996) Effects of valproate on amino acid and monoamine concentrations in striatum of audiogenic seizure-prone balb/c mice. *Mol Chem Neuropathol* 27: 307–324
168 Baf MHM, Subhash MN, Lakshmana KM, Rao BSSR (1994) Sodium valproate induced alterations in monoamine levels in different regions of the rat brain. *Neurochem Int* 24: 67–72
169 Biggs CS, Pearce BR, Fowler LJ, Whitton PS (1992a) Regional effects of sodium valproate on extracellular concentrations of 5-hydroxytryptamine, dopamine, and their metabolites in the rat brain: an in vivo microdialysis study. *J Neurochem* 59: 1702–1708
170 Nagao T, Oshimo T, Mitsunobo K, Sato M, Otsuki S (1979) Cerebrospinal fluid monoamine metabolites and cyclic nucleotides in chronic schizophrenic patients witgh tardive dyskinesia or drug-induced tremor. *Biol Psychiatr* 14: 509–523
171 Lust WD, Kupferberg HJ, Yonekawa WD, Penry JK, Passoneau JV, Wheaton AB (1978) Changes in brain metabolites induced by convulsants or electroshock: effects of anticonvulsant agents. *Mol Pharmacol* 14: 347–356
172 McCandless DW, Feussner GK, Lust WD, Passoneau JV (1979) Metabolite levels in brain following experimental seizures: the effects of isoniazid and sodium valproate in cerebellar and cerebral cortical layers. *J Neurochem* 32: 755–760
173 Maes M, Calabrese J, Jayathilake K, Meltzer HY (1997) Effects of subchronic treatment with valproate on L-5-HTP-induced cortisol responses in mania: Evidence for increased central serotonergic neurotransmission. *Psychiatry Res* 71: 67–76

174 Nathanson JA (1977) Cyclic nucleotides and nervous system function. *Physiol Rev* 57: 157–256
175 Trabucchi M, Cerri C, Spano PF, Kumakura K (1977) Guanosine 3′,5′-monophosphate in the CSF of neurological patients. *Arch Neurol* 34: 12–13
176 Nau H, Löscher W (1984) Valproic acid and metabolites: Pharmacological and toxicological studies. *Epilepsia* 25(1): 14–22
177 Semmes RLO, Shen DD (1991) Comparative pharmacodynamics and brain distribution of E-Δ^2-valproate and valproate in rats. *Epilepsia* 32: 232–241
178 Löscher W (1992) Pharmacological, toxicological and neurochemical effects of $\Delta^{2(E)}$-valproate in animals. *Pharm Weekblad* 14: 139–143
179 Löscher W, Böhme G, Schäfer H, Kochen W (1981) Effect of metabolites of valproic acid on the metabolism of GABA in brain and brain nerve endings. *Neuropharmacology* 20: 1187–1192
180 Wamil AW, Löscher W, Mclean MJ (1997) Trans-2-en-valproic acid limits action potential firing frequency in mouse central neurons in cell culture. *J Pharmacol Exp Ther* 280: 1349–1356
180 a Löscher W, Hönack D (1996) Valproate and its major metabolite E-2-en-valproate induce different effects on behaviour and brain monoamine metabolism in rats. *Eur J Pharmacol* 299: 61–67
181 Löscher W, Nau H (1982) Valproic acid: metabolite concentrations in plasma and brain, anticonvulsant activity, and effects on GABA metabolism during subacute treatment in mice. *Arch Int Pharmacodyn Ther* 257: 20–31
182 Frey H-H, Löscher W (1978) Distribution of valproate across the interface between blood and cerebrospinal fluid. *Neuropharmacology* 17: 637–642
183 Adkison KDK, Shen DD (1996) Uptake of valproic acid into rat brain is mediated by a medium-chain fatty acid transporter. *J Pharmacol Exp Ther* 276: 1189–1200
184 Cutrer FM, Limmroth V, Moskowitz MA (1997) Possible mechanisms of valproate in migraine prophylaxis. *Cephalalgia* 17: 93–100

Milestones in Drug Therapy
Valproate
ed. by W. Löscher

Chemistry and biotransformation

Frank S. Abbott and M. Reza Anari

Faculty of Pharmaceutical Sciences, The University of British Columbia, 2146 East Mall, Vancouver, B.C., Canada V6T 1Z3

Introduction

Since the serendipitous discovery of the anticonvulsant activity of valproic acid (VPA) by Munier in 1963, and its introduction to the clinic in 1967, the study of the metabolic fate and diverse pharmacological activity of VPA metabolites have progressively continued for three decades. Up to fifty VPA metabolites were identified by 1987 [1], with a few of these demonstrating significant biological activity. As a result, VPA metabolites have been studied in detail not only with respect to their antiepileptic potential ([2] and see below), but also as model compounds to characterize the structural requirements for induction of teratogenic or hepatotoxic effects (see Radatz and Nau, this volume) and also to elucidate the biochemical nature of novel pathways of biotransformation, e.g. cytochrome P450-mediated desaturation and (ω-2)-hydroxylation [3]. In light of the need for rational drug-design of new efficacious antiepileptic drugs having minimal side-effects and free of the undesired interactions with other antiepileptic agents, structure-activity relationship studies of VPA analogues and metabolites have played a pivotal role towards the development of new antiepileptic agents [4–7]. The current chapter describes the chemistry and structure-activity relationships (SAR) of VPA analogues and related metabolites, followed by an overview on the metabolic pathways involved in VPA biotransformation.

Chemistry

Valproic acid (2-propylpentanoic acid, dipropylacetic acid) (Fig. 1) is a C-8 branched chain fatty acid having a molecular weight of 144.2 g mol^{-1}. The pure acid is a colorless liquid (bp 221–222 °C) with a characteristic odor that is only slightly soluble in water but is highly soluble in organic solvents (log P = 2.72–2.75) [8, 9]. The sodium salt is very soluble in water and in some organic solvents (e.g. methanol and acetone). A common therapeutic form of the drug in the US is divalproex sodium (Depakote, Epival), a stable coordination compound derived from sodium valproate and

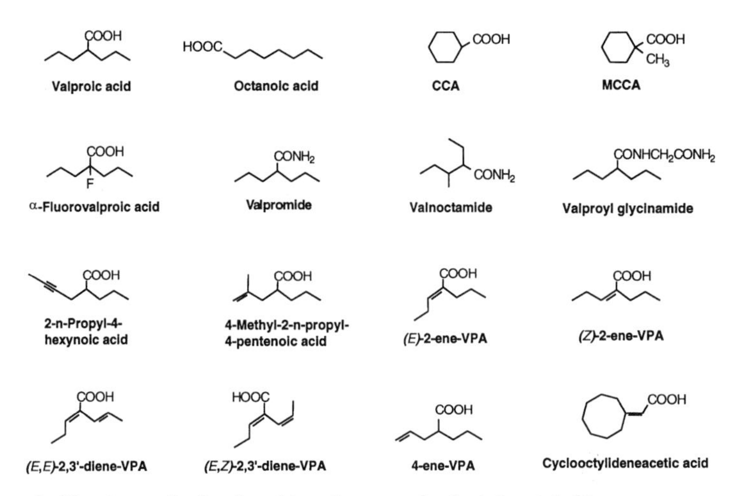

Figure 1. Structures of valproic acid analogues and selected metabolites.

valproic acid in a 1:1 molar ratio, the chemical form being described as sodium hydrogen-bis-(2-propylpentanoate). The free acid (pKa = 4.56–4.8) [9] and sodium salt forms are stable compounds. Further details of physical constants and spectroscopic details of valproic acid can be found in the monograph by Chang [10]. Because valproic acid has no significant chromophore (UV_{max} = 213 nm, ε = 86) it is not a good candidate for sensitive high performance liquid chromatography (HPLC) detection. Most of the analytical methods used to effectively assay valproic acid for research purposes and to study its metabolism have centered around gas chromatography (GC) [11, 12], gas chromatography-mass spectrometry (GC-MS) [13–22] and liquid chromatography-mass spectrometry (LC-MS) [23, 24] techniques.

The chemical properties of the phase I metabolites of valproic acid are not as well documented and the characterization of metabolites frequently relies on chromatographic and mass spectral data. Where the chemistry is described the metabolites are typically more polar as indicated by their measured log P values and the acidic compounds have pKa values similar to that of valproic acid [25]. Like valproic acid, the metabolites are liquids or low melting solids [9, 12, 26]. In contrast to valproic acid, the lack of stability by some of the metabolites presents interesting problems with respect to their analysis. The 4-hydroxy-VPA metabolite (4-HO-VPA) readily forms a γ-lactone [12] and some labs find it difficult to obtain a proper derivative of 4-HO-VPA [14, 22]. The 5-hydroxy-VPA (5-HO-VPA) also forms a lactone, but only in strong acid. One must be careful to control conditions when extracting and derivatizing the 3-keto-VPA metabolite that readily undergoes heat and acid catalyzed decarboxylation to 3-heptanone. A stable isotope labeled internal standard of 3-keto-VPA is a valuable asset to reliable analysis [17, 22] and earlier measurements of this metabolite in blood and urine likely underestimated the true values [27]. Attempts to convert the keto function to an oxime may lead to the formation of artifacts because under the derivatization conditions, valproic acid partially converts to a hydroxamic acid [28]. A recent assay claims good precision for 3-keto-VPA under controlled conditions without the use of a labeled internal standard [14].

Structure activity relationships

A discussion of the chemistry of this unique drug and its metabolites would not be complete without addressing the effects of even subtle changes on the structure with respect to the pharmacological effects. In this context, the *in vivo* metabolism of VPA to numerous metabolites, bearing various functional groups at different positions on the side chains together with configurational consideration of these molecules has provided a rich source of compounds for SAR studies, which will be addressed following the parent drug.

Structure activity relationships of valproic acid

Side-chain length and substitution

Anticonvulsant activity: In an early study of anticonvulsant activity against pentylenetetrazole-induced seizures in mice, the side-chain length of pentanoate derivatives branched at the 2-position was significantly correlated with the anticonvulsant activity [29]. Analogues with side chains larger than valproic acid, i.e. C-9, were more potent than VPA but neurotoxic effects such as sedation and ataxia were more frequent. Straight chain analogues were poorly active. These results and the balance of activity over the side-effects appeared to establish valproic acid as the optimal structure [30, 31]. Similar results were obtained for valproate analogues tested for anticonvulsant activity against audiogenic seizures in DBA/2 mice, larger molecules showing greater potency [32]. In a subsequent study where systemic and intracerebroventricular administration were compared, the effect of side chain length on potency no longer held [33]. This suggested that the larger molecules simply enjoyed a more efficient cerebral uptake rather than exhibiting some mechanistic advantage intracellularly or by acting on lipophilic membranes. Similar results have been obtained by Perlman and Goldstein [34] who found that on the basis of brain concentrations, the straight chain heptanoic and octanoic acids were both more potent than VPA. This further emphasizes the influence of drug distribution and disposition in the whole animal even when comparing compounds of closely related structures. A good example of this can be found in the series of articles by Liu and Pollack [4, 35, 36] in which the pharmacokinetics and pharmacodynamics of valproic (VPA), octanoic (OA), cyclohexanecarboxylic (CCA), and 1-methyl-1-cyclohexanecarboxylic (MCCA) acids were compared in rats (see Fig. 1 for structures). All four compounds demonstrated nonlinear disposition but apparently by different mechanisms. VPA and MCCA, which share α-branching in their structures, exhibit enterohepatic recirculation and are excreted into urine as their glucuronides. OA and CCA displayed much higher clearances but did not exhibit enterohepatic recirculation nor were they excreted as base labile conjugates. Watkins and Klaassen [37] had earlier shown that the choleretic activity of close structural analogues of valproate appears to parallel their ability as anticonvulsants. Structures that did not have α-branching were not choleretic. Returning to VPA, CCA, and MCCA, only VPA was found to exhibit a significant dissociation between its pharmacokinetics and pharmacodynamics by demonstrating a marked delay in the production of maximal anticonvulsant activity [36].

Teratogenicity: Several papers have described the structure activity relationships of valproate analogues and metabolites with respect to their teratogenic potential [1, 38–43] (see Radatz and Nau, this volume, for details). The structural features required for teratogenic properties were a free carboxyl group, branching at C-2 and the presence of a proton on C-2. For example, the metabolite of VPA with a double bond at C-4 retained the teratogenicity while 2-ene-VPA was free of this side-effect [40, 42]. Anticonvulsant and neurotoxic properties of VPA analogues were unrelated to their teratogenic potential [44]. Stereoselectivity was evident in the teratogenic properties of individual isomers but was absent for anticonvulsant and neurotoxic effects. Valproate analogues substituted with an additional methyl group at positions 2, 3, or 4 exhibit greatly reduced teratogenicity and embryolethality but retain high anticonvulsant activity. An example is 4-methyl-2-n-propyl-4-pentenoic acid (Fig. 1) [38].

Quantitative structure activity relationship (QSAR) studies: QSAR studies of valproate derivatives have been limited. Abbott and Acheampong [25] examined a broad array of carboxylic acids and tetrazoles for anti-PTZ activity in mice using Taft's steric parameter, log P, $(\log P)^2$ and pKa as physicochemical descriptors. The best correlation was found for log P and pKa, indicating optimal activity with increasing lipophilicity and decreasing acidity. Others have found this trend of lipophilicity and potency but it is not a strict correlation [38, 45]. For example 2-n-propyl-4-hexynoic acid (Fig. 1) was equivalent to valproate in potency but had a much lower log P (1.84) [45].

Fluoride substitution
Alpha-fluorovalproic acid (Fig. 1) showed interesting differences from VPA when studied in mice [7]. The α-fluoro group has a marked effect on the acidity of the carboxyl group (pKa = 3.55) and metabolism and distribution characteristics of α-fluoro-VPA were quite dissimilar to VPA. Very little of the α-fluoro-VPA is excreted as the glucuronide and with normal β-oxidation blocked by the fluoro group a substantial amount of the drug is recovered in the urine as the glutamine conjugate. Like VPA, the α-fluoro analogue appears to share the properties of asymmetric transport to brain but differs in achieving peak brain levels 45 min later than in the serum while VPA peaks in both serum and brain within 15 min. The α-fluoro-VPA is not as potent as VPA but is sufficiently active to be of future interest as a potential anticonvulsant drug [7].

Amide substitution
Of 32 analogues and metabolites of VPA tested, Löscher and Nau [30] showed that valpromide (Fig. 1) was the most active. Amides tend to be more potent than their corresponding acids having less serum protein binding and greater CNS penetration through, presumably, a passive process. In man valpromide is extensively hydrolyzed to VPA and is thus a potential teratogen. Valpromide is also a selective inhibitor of human microsomal epoxide hydrolase [46]. In seeking to prepare more stable amide analogues of VPA, Haj-Yehia et al.[47] found that branching in both the α- and β-positions of pentanamide (e.g. valnoctamide, Fig. 1) was necessary to guarantee stability to hydrolysis *in vivo*. More recently valproyl glycinamide (Fig. 1) has been reported to be more potent than VPA by virtue of better brain partitioning [48, 49]. Some biotransformation of the glycinamide derivative to the active metabolite, valproyl glycine occurred but no VPA was detected. Other variations on the valproate structure have been spiro analogues [50] and esters of the 3- and 4-amino derivatives [51].

Structure activity relationships of valproic acid metabolites

The investigation of metabolites for activity is a natural outcome for most drugs that are extensively metabolized and VPA is no exception. The promise of an alternative drug to VPA that retains the broad spectrum, potency and low neurotoxicity of VPA but is free of the teratogenic and hepatotoxic potential was a worthy goal for investigating the activity of VPA metabolites. Metabolite activity was also thought to explain the biphasic onset and prolonged effects that are characteristic of the pharmacodynamics of VPA anticonvulsant activity. A likely candidate to possess the sought after ideal properties was the major serum metabolite (*E*)-2-ene-

VPA, the investigation of which fuelled the research activities of a number of laboratories for several years. The apparent failure to bring 2-ene-VPA to fruition means that the quest for the consummate "valproate alternative" continues today. The first extensive study of metabolite activity was reported by Löscher in 1981 [2] followed by a more systematic study in 1985 [30]. Depending on the test, all 11 metabolites showed some degree of anticonvulsant activity although none was as potent as VPA. Of most interest were the mono-unsaturated metabolites (2-, 3- and 4-ene-VPA), the position and configuration of the double bond influencing the level of activity. The position of the double bond was reported to influence the serum protein binding of these metabolites with the 2-position most highly favored for binding of the unsaturated compound [52]. Double bond position also affected the asymmetric transport across the blood brain barrier, a factor significant to potency considerations. Di-unsaturated metabolites (2,4′- and 4,4′-diene-VPA) tend to be less potent [30] although the 2,3′-diene-VPA isomers have potencies comparable to 2-ene-VPA [9, 25]. Considering that only (*E*)-2-ene-VPA was non-teratogenic [53] and was reported to lack the hepatotoxic potential of 4-ene-VPA [54, 55], significant attention was directed to this metabolite. Interestingly, the anticonvulsant activity of (*E*)-2-ene-VPA has a somewhat mixed history with reported potencies ranging from 50% to more than 2 fold that of VPA. Testing results from various investigators are summarized in Table 1. The variability in observed potencies was difficult to explain although the purity of the compound, the strain of the rodent model used, the method of administration (ip vs. oral) [56] and the timing of the test [57] may partially explain the discrepancies. Even though (*E*)-2-ene-VPA was more neurotoxic than VPA in most of the animal models evaluated, the protective index was determined to be comparable to VPA [58].

The metabolic profile of (*E*)-2-ene-VPA in human microsomes [59] indicated that substantial amounts of 4-ene-VPA and 2,4-diene-VPA were formed, both of which are known to be hepatotoxic in rats [54]. This result alone may have deterred further clinical investigations of (*E*)-2-ene-VPA. The interest in (*E*)-2-ene-VPA as a potential alternative to valproate prompted a quantitative structure activity study of structurally related unsaturated analogues [9]. Although shape parameters were calculated by molecular modeling techniques, only volume and lipophilicity were found to strongly correlate with potency. A C-10 analogue, cyclooctylideneacetic acid (Fig. 1) was significantly more potent than (*E*)-2-ene-VPA and exhibited but modest sedation. Thus there would appear to be further opportunities for the development of 2-ene-VPA analogues that dissociate neurotoxicity from the anticonvulsant action.

Stereochemistry

Valproic acid itself is a symmetrical molecule and does not contain a chiral carbon. The metabolic introduction of a hydroxyl group at C-3, 4, or 5 or a

Table 1. Comparisons of the anticonvulsant activity of VPA and its major serum metabolite (*E*)-2-ene-VPA in various animal models

Species	Test [a]	Potency (ED50)		Potency (*E*)-2-ene VPA/VPA	Ref.
		VPA	2-ene-VPA		
Mice	scPTZ	70% Protection	65% Protection	1.0	[169]
Gerbils	Air Blast	73 [b]	90 [b]	0.8	[53]
Rats	Spontaneous	81 [b]	66 [b]	1.2	[53]
Mice	scPTZ	325 [b]	225 [b]	1.4	[53]
Mice	MES	250 [b]	220 [b]	1.1	[53]
Mice	scPTZ	1.03 [c]	2.05 [c]	0.5	[8]
Mice	MP	0.71 [c]	1.69 [c]	0.4	[8]
Mice	Bic	1.06 [c]	2.16 [c]	0.5	[8]
Mice	Pic	0.78 [c]	1.67 [c]	0.5	[8]
Mice	MES	4.35 [c]	5.3 [c]	0.8	[8]
Mice	scPTZ	240 [b]	220 [b]	1.1	[30]
Mice	MES	210 [b]	230 [b]	1.1	[30]
Mice	scPTZ	0.7 [c]	1.46 [c]	0.5	[25]
Mice	i.v. PTZthres	113 [b]	80 [b]	1.4	[58]
Mice	MESthres	69 [b]	54 [b]	1.3	[58]
Mice	MES	320 [b]	265 [b]	1.2	[58]
Mice	scPTZ	160 [b]	110 [b]	1.5	[58]
Rats	MES	140 [b]	140 [b]	1.0	[58]
Rats	scPTZ	195 [b]	153 [b]	1.3	[58]
Dogs	PTZthres			1.3	[58]
Rats	PTZthres	167 [b]	109 [b]	1.5	[56]
			Intrinsic potency [d]	2.4	[56]
Rats	A-K Generalized	190 [b]	85 [b]	2.2	[170]
Rats	A-K Afterdischarge	300 [b]	210 [b]	1.4	[170]
Mice	scPTZ	0.83 [c]	0.84 [c]	1.0	[9]
			Intrinsic potency [d]	2.3	[9]

[a] scPTZ, subcutaneous pentylenetetrazole; PTZthres, pentylenetetrazole threshold; MES, maximal electroshock seizures; MESthres, maximal electroshock seizures threshold; MP, mercaptopropionic acid; Bic, bicuculline; Pic, picrotoxin; A-K, Amygdala-Kindled.
[b] mg/kg.
[c] mmol/kg.
[d] Intrinsic potency is based on the whole brain drug concentration at ED50.

double bond at C-3 or C-4 creates a chiral center at C-2. In the case of 3- and 4-hydroxy VPA additional chirality is introduced to produce diastereomers that can be separated readily by traditional GC [12, 22, 60]. The configuration, *E* or *Z*, of the double bonds at C-2 or C-3 needs also to be considered. Teratogenic effects show a distinct stereoselectivity with the metabolite 4-ene-VPA and its analogue 4-yne-VPA demonstrating greater potency of the S-isomer that is not related to pharmacokinetic differences [61] and for 4-yne-VPA is evident across species [62]. For anticonvulsant activity, Löscher and Nau [30] demonstrated that (*E*)-2-ene-VPA was 50% more potent than its Z-isomer (Fig. 1) and such a configurational preference for potency was confirmed with (*E*)- and (*Z*)-2-pentenoic acid [9]. Striking differences in potency and neurotoxicity were evident for the (*E,E*)- and

(*E,Z*)-isomers of the 2,3′-diene-VPA metabolites (Fig. 1) where the configuration differs about the C-3 double bond. The (*E,Z*)-isomer was slightly more potent than VPA and showed no sedative effects. By contrast the (*E,E*)-isomer was poorly active and displayed extreme muscle rigidity in the animals tested [9]. Such differences in both the qualitative and quantitative profiles of the isomers might be explained by the degree of conformational flexibility of the isomers or require some form of specific receptor interaction.

Metabolic pathways of valproic acid

Unlike its simple molecular structure, valproic acid metabolism is extremely complex and a variety of overlapping phase I and II metabolic pathways are known to catalyze its biotransformation. The small molecular size and structural similarity of various VPA metabolites along with the presence of several stereo- and geometric-isomers of many metabolites have added a significant level of complexity to the identification and profiling of numerous VPA metabolites *in vivo*. A milestone in the characterization of VPA metabolites was made largely as a result of the introduction of sensitive and specific analytical methods based on combined gas chromatography-mass spectrometry systems in the 1970s. The advent of GC-MS analytical methodologies [12, 13, 16, 18, 19, 21, 22, 60, 63, 64] along with the application of stable isotope techniques [60, 65–67] has greatly facilitated the identification of numerous VPA metabolites and their diverse metabolic pathways in various animal models, e.g. rats [67–78], mice [69, 75], rabbits [68, 75, 79], dogs [68, 69], monkeys [80–83], and also humans [27, 68–70, 75, 84]. While an earlier literature review by Gugler and von Unruh [85] covers the relevant metabolic studies of VPA published before 1980, recent reviews by Zaccara [86] and Baillie [87] deal with the literature until 1990, and the most recent comprehensive reviews by Baillie [88, 89] and other investigators [90, 91] cover research prior to 1995. Based on these investigations of VPA metabolism, it is safe to say for phase I biotransformation that VPA is primarily metabolized by the enzyme systems normally reserved for the biotransformation of endogenous fatty acids, e.g. the β-oxidative and ω- and (ω-1)-oxidative pathways. Metabolites formed in these pathways are found predominantly in their free forms in plasma, whereas the glucuronide and mercapturic acid conjugates of VPA metabolites and of the parent drug are found mainly in urine and bile.

Advances in molecular biological techniques along with the availability of sensitive LC-MS analytical systems have recently provided an extra level of insight into the identification of polar VPA conjugates [23, 24, 78, 92–97] and the characterization of specific enzymes involved in the metabolic fate of VPA [59, 77–79, 84, 92, 96]. The discussion that follows will address the most recent literature on VPA metabolites according to the biochemical pathways from which they are most likely derived, starting with the phase I oxidative metabolic pathways followed by phase II conjugation reactions.

Phase I biotransformation of VPA

β-Oxidation: β-Oxidation takes place primarily in mitochondria and plays a key role in the oxidative catabolism of fatty acids to acetyl-coenzyme A (acetyl-CoA) derivatives [72]. Being a medium-chain fatty acid in nature, VPA has long been considered to be a substrate for the fatty acid β-oxidation pathway. Thus, metabolites of VPA such as 2-ene-VPA [12, 19, 98–102], 3-hydroxy-VPA (3-HO-VPA) [12, 69, 72, 98–100, 103], and 3-keto-VPA [12, 27, 70, 72, 98, 99, 103] were assumed to result from the mitochondrial metabolism of VPA based on their structural resemblance to those of the fatty acid metabolites (Fig. 2). In this context, enhanced or reduced formation and excretion of 2-ene-VPA and 3-keto-VPA resulting from the use of various β-oxidation inhibitors or inducers or under a different nutritional status [72, 102, 104–106] were considered as further support for the β-oxidation pathway in VPA biotransformation. On these grounds VPA was also shown to compete with endogenous long-chain fatty acids for β-oxidation enzymes [106, 107] and inhibit fatty acid oxidation in the liver [108, 109]. Such an interaction was suggested to account for the increased excretion of C6-C10 dicarboxylic acids in the urine of epileptic patients treated with VPA [110].

Stable isotope-labeling techniques [67] have demonstrated that the 3-HO-VPA is not an exclusive product of β-oxidation of VPA but has a dual origin *in vivo,* being derived in part by direct cytochrome P450-dependent hydroxylation of the parent drug [111]. The relative contributions of β-oxidation enzymes and cytochrome P450 to the formation of 3-HO-VPA in humans is yet to be established (see below). Stable isotope-labeling studies also demonstrated that 3-ene-VPA is formed reversibly from 2-ene-VPA by isomerization and that further desaturation of 3-ene-VPA during β-oxidation gives rise to the diene structure [67, 76, 112]. The experiments with authentic metabolites incubated with isolated mitochondrial preparations showed that (*E*)-2-ene-VPA, 3-ene-VPA, and (*E,E*)-2,3′-diene-VPA are interconverted by isomerization and reduction processes and that all three serve as precursors of 3-keto-VPA (Fig. 2) [76]. The stereospecificity of VPA β-oxidation with respect to the geometric isomers was similar to that of the fatty acid β-oxidation and favored the formation of the *E* isomer (previously named as *trans*). Similarly, the desaturation of the chiral ^{13}C-labeled analogs of VPA exhibited a slight preference for oxidation on the *pro*-S side-chain which was evident for (*E*)-2-ene-VPA, 3-ene-VPA, 3-HO-VPA, and 3-keto-VPA [112]. These observations confirm a common origin of the above-mentioned metabolites and their interconversion.

A milestone in the characterization of the complex enzymatic process involved in VPA β-oxidation has been achieved recently by utilizing mitochondrial subcellular fractions and purified β-oxidation enzymes [76, 90, 92, 112, 113]. These studies provided direct evidence for the functional role of individual enzymes involved in the process of VPA β-oxidation, which appeared to be different in some steps from those reported for endogenous straight-chain fatty acids [92, 113]. Incubation of VPA with coupled rat liver mitochondria fortified with appropriate cofactors resulted in the formation of (*E*)-2-ene-VPA and 3-keto-VPA as major metabolites along with the unsaturated derivatives 3-ene-VPA and (*E,E*)-2,3′-diene-VPA [76]. Furthermore, VPA-CoA, (*E*)-2-ene-VPA-CoA, 3-HO-VPA-CoA, and 3-keto-VPA-CoA were separated from the incubations of freshly isolated rat liver mitochondria [92]. With the aid of purified mitochondrial enzymes, the first step of the β-oxidation reaction, i.e. the conversion of VPA-CoA to (*E*)-2-ene-VPA-CoA, was shown to be mediated selectively by 2-methyl-branched-chain acyl-CoA dehydrogenase [113]. This reaction, which appeared to be highly stereoselective for endogenous fatty acids, showed a low degree of stereoselectivity for the conversion of VPA to (*E*)-2-ene-VPA, (*E*)-3-ene-VPA, 3-HO-VPA and 3-keto-VPA [112]. The subsequent step which involves hydration of (*E*)-2-ene-VPA-CoA to 3-hydroxy-VPA-CoA has been shown to be catalyzed by enoyl-CoA hydratase (crotonase) [92]. In the third step of β-oxidation, unlike the 3-hydroxy-fatty acid intermediates, 3-hydroxy-VPA-CoA was not a substrate for soluble mitochondrial or peroxisomal L-3-hydroxyacyl-CoA dehydrogenases, but surprisingly was oxidized to 3-keto-VPA-CoA by a novel membrane-bound NAD^+-specific 3-hydroxyacyl-CoA dehydrogenase [92]. Finally, 3-keto-VPA-CoA appeared to be resistant to cleavage by 3-keto-acyl-CoA thiolase, an enzyme that catalyzes the final step of fatty acyl-CoA hydrolysis in the β-oxidation pathway [92]. The slow degradation of acyl-CoA metabolites of VPA in the mitochondrial matrix, presumably by hydrolysis [92], was suggested to result in the depletion of free CoA, thereby inhibiting β-oxidation [114]. The reversible binding of valproate metabolites, e.g. 3-keto-2-VPA-CoA, to one or several of the β-oxidation enzymes has also been suggested as an alternate mechanism to explain the inhibition of β-oxidation by VPA [114]. In this context, valproate and other substituted branched-chain carboxylic acids have not only been used recently as tools to investigate the regulation of the β-oxidation pathway, but also as potential hypoglycemic drugs [114, 115].

It has been claimed that the administration of VPA to rats results in the proliferation of peroxisomes [116, 117] that would lead to the enhanced peroxisomal β-oxidation of VPA [118]. More recent evidence appears to confirm that peroxisomes are not significantly involved in the oxidation of either VPA or its metabolites [116, 119, 120].

The β-oxidation pathway may play a primary role in the toxicological properties of VPA, where 4-ene-VPA, a known hepatotoxin [54], is further meta-

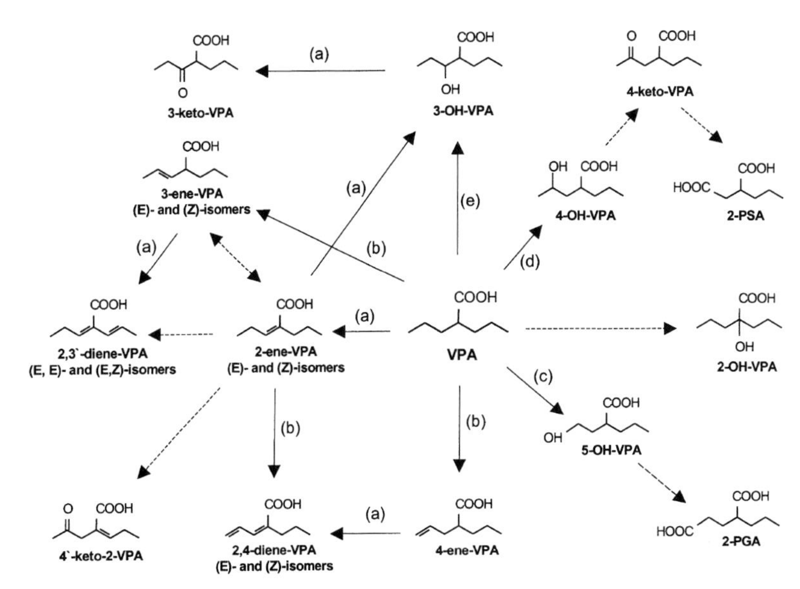

Figure 2. Summary of phase I metabolic pathways for frequently observed valproate metabolites in human. The putative characterized enzymatic pathways are as follows: (a) β-oxidation; (b) P450-dependent desaturation; (c) P450-dependent ω-hydroxylation; (d) P450-dependent (ω-1)-hydroxylation; (e) P450-dependent (ω-2)-hydroxylation. The broken lines indicate a metabolic route in which the details are not yet confirmed.

bolized by β-oxidation [76], the resulting diunsaturated acyl-CoA derivative and the α,β-unsaturated ketone acyl-CoA compound are both highly reactive products [88, 94, 121, 122]. The primary route of biotransformation for 4-ene-VPA, in fact, was shown to be β-oxidation [83, 121] and considerable levels of di- and keto-unsaturated reactive metabolites were found to be formed through the β-oxidation of 4-ene-VPA [23, 94, 95] (Fig. 2). To illustrate the significance of β-oxidation to the metabolic activation of 4-ene-VPA, the metabolism of fluorinated analogues of 4-ene-VPA, known to be resistant to β-oxidation due to the lack of an α-hydrogen, was studied in rats [7, 95, 97]. The α-fluoro-4-ene-VPA did not undergo biotransformation to unsaturated reactive metabolites, nor did it show any microvesicular steatosis or glutathione depletion that is common to 4-ene-VPA [7, 95, 97].

Desaturation: Mono-unsaturated metabolites have been identified in the plasma and urine of patents and animals given VPA since early metabolic studies [98, 99]. The first unsaturated metabolite was characterized as 2-ene-VPA [98], which is known to be the precursor of other desaturated metabolites such as 3-ene-VPA, and 2,3′-diene-VPA [76] (Fig. 2). As discussed earlier, the *E* (major) geometric isomers of the above-mentioned metabolites are known to be the product of VPA β-oxidation, although it is not clear whether the *Z* (minor) geometric isomers are also formed through this pathway.

Further profiling of VPA metabolites in epileptic children resulted in the identification of a terminal desaturated olefin, 4-ene-VPA [99] (Fig. 2). Even though this desaturated metabolite was found to be a minor metabolite in the plasma and urine of experimental animals and humans [73, 83], 4-ene-VPA has received considerable attention because terminal olefins are generally known as precursors to reactive toxic species, capable of binding covalently to cellular nucleophiles and inactivating various enzymes [123]. Indeed, 4-ene-VPA was found to be the most toxic metabolite of VPA in isolated hepatocytes [124], a potent inducer of hepatic microvesicular steatosis in rats [54, 125], and clearly a teratogenic compound in mice [42]. *In vitro* studies showed that 4-ene-VPA gives rise to covalent binding in liver tissues [126] and inhibits both hepatic cytochrome P450 [74] and fatty acid β-oxidation [109]. Thus, the possible involvement of the terminal desaturation pathway, termed the 4-ene pathway, in the etiology of VPA-induced idiosyncratic hepatotoxic reaction [127, 128] has stimulated many researchers to investigate the mechanism of this bioactivation reaction as well as the nature of the enzymes involved in side-chain terminal desaturation [88, 89, 122].

A novel discovery in VPA biotransformation was the finding that terminal desaturation of VPA to give 4-ene-VPA is a unique reaction catalyzed by cytochrome P450 (P450) [3]. This bioactivation reaction was shown to be catalyzed by rat, rabbit, mouse, and human liver microsomes and by purified cytochrome P450 2B1 in a reaction that was dependent on NADPH and oxygen and was suppressed by cytochrome P450 inhibitors [3, 75]. On the other hand, phenobarbital and a few other anticonvulsants with known cytochrome P450 inducing capabilities, were shown to highly enhance the hepatic microsomal formation of 4-ene-VPA [75]. The desaturation of alkyl hydrocarbons is a rare type of oxidation reaction for cytochrome P450 mixed function oxidase [123] that was initially identified for VPA and testosterone [129], and more recently for lauric acid [130]. The novelty of the reaction in the case of VPA is that, unlike testosterone, the desaturation takes place at a nonactivated position, suggesting its wide applicability to cover a broad spectrum of aliphatic substrates [3, 130].

Investigations of the mechanism of P450-dependent desaturation using inter- and intramolecular deuterium isotope effects have revealed that an initial hydrogen atom abstraction from the (ω-1)-position to form a carbon-centered free radical at C-4 is an essential common rate limiting step for the formation of both 4-HO-VPA and 4-ene-VPA products [75]. The subsequent balance between oxygen rebound or hydrogen atom removal would then determine the relative formation of 4-HO-VPA or 4-ene-VPA, respectively [75]. Studies on the stereospecificity of the reaction by utilizing (R)- and (S)-[3-^{13}C]-VPA enantiomers demonstrated that the *pro*-(R)-side chain is preferentially desaturated in intact hepatocytes [131], unlike the β-oxidation desaturated products that showed only slight preference for the *pro*-(S)-side chain [112].

With respect to the nature of the P450 enzymes that catalyze 4-ene-VPA formation, initial studies on subcellular fractions demonstrated that rat [3] and rabbit [75] CYP2B isoforms as well as rabbit CYP4B1 [79] showed significant catalytic turnover for VPA terminal desaturation. In contrast, the lauric acid ω-hydroxylases, CYP4A1 and CYP4A3, did not give rise to any detectable level of 4-ene-VPA [79]. In light of the recent advances in molecular biology techniques and the availability of individual human cDNA-expressed P450 isoforms, it has become possible to identify human cytochrome P450 enzymes capable of bioactivating VPA to its terminal desaturated olefin [84]. These experiments demonstrated that multiple human P450 enzymes are involved in desaturating VPA and include CYP2C9, and CYP2A6 [84] as well as CYP2B6 (Anari and Abbott, unpublished observations).

Besides the monounsaturated metabolites discussed above, a few diene derivatives of VPA have been detected in the plasma and urine of patients under VPA therapy [132, 133]. (*E,E*)-2,3′-VPA was shown to be the major diene metabolite in both humans and animals [67, 73, 132, 133] and is formed largely by further β-oxidation of 3-ene-VPA [67, 76] (Fig. 2). Significant amounts of (*E*)-2,4-diene-VPA have also been found, which is known to arise as a secondary metabolite of 4-ene-VPA in rat [23, 24, 54, 63, 73, 121] and monkey [82] (Fig. 2). Two independent metabolic pathways appear to contribute to the formation of (*E*)-2,4-diene-VPA; the microsomal cytochrome P450 system that catalyzes the dehydrogenation of 2-ene-VPA to the free acid form of (*E*)-2,4-diene-VPA [77], and the mitochondrial β-oxidation pathway that converts 4-ene-VPA-CoA to the corresponding diene-VPA-CoA thioester [23, 93, 94] (see β-oxidation and GSH conjugation sections). At last, two triene metabolites of VPA have been detected in studies in rats, but these desaturated trienes have yet to be characterized in human [54, 73].

Epoxidation: Theoretically, all the unsaturated metabolites of VPA could undergo epoxidation as a second step of biotransformation, but so far only 4-ene-VPA was shown to be converted by a cytochrome P450-mediated reaction to 4,5-epoxide-VPA [73, 121] (Fig. 3). Using ^{18}O as a stable isotope metabolic tracer, 4,5-epoxide-VPA was shown to undergo an intramolecular rearrangement to yield the stable cyclized product, 4,5-dihydroxy-VPA-γ-lactone [74]. Further proof for the formation of an intermediate epoxide metabolite has been obtained from a recent study that characterized corresponding glutathione conjugates of 4,5-epoxide-VPA in rat bile [94] (see below). Evidence for the *in vivo* formation of GSH adducts of 4,5-epoxide-VPA has raised some concern about the half-life and reactivity of the epoxides which might diffuse away from their site of formation in the endoplasmic reticulum and alkylate critical biomacromolecules [89, 94]. While GSH adducts of 4,5-epoxide-VPA have been formed in the rat, similar NAC metabolites have not been detected in the urine of patients on VPA therapy [134].

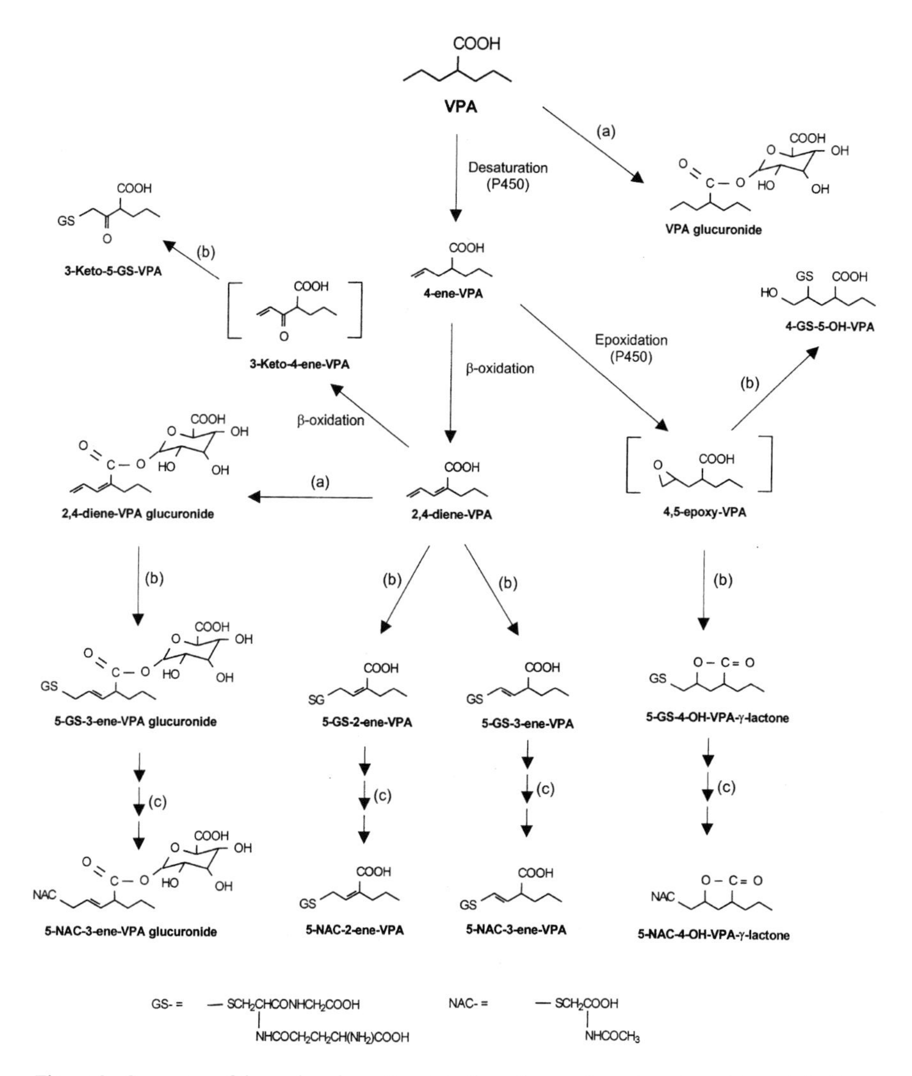

Figure 3. Summary of the major phase II metabolic pathways for valproate and its metabolites. The putative enzymatic pathways are as follows: (a) glucuronidation; (b) glutathione conjugation; (c) mercapturic acid pathway. Compounds shown in brackets have not been isolated but represent proposed intermediates of phase II metabolites. The thioester CoA intermediates were shown in their free acid forms for the sake of simplicity.

Hydroxylations: Valproate is hydroxylated mainly at positions 4 and 5 by the action of cytochrome P450 enzymes [111], which correspond to the ω- and (ω-1) hydroxylation in endogenous fatty acids [68–71, 80, 99, 100] (Fig. 2). These hydroxylated metabolites are known to be formed mainly in liver but other human tissues such as lung, brain, and adrenal glands also participate in valproate hydroxylation [135]. Similar to the 4-ene pathway, the coadministration of cytochrome P450 inducers influences the P450-

dependent hydroxylation of valproate and in fact, the ω-hydroxylated product, 5-HO-VPA, was found in a markedly increased amount in a patient who developed a fulminant hepatitis and was under concomitant phenobarbital-VPA therapy [136]. The occurrence of 3-HO-VPA, may be mediated by either β-oxidation or cytochrome P450-catalyzed (ω-2)-hydroxylation as noted earlier [67, 111, 135]. The origin of 2-hydroxy-VPA, a recently identified hydroxylated metabolite of VPA in human, has yet to be determined [21].

Ketone metabolite formation: Further oxidation of 4-and 5-hydroxy-VPA most likely accounts for the formation of the secondary metabolites 4-keto-VPA [60] and 2-propylglutaric acid (PGA) [27, 68, 70, 98], respectively (Fig. 2). The metabolic source of 3-keto-VPA, a major urinary metabolite of VPA in most species [137], is most probably through the β-oxidation of (*E*)-2-ene-VPA [67, 92], although an alternative source that involves oxidation by cytosolic dehydrogenases cannot be excluded [89].

Dicarboxylic acid metabolite formation: 2-n-Propylglutaric acid (PGA) [71] and 2-n-propylsuccinic acid (PSA) [73] are the two major dicarboxylic acids metabolites formed by further oxidation of 5-hydroxy- and 4-hydroxy-VPA metabolites, respectively [73]. The finding that the terminal olefin metabolite, 4-ene-VPA can also serve as a precursor to PGA in rat liver [121] raises the possibility that this dicarboxylic acid may, in fact, have a dual origin [89].

Miscellaneous pathways: Studies of the biotransformation of mono-desaturated VPA metabolites in animals have revealed the possible involvement of a few uncommon pathways of biotransformation, e.g. hydration and reduction. The hydration of the 4-ene-VPA double bond was suggested as a possible metabolic route to convert this olefin to 5-HO-VPA in rats [73, 121] and monkeys [82]. Sequential reduction-oxidation would also result in the formation of the same metabolite [121]. Reduction of the mono-unsaturated metabolites, i.e. 2-ene- and 3-ene-VPA, to yield the parent drug was reported to be significant in rats as up to 2% of an administered dose of these compounds was recovered in the urine as VPA [73].

The study of the metabolism of 4-ene-VPA in isolated perfused rat liver showed that metabolism occurred on both the unsaturated and saturated side chains of 4-ene-VPA to yield a variety of products, the most significant of which were the unconjugated diene (*E*)-2,4-diene-VPA and the allylic alcohol 3-hydroxy-4-ene-VPA [121]. These metabolites correspond to the first and second intermediates, respectively, of the proposed β-oxidation pathway leading to 3-keto-4-ene-VPA, the putative toxic alkylating species found in samples of either liver perfusate or bile (see GSH conjugation, Fig. 3). The chemical properties of synthetic 3-keto-4-ene-VPA indicate that this α,β-unsaturated ketone is indeed a highly reactive electrophile that readily undergoes Michael-type addition reactions through nucleophilic attack at the olefinic terminus [121]. Studies on the metabolism of 4-ene-VPA in the rhesus monkey, as a close animal model to human, showed that unlike VPA where metabolism *via* the β-oxidation pathway accounts for only 6% of the administered dose, 4-ene-VPA β-oxidation metabolism is enhanced up to 22% of the administered dose [82]. The 4-ene-VPA, therefore, seems to exhibit a marked preference for metabolism by β-oxidation as compared with the parent drug, a property that may contribute (through the formation of 3-keto-4-ene-VPA) to the high potency of this terminal olefin as an inhibitor of fatty acid β-oxidation *in vitro* [109] and *in vivo* [54, 125].

Phase II Biotransformation of VPA and Its Metabolites

Glucuronidation

Conjugation of VPA with D-glucuronic acid results in the formation of the corresponding ester glucuronide [72] (Fig. 3), a major VPA metabolite in most animals [15] and human [98, 138]. The proportion of a dose of VPA in humans metabolized by the glucuronidation pathway appears to increase with increasing dose at the expense of the oxidative pathways, particularly, β-oxidation [138, 139].

The glucuronide conjugate is excreted into urine but is also found at high concentrations in the bile of rats given VPA [15, 140]. The biliary excretion of VPA glucuronide in the rat is consistent with the enterohepatic recycling of VPA in this species [141, 142] and may also explain the marked choleretic activity of VPA due to the osmotic properties of VPA-glucuronide [143, 144]. Inhibitors of glucuronide conjugation *in vivo* prevented VPA-induced choleresis [143–145]. Conversely, inducers of glucuronide conjugation such as phenobarbital [146, 147] did not increase bile flow or the excretion of VPA-glucuronide [143].

The valproate conjugation with glucuronic acid has been shown to occur from the action of hepatic microsomal UDP-glucuronosyltransferase enzymes [147]. A high metabolic capacity was observed in this *in vitro* system for valproate glucuronidation, which in fact inhibited the conjugation of other drugs in a competitive fashion [147]. The nature of specific UDP-glucuronosyltransferase isoforms involved in valproate conjugation has yet to be reported.

VPA-glucuronide was characterized directly by GC-MS, which indicated the presence of the 1-*O*-acyl-β-linked ester structure [72]. VPA-glucuronide is like other acyl glucuronide conjugates, which are intrinsically reactive molecules, capable of undergoing a number of reactions including hydrolysis, rearrangement, and covalent binding to proteins [148]. Hydrolysis, catalyzed by hydroxide ion, β-glucuronidases, esterases, or serum albumin, leads to regeneration of the pharmacologically active drug. Rearrangement of the acyl moiety of valproate glucuronide occurs in a pH-dependent fashion to yield the β-glucuronidase-resistant 2-, 3-, and 4-*O* positional isomers [149]. Consequently, care is necessary to delineate appropriate conditions of handling and storage to prevent such rearrangements when β-glucuronidase is used to cleave the conjugate [149]. The valproate rearranged isomers, unlike the acyl glucuronide itself, may exist transiently in the open chain form of the sugar ring and thereby may react non-covalently with protein amino groups *via* the exposed aldehyde group [140, 148]. Alternatively, the acyl glucuronide itself may interact directly with nucleophilic –SH, –OH, and $–NH_2$ groups on protein, with loss of the glucuronic acid moiety [148]. Recent studies have shown that VPA-glucuronides give rise to covalent adducts with proteins [150], although these conjugates appeared to be only weakly immunogenic in humans [151]. The toxicological significance of β-glucuronidase resistant forms that show different disposition in rats [152] and were found at high levels in the plasma of a patient with hepatobiliary and renal dysfunction [153] remains to be established.

It should be noted that glucuronidation also represents an important pathway of biotransformation for primary metabolites of VPA which have been formed through a variety of oxidative phase I pathways [21, 73, 82, 83, 121, 143, 154, 155]. Recent LC/MS/MS analysis of (E)-2,4-diene-VPA metabolites in rats indicated the presence of novel diconjugates characterized as 5-GS-3-ene-VPA-glucuronide and the corresponding 5-NAC-3-ene-VPA-glucuronide [96] (Fig. 3). The amount of biliary 5-GS-3-ene-VPA-glucuronide was seven-fold greater than the corresponding 5-GS-3-ene-VPA, the sum of the two metabolites accounting for 6.6% of the dose [96]. *In vitro* incubation of 2,4-diene-VPA-glucuronide with GSH also led to the formation of 5-GS-3-ene-VPA-glucuronide. This direct evidence of the reactivity of a valproate acyl glucuronide with GSH through a Michael addition mechanism supports the bioactivation potential of the glucuronidation pathway in valproate biotransformation [96]. The toxicological significance of this pathway and its occurrence in human where such a metabolite could potentially deplete cellular GSH pools or react with macromolecules has yet to be investigated.

GSH conjugation

The tripeptide thiol, glutathione (GSH), is known to react with a variety of electrophilic compounds, thus preventing their reaction with essential components of the cell. Considering the hepatotoxic [127, 128] and teratogenic side-effects [42] of VPA, the identification of VPA metabolite: GSH conjugates was of great toxicological interest. Hence, the detection and characterization of GSH adducts of VPA metabolites helped gain significant insight into the structure, mechanism of bioactivation, subcellular site of formation, and the potential target sites of VPA reactive metabolites [23, 24, 78, 93–97, 156, 157].

Although, covalent binding to cellular fractions has been reported to occur during the course of VPA metabolic bioactivation [74, 126], it was not until 1991, when the structure of the first VPA metabolite: GSH adduct was elucidated [23]. Thereafter, the availability of combined liquid chromatography-tandem mass spectrometry (LC-MS/MS), as a powerful analytical technique to characterize polar drug metabolites, has greatly facilitated the study of phase II VPA metabolites [23, 24, 78, 93-97]. The first thiol adduct characterized in the urine of humans given VPA was the N-acetylcysteine (NAC) conjugate, 5-NAC-3-ene-VPA [23] (Fig. 3). NAC-conjugates are formed through the mercapturic acid pathway and are the end products of GSH adduct catabolism. Attempts to characterize the precursor GSH adduct led to identification of the corresponding 5-GS-3-ene-VPA conjugate in the bile of rats treated with 4-ene-VPA, along with its metabolite 5-NAC-3-ene-VPA in the urine [23]. Unlike the ester form, the free acid of 2,4-diene-VPA showed very low reactivity towards GSH *in vitro* [23]. Therefore, considering the high urinary recovery of the NAC conjugate of 2,4-diene-VPA *in vivo*, Kassahun et al. [23] proposed that metabolic activation of this olefin in mitochondria yields the CoA thioester of the conjugated diene, an electrophilic intermediate possibly responsible for the hepatotoxicity associated with 4-ene-VPA.

In this context, a further study of GSH and NAC adducts supported the role of mitochondrial β-oxidation in the metabolic activation of 4-ene-VPA to reactive intermediates [94]. With the aid of ionspray LC-MS/MS, three additional GSH adducts were detected in the bile of 4-ene-VPA treated rats, two of which were characterized as 4-HO-5-GS-VPA-γ-lactone and 3-keto-5-GS-VPA, with the thiol identified tentatively as 4-GS-5-HO-VPA [94](Fig. 3). The most abundant conjugate, 4-HO-5-GS-VPA-γ-lactone, and the minor adduct, 4-GS-5-HO-VPA, were representative products of the unstable 4,5-epoxide of 4-ene-VPA (Fig. 3). The novel 3-keto-5-GS-VPA adduct and the previously identified metabolite in both rats and humans, 5-GS-3-ene-VPA [23], were most likely representative of the products of metabolic activation of 4-ene-VPA by the mitochondrial β-oxidation pathway [94] (Fig. 3). In contrast, when rats were given an equivalent dose of 2-ene-VPA, only 5-GS-3-ene-VPA was detected at low concentrations in bile, consistent with the result of GSH depletion in rat liver homogenates [94]. Therefore, it would appear that β-oxidation of 4-ene-VPA leads to the generation of two reactive intermediates, 2,4-diene-VPA-CoA and 3-keto-4-ene-VPA-CoA, both of which are formed within the mitochondrion as CoA thioester derivatives [23, 89, 94]. When 4-ene-VPA was substituted with a fluoro atom at the α-position to block β-oxidation, no thiol conjugates from the biotransformation of α-fluoro-4-ene-VPA could be detected in the bile and urine of rats [97].

Quantitative analysis of the corresponding valproate NAC conjugates in rat urine indicated that metabolism of 4-ene-VPA through the GSH-dependent pathways accounts for approximately 20% of the administered dose

[94] (Fig. 3). GSH- and NAC-conjugates of (E)-2,4-diene-VPA were identified as the biliary metabolites in 2,4-diene-VPA treated rats along with the GSH-glucuronide di-conjugates, formed at 10% of the level of the corresponding mono-GSH conjugate [24]. Metabolic profiling of VPA patients urines for 5-NAC-3-ene-VPA and 5-NAC-2-ene-VPA using GC-MS has recently been carried out [134]. Age and polytherapy influence the excretion of these thiol conjugates and perhaps this method will be useful to monitor the degree of exposure by a patient to reactive VPA metabolites [134].

In light of possible enzymes involved in the formation of GSH conjugates of reactive VPA metabolites, glutathione-S-transferase (GST) was suspected of being the catalyst for this conjugation reaction [78]. The addition of rat hepatic cytosol, a rich source of GST enzymes, was shown to enhance up to 23 fold the *in vitro* conjugation of GSH with N-acetyl-S-((*E*)-2-propyl-2,4-pentadienoyl)cysteamine, a structural mimic of (*E*)-2,4-diene-VPA-CoA thioester [78]. Partially purified GST was also found to catalyze the conjugation reaction similar to that of the isolated subcellular fractions [78]. Again, no reaction with GSH could be detected for the free acid form of (*E*)-2,4-diene-VPA indicating that the *in vivo* production of the GSH conjugates of (*E*)-2,4-diene-VPA is most likely catalyzed by GST enzymes, with the esterified diene being essential for the conjugation reaction.

Amino acid conjugation

Early studies of the metabolic fate of VPA in human subjects failed to detect the presence of valproyl glycine [27]. However, small amounts of the glycine conjugate have been identified together with significantly greater quantities of the glycine conjugates of unsaturated VPA metabolites in rat urine [73]. Moreover, the glycine conjugate of (*E*)-2,4-diene-VPA was identified in the urine of rhesus monkeys given 4-ene-VPA [82].

In recent human studies, LC-MS was used to identify the glycine, glutamate, and glutamine conjugates of VPA in the urine of patients on VPA therapy [158]. The VPA-glutamate conjugate, the most abundant of the amino acid conjugates found, is unusual for a xenobiotic and the significance of this finding needs to be explored.

Miscellaneous pathways

As noted earlier, VPA biotransformation highly resembles oxidative catabolism of fatty acids *in vivo*. Therefore it is not surprising to see other metabolites similar to those formed during the mitochondrial fatty acid translocation and activation. Recent advances in analytical techniques based on LC-MS systems have enabled the characterization of a few novel valproate metabolites generated by the action of the β-oxidation enzyme complex, i.e. valproyl-carnitine, VPA-CoA, and VPA-adenylate.

Millington et al. first identified a novel valproyl-L-carnitine conjugate in the urine of pediatric patients treated chronically with VPA [159, 160].

L-Carnitine ester formation which is an essential reaction for the entry of long-chain fatty acids into mitochondria was not an obligatory step for the β-oxidation of VPA by isolated rat liver mitochondria [76]. Although L-carnitine ester conjugation accounted for less than 1% of the VPA dose in pediatric patients [159], this metabolic pathway has gained considerable toxicological interest due to reports on carnitine depletion secondary to VPA administration and also the association of carnitine deficiency with VPA-induced Reye-like syndrome [161–164]. Of even greater toxicological interest is the reported hepatoprotective effects of L-carnitine in experimental animals treated with VPA or 4-ene-VPA [165–168], even though the beneficial effect of carnitine remains to be established in humans [161, 162, 164].

The coenzyme A thioester derivative of VPA (VPA-CoA) and its metabolites have been identified through *in vitro* studies of VPA mitochondrial β-oxidation [76, 92, 107] as described earlier. The fact that both carnitine and glycine conjugates have been detected as urinary metabolites of VPA provides indirect evidence for the existence of CoA derivatives of VPA. The glutamine conjugate has been identified recently as a metabolite of 2-fluoro-VPA [7, 97] and the formation of carnitine and amino acid conjugates is known to proceed *via* CoA thioesters.

Recently, a novel acyl adenylate mono phosphate derivative of VPA (VPA-AMP) was identified by Schulz and co-workers [107] from the incubation of VPA with soluble extracts of rat liver mitochondria fortified with ATP, CoA, and $MgCl_2$. VPA-AMP was found to exist, at least partially, in a free form and was enzymatically converted to valproyl-CoA in the presence of CoASH. Formation of VPA-AMP, which was also shown by the authors to occur during the metabolism of VPA by rat liver mitochondria in hepatocytes, appeared to be catalyzed by medium-chain acyl-CoA synthetase [107]. The formation of VPA-AMP, which serves as the metabolic intermediate between VPA and VPA-CoA, supports the VPA-CoA thioester formation as an important pathway of cellular VPA metabolism.

Conclusions

Valproic acid metabolism is extremely complex and a variety of overlapping phase I and II metabolic pathways catalyze its biotransformation to over 50 known metabolites. Based on decades of investigation into the metabolism of VPA in various experimental animals and human subjects, it is safe to say for phase I biotransformation that VPA metabolism generally follows that of the endogenous fatty acids, e.g. being primarily metabolized by the β-oxidation and ω- and (ω-1)-oxidative pathways. For the phase II biotransformation of VPA, conjugation with glucuronic acid is the major metabolic pathway in most species and accounts for the majority of the urinary excreted forms of the drug. Interestingly, conjugates of glutathione and N-acetylcysteine have recently been characterized in rats and humans indicating the formation of intermediate reactive metabolites during the process of VPA biotransformation. Whether these intermediate reactive metabolites are involved in certain serious adverse reactions to the drug is an issue that requires further investigation.

Although much remains to be learned about the diverse metabolic pathways of valproate in human, a significant body of research has recently addressed the nature of enzymatic pathways involved in VPA biotransformation. In this context, specific human cytochrome P450 enzymes appear to catalyze the novel terminal desaturation of VPA in human. Moreover, a number of recent studies have demonstrated the structure-activity relationships of VPA metabolites with respect to both their pharmacological activity and cytotoxic potential. The significance of geometric and enantiomeric configurations to the dynamics of these responses provides opportunity for further discovery. The knowledge of structure-activity relationships and the metabolic fate of valproate is pivotal to the rational design of new antiepileptic drugs with enhanced efficacy and less toxic side-effects.

With recent advances in molecular biology techniques and the availability of increasingly sensitive analytical methodologies, we envision further discoveries regarding the biological fate and structure-activity relationships of valproate/valproate analogues to be disclosed in the near future.

References

1 Di Carlo FJ, Bickart P, Auer CM (1986) Structure-metabolism relationships (SMR) for the prediction of health hazards by the Environmental Protection Agency. II. Application to teratogenicity and other toxic effects caused by aliphatic acids. *Drug Metab Rev* 17: 187–220

2 Loscher W (1981) Anticonvulsant activity of metabolites of valproic acid. *Arch Int Pharmacodyn Ther* 249: 158–63

3 Rettie AE, Rettenmeier AW, Howald WN, Baillie TA (1987) Cytochrome P-450-catalyzed formation of delta 4-VPA, a toxic metabolite of valproic acid. *Science* 235: 890–3

4 Liu MJ, Brouwer KL, Pollack GM (1992) Pharmacokinetics and pharmacodynamics of valproate analogs in rats. III. Pharmacokinetics of valproic acid, cyclohexanecarboxylic acid, and 1-methyl-1-cyclohexanecarboxylic acid in the bile-exteriorized rat. *Drug Metab Dispos* 20: 810–5

5 Bialer M, Hadad S, Kadry B, Abdul-Hai A, Haj-Yehia A, Sterling J, Herzig Y, Yagen B (1996) Pharmacokinetic analysis and antiepileptic activity of tetra-methylcyclopropane analogues of valpromide. *Pharm Res* 13: 284–9

6 Bialer M (1991) Clinical pharmacology of valpromide. *Clin Pharmacokinet* 20: 114–22

7 Tang W, Palaty J, Abbott FS (1997) Time course of alpha-fluorinated valproic acid in mouse brain and serum and its effect on synaptosomal gamma-aminobutyric acid levels in comparison to valproic acid. *J Pharmacol Exp Ther* 282: 1163–72

8 Keane PE, Simiand J, Morre M (1985) Comparison of the pharmacological and biochemical profiles of valproic acid (VPA) and its cerebral metabolite (2-en-VPA) after oral administration in mice. *Methods Find Exp Clin Pharmacol* 7: 83–6

9 Palaty J, Abbott FS (1995) Structure-activity relationships of unsaturated analogues of valproic acid. *J Med Chem* 38: 3398-406

10 Chang ZL (1979) Sodium valproate and valproic acid. In: K Florey (ed): *Analytical profiles of drug substances*. Academic Press, New York, 529–556

11 Löscher W (1981) Concentration of metabolites of valproic acid in plasma of epileptic patients. *Epilepsia* 22: 169–78
12 Schafer H, Luhrs R (1978) Metabolite pattern of valproic acid. Part I: gas chromatographic determination of the valproic acid metabolite artifacts, heptanone-3, 4- and 5-hydroxy-valproic acid lactone. *Arzneimittelforschung* 28: 657–62
13 Abbott FS, Kassam J, Acheampong A, Ferguson S, Panesar S, Burton R, Farrell K, Orr J (1986) Capillary gas chromatography-mass spectrometry of valproic acid metabolites in serum and urine using tert.-butyldimethylsilyl derivatives. *J Chromatogr* 375: 285–98
14 Darius J (1996) On-column gas chromatographic-mass spectrometric assay for metabolic profiling of valproate in brain tissue and serum. *J Chromatogr B Biomed Appl* 682: 67–72
15 Dickinson RG, Harland RC, Ilias AM, Rodgers RM, Kaufman SN, Lynn RK, Gerber N (1979) Disposition of valproic acid in the rat: dose-dependent metabolism, distribution, enterohepatic recirculation and choleretic effect. *J Pharmacol Exp Ther* 211: 583–95
16 Fisher E, Wittfoht W, Nau H (1992) Quantitative determination of valproic acid and 14 metabolites in serum and urine by gas chromatography/mass spectrometry. *Biomed Chromatogr* 6: 24–9
17 Kassahun K, Farrell K, Zheng JJ, Abbott F (1990) Metabolic profiling of valproic acid in patients using negative-ion chemical ionization gas chromatography-mass spectrometry. *J Chromatogr* 527: 327–41
18 Nau H, Wittfoht W, Schafer H, Jakobs C, Rating D, Helge H (1981) Valproic acid and several metabolites: quantitative determination in serum, urine, breast milk and tissues by gas chromatography-mass spectrometry using selected ion monitoring. *J Chromatogr* 226: 69–78
19 Rettenmeier AW, Howald WN, Levy RH, Witek DJ, Gordon WP, Porubek DJ, Baillie TA (1989) Quantitative metabolic profiling of valproic acid in humans using automated gas chromatographic/mass spectrometric techniques. *Biomed Environ Mass Spectrom* 18: 192–9
20 Sugimoto T, Muro H, Woo M, Nishida N, Murakami K (1996) Metabolite profiles in patients on high-dose valproate monotherapy. *Epilepsy Res* 25: 107–12
21 Tatsuhara T, Muro H, Matsuda Y, Imai Y (1987) Determination of valproic acid and its metabolites by gas chromatography-mass spectrometry with selected ion monitoring. *J Chromatogr* 399: 183–95
22 Yu D, Gordon JD, Zheng J, Panesar SK, Riggs KW, Rurak DW, Abbott FS (1995) Determination of valproic acid and its metabolites using gas chromatography with mass-selective detection: application to serum and urine samples from sheep. *J Chromatogr B Biomed Appl* 666: 269–81
23 Kassahun K, Farrell K, Abbott F (1991) Identification and characterization of the glutathione and N-acetylcysteine conjugates of (*E*)-2-propyl-2,4-pentadienoic acid, a toxic metabolite of valproic acid, in rats and humans. *Drug Metab Dispos* 19: 525–35
24 Tang W, Abbott FS (1996) Characterization of thiol-conjugated metabolites of 2-propyl-pent-4-enoic acid (4-ene VPA), a toxic metabolite of valproic acid, by electrospray tandem mass spectrometry. *J Mass Spectrom* 31: 926–36
25 Abbott FS, Acheampong AA (1988) Quantitative structure-anticonvulsant activity relationships of valproic acid, related carboxylic acids and tetrazoles. *Neuropharmacology* 27: 287–94
26 Lee RD, Kassahun K, Abbott FS (1989) Stereoselective synthesis of the diunsaturated metabolites of valproic acid. *J Pharm Sci* 78: 667–71
27 Gompertz D, Tippett P, Bartlett K, Baillie T (1977) Identification of urinary metabolites of sodium dipropylacetate in man; potential sources of interference in organic acid screening procedures. *Clin Chim Acta* 74: 153–60
28 Libert R, Van Hoof F, Schanck A, Hoffmann E (1986) The hydroxamate of valproic acid, a compound produced by oximation of urine from patients under valproic therapy. *Biomed Environ Mass Spectrom* 13: 599–603
29 Keane PE, Simiand J, Mendes E, Santucci V, Morre M (1983) The effects of analogues of valproic acid on seizures induced by pentylenetetrazol and GABA content in brain of mice. *Neuropharmacology* 22: 875–9

30 Löscher W, Nau H (1985) Pharmacological evaluation of various metabolites and analogues of valproic acid. Anticonvulsant and toxic potencies in mice. *Neuropharmacology* 24: 427–35
31 Morre M, Keane PE, Vernieres JC, Simiand J, Roncucci R (1984) Valproate: recent findings and perspectives. *Epilepsia* 25: S5-9
32 Chapman AG, Meldrum BS, Mendes E (1983) Acute anticonvulsant activity of structural analogues of valproic acid and changes in brain GABA and aspartate content. *Life Sci* 32: 2023–31
33 Chapman AG, Croucher MJ, Meldrum BS (1984) Anticonvulsant activity of intracerebroventricularly administered valproate and valproate analogues. A dose-dependent correlation with changes in brain aspartate and GABA levels in DBA/2 mice. *Biochem Pharmacol* 33: 1459–63
34 Perlman BJ, Goldstein DB (1984) Membrane-disordering potency and anticonvulsant action of valproic acid and other short-chain fatty acids. *Mol Pharmacol* 26: 83–9
35 Liu MJ, Pollack GM (1993) Pharmacokinetics and pharmacodynamics of valproate analogs in rats. II. Pharmacokinetics of octanoic acid, cyclohexanecarboxylic acid, and 1-methyl-1-cyclohexanecarboxylic acid. *Biopharm Drug Dispos* 14: 325–39
36 Liu MJ, Pollack GM (1994) Pharmacokinetics and pharmacodynamics of valproate analogues in rats. IV. Anticonvulsant action and neurotoxicity of octanoic acid, cyclohexanecarboxylic acid, and 1-methyl-1-cyclohexanecarboxylic acid. *Epilepsia* 35: 234–43
37 Watkins JB, Klaassen CD (1983) Choleretic effect of structural analogs of valproic acid in the rat. *Res Commun Chem Pathol Pharmacol* 39: 355–66
38 Bojic U, Elmazar MM, Hauck RS, Nau H (1996) Further branching of valproate-related carboxylic acids reduces the teratogenic activity, but not the anticonvulsant effect. *Chem Res Toxicol* 9: 866–70
39 Courage-Maguire C, Bacon CL, Nau H, Regan CM (1997) Correlation of *in vitro* anti-proliferative potential with *in vivo* teratogenicity in a series of valproate analogues. *Int J Dev Neurosci* 15: 37–43
40 Klug S, Lewandowski C, Zappel F, Merker HJ, Nau H, Neubert D (1990) Effects of valproic acid, some of its metabolites and analogues on prenatal development of rats *in vitro* and comparison with effects *in vivo*. *Arch Toxicol* 64: 545–53
41 Maar TE, Ellerbeck U, Bock E, Nau H, Schousboe A, Berezin V (1997) Prediction of teratogenic potency of valproate analogues using cerebellar aggregation cultures. *Toxicology* 116: 159–68
42 Nau H, Loscher W (1986) Pharmacologic evaluation of various metabolites and analogs of valproic acid: teratogenic potencies in mice. *Fundam Appl Toxicol* 6: 669–76
43 Vorhees CV, Acuff-Smith KD, Weisenburger WP, Minck DR, Berry JS, Setchell KD, Nau H (1991) Lack of teratogenicity of trans-2-ene-valproic acid compared to valproic acid in rats. *Teratology* 43: 583–90
44 Hauck RS, Nau H, Elmazar MM (1991) On the development of alternative antiepileptic drugs. Lack of enantioselectivity of the anticonvulsant activity, in contrast to teratogenicity, of 2-n-propyl-4-pentenoic acid and 2-n-propyl-4-pentynoic acid, analogues of the anticonvulsant drug valproic acid. *Naturwissenschaften* 78: 272–4
45 Elmazar MM, Hauck RS, Nau H (1993) Anticonvulsant and neurotoxic activities of twelve analogues of valproic acid. *J Pharm Sci* 82: 1255–8
46 Robbins DK, Wedlund PJ, Elsberg S, Oesch F, Thomas H (1992) Interaction of valproic acid and some analogues with microsomal epoxide hydrolase. *Biochem Pharmacol* 43: 775–83
47 Haj-Yehia A, Hadad S, Bialer M (1992) Pharmacokinetic analysis of the structural requirements for forming "stable" analogues of valpromide. *Pharm Res* 9: 1058–63
48 Hadad S, Bialer M (1995) Pharmacokinetic analysis and antiepileptic activity of N-valproyl derivatives of GABA and glycine. *Pharm Res* 12: 905–1010
49 Blotnik S, Bergman F, Bialer M (1997) The disposition of valproyl glycinamide and valproyl glycine in rats. *Pharmaceutical Research* 14: 873–878
50 Scott KR, Moore JA, Zalucky TB, Nicholson JM, Lee JA, Hinko CN (1985) Spiro [4.5] and spiro [4.6] carboxylic acids: cyclic analogues of valproic acid. Synthesis and anticonvulsant evaluation. *J Med Chem* 28: 413–7
51 Scott KR, Adesioye S, Ayuk PB, Edafiogho IO, John D, Kodwin P, Maxwell-Irving T, Moore JA, Nicholson JM (1994) Synthesis and evaluation of amino analogues of valproic acid. *Pharm Res* 11: 571–4

52 Adkison KD, Ojemann GA, Rapport RL, Dills RL, Shen DD (1995) Distribution of unsaturated metabolites of valproate in human and rat brain–pharmacologic relevance? *Epilepsia* 36: 772–82
53 Löscher W, Nau H, Marescaux C, Vergnes M (1984) Comparative evaluation of anticonvulsant and toxic potencies of valproic acid and 2-en-valproic acid in different animal models of epilepsy. *Eur J Pharmacol* 99: 211–8
54 Kesterson JW, Granneman GR, Machinist JM (1984) The hepatotoxicity of valproic acid and its metabolites in rats. I. Toxicologic, biochemical and histopathologic studies. *Hepatology* 4: 1143–52
55 Löscher W, Nau H, Wahnschaffe U, Honack D, Rundfeldt C, Wittfoht W, Bojic U (1993) Effects of valproate and E-2-en-valproate on functional and morphological parameters of rat liver. II. Influence of phenobarbital comedication. *Epilepsy Res* 15: 113–31
56 Semmes RL, Shen DD (1991) Comparative pharmacodynamics and brain distribution of *E*-delta 2-valproate and valproate in rats. *Epilepsia* 32: 232–41
57 Löscher W (1992) Pharmacological, toxicological and neurochemical effects of delta 2(*E*)-valproate in animals [see comments]. *Pharm Weekbl Sci* 14: 139–43
58 Löscher W, Honack D, Nolting B, Fassbender CP (1991) Trans-2-en-valproate: reevaluation of its anticonvulsant efficacy in standardized seizure models in mice, rats and dogs. *Epilepsy Res.* 9: 195–210
59 Fabre G, Briot C, Marti E, Montseny JP, Bourrie M, Masse D, Berger Y, Cano JP (1992) Delta 2-valproate biotransformation using human liver microsomal fractions. *Pharm Weekbl Sci* 14: 146–51
60 Acheampong A, Abbott F, Burton R (1983) Identification of valproic acid metabolites in human serum and urine using hexadeuterated valproic acid and gas chromatographic mass spectrometric analysis. *Biomed Mass Spectrom* 10: 586–95
61 Nau H, Hauck RS, Ehlers K (1991) Valproic acid-induced neural tube defects in mouse and human: aspects of chirality, alternative drug development, pharmacokinetics and possible mechanisms. *Pharmacol Toxicol* 69: 310–21
62 Andrews JE, Ebron-McCoy MT, Bojic U, Nau H, Kavlock RJ (1997) Stereoselective dysmorphogenicity of the enantiomers of the valproic acid analogue 2-N-propyl-4-pentynoic acid (4-yn-VPA): cross-species evaluation in whole embryo culture. *Br J Psychiatry* 170: 484–5
63 Singh K, Abbott FS, Orr JM (1987) Capillary GCMS assay of 2-n-propyl-4-pentenoic acid (4-ene VPA) in rat plasma and urine. *Res Commun Chem Pathol Pharmacol* 56: 211–23
64 Kassahun K, Burton R, Abbott FS (1989) Negative ion chemical ionization gas chromatography/mass spectrometry of valproic acid metabolites. *Biomed Environ Mass Spectrom* 18: 918–26
65 Acheampong AA, Abbott FS, Orr JM, Ferguson SM, Burton RW (1984) Use of hexadeuterated valproic acid and gas chromatography-mass spectrometry to determine the pharmacokinetics of valproic acid. *J Pharm Sci* 73: 489–94
66 Baillie TA (1985) Mass spectrometry and stable isotopes in the study of mechanistic aspects of metabolic pathways. In: AL Burlingame, N Castagnoli Jr. (eds): *Mass spectrometry in the health and life sciences*. Elsevier Science Publishers, Amsterdam, 349–362
67 Rettenmeier AW, Gordon WP, Barnes H, Baillie TA (1987) Studies on the metabolic fate of valproic acid in the rat using stable isotope techniques. *Xenobiotica* 17: 1147–57
68 Ferrandes B, Eymard P (1977) Metabolism of valproate sodium in rabbit, rat, dog, and man. *Epilepsia* 18: 169–82
69 Jakobs C, Löscher W (1978) Identification of metabolites of valproic acid in serum of humans, dog, rat, and mouse. *Epilepsia* 19: 591–602
70 Kochen W, Imbeck H, Jakobs C (1977) Studies on the urinary excretion of metabolites of valproic acid (dipropylacetic acid) in rats and humans (in German). *Arzneimittelforsch* 27: 1090–1099
71 Kuhara T, Matsumoto I (1974) Metabolism of branched medium chain length fatty acid. I. Omega- oxidation of sodium dipropylacetate in rats. *Biomed Mass Spectrom* 1: 291–4
72 Matsumoto I, Kuhara T, Yoshino M (1976) Metabolism of branched medium chain length fatty acid. II–beta-oxidation of sodium dipropylacetate in rats. *Biomed Mass Spectrom* 3: 235–40
73 Granneman GR, Wang SI, Machinist JM, Kesterson JW (1984) Aspects of the metabolism of valproic acid. *Xenobiotica* 14: 375–87

74 Prickett KS, Baillie TA (1986) Metabolism of unsaturated derivatives of valproic acid in rat liver microsomes and destruction of cytochrome P-450. *Drug Metab Dispos* 14: 221–9
75 Rettie AE, Boberg M, Rettenmeier AW, Baillie TA (1988) Cytochrome P-450-catalyzed desaturation of valproic acid *in vitro*. Species differences, induction effects, and mechanistic studies. *J Biol Chem* 263: 13733–8
76 Bjorge SM, Baillie TA (1991) Studies on the beta-oxidation of valproic acid in rat liver mitochondrial preparations. *Drug Metab Dispos* 19: 823–9
77 Kassahun K, Baillie TA (1993) Cytochrome P-450-mediated dehydrogenation of 2-n-propyl-2(*E*)-pentenoic acid, a pharmacologically-active metabolite of valproic acid, in rat liver microsomal preparations. *Drug Metab Dispos* 21: 242–8
78 Tang W, Borel AG, Abbott FS (1996) Conjugation of glutathione with a toxic metabolite of valproic acid, (*E*)-2-propyl-2,4-pentadienoic acid, catalyzed by rat hepatic glutathione-S-transferases. *Drug Metab Dispos* 24: 436–46
79 Rettie AE, Sheffels PR, Korzekwa KR, Gonzalez FJ, Philpot RM, Baillie TA (1995) CYP4 isozyme specificity and the relationship between omega-hydroxylation and terminal desaturation of valproic acid. *Biochemistry* 34: 7889–95
80 Schobben F, Vree TB, van der Kleijn E, Claessens R, Renier WO (1980) Metabolism of Valproic acid in monkey and man. In: SI Johannessen, PL Morselli, CE Pippinger, A Richens, D Schmidt, H Meinardi (eds): *Antiepileptic therapy: Advances in drug monitoring*. Raven Press, New York, 91–102
81 Koch KM, Ludwick BT, Levy RH (1981) Phenytoin–valproic acid interaction in rhesus monkey. *Epilepsia* 22: 19–25
82 Rettenmeier AW, Gordon WP, Prickett KS, Levy RH, Baillie TA (1986) Biotransformation and pharmacokinetics in the rhesus monkey of 2-n-propyl-4-pentenoic acid, a toxic metabolite of valproic acid. *Drug Metab Dispos* 14: 454–64
83 Rettenmeier AW, Gordon WP, Prickett KS, Levy RH, Lockard JS, Thummel KE, Baillie TA (1986) Metabolic fate of valproic acid in the rhesus monkey. Formation of a toxic metabolite, 2-n-propyl-4-pentenoic acid. *Drug Metab Dispos* 14: 443–53
84 Sadeque AJM, Fisher MB, Korzekwa KR, Gonzalez FJ, Rettie AE (1997) Human CYP2C9 and CYP2A6 mediate formation of the hepatotoxin 4-ene- valproic acid. *J Pharmacol Exp Ther* 283: 698–703
85 Gugler R, von Unruh GE (1980) Clinical pharmacokinetics of valproic acid. *Clin Pharmacokinet* 5: 67–83
86 Zaccara G, Messori A, Moroni F (1988) Clinical pharmacokinetics of valproic acid – 1988. *Clin Pharmacokinet* 15: 367–89
87 Baillie TA, Rettenmeier AW (1989) Valproate: biotransformation. In: R Levy, R Mattson, B Meldrum, JK Penty, FE Dreifuss (eds): *Antiepileptic drugs*. Raven Press, New York, 601–619
88 Baillie TA (1992) Metabolism of valproate to hepatotoxic intermediates. *Pharm Weekbl Sci* 14: 122–5
89 Baillie TA, Sheffels PA (1995) Valproic acid: chemistry and biotransformation. In: RH Levy, RH Mattson, BS Meldrum (eds): *Antiepileptic drugs*. Raven Press, New York, 589–604
90 Ponchaut S, Veitch K (1993) Valproate and mitochondria. *Biochem Pharmacol* 46: 199–204
91 Van der Kleijn E, Meinardi H, Plomp-Van Heest M, Polderman AKS, Vinks P (1992) Recent developments on valproate and its metabolites. *Pharm. Weekl. Sci.* 14: 97–108
92 Li J, Norwood DL, Mao LF, Schulz H (1991) Mitochondrial metabolism of valproic acid. *Biochemistry* 30: 388–94
93 Kassahun K, Abbott F (1993) In vivo formation of the thiol conjugates of reactive metabolites of 4-ene VPA and its analog 4-pentenoic acid. *Drug Metab Dispos* 21: 1098–106
94 Kassahun K, Hu P, Grillo MP, Davis MR, Jin L, Baillie TA (1994) Metabolic activation of unsaturated derivatives of valproic acid. Identification of novel glutathione adducts formed through coenzyme A-dependent and -independent processes. *Chem Biol Interact* 90: 253–75
95 Tang W, Borel AG, Fujimiya T, Abbott FS (1995) Fluorinated analogues as mechanistic probes in valproic acid hepatotoxicity: hepatic microvesicular steatosis and glutathione status. *Chem Res Toxicol* 8: 671–82

96 Tang W, Abbott FS (1996) Bioactivation of a toxic metabolite of valproic acid, (*E*)-2-propyl-2,4-pentadienoic acid, *via* glucuronidation. LC/MS/MS characterization of the GSH-glucuronide diconjugates. *Chem Res Toxicol* 9: 517–26
97 Tang W, Abbott FS (1997) A comparative investigation of 2-propyl-4-pentenoic acid (4-ene VPA) and its alpha-fluorinated analogue: phase II metabolism and pharmacokinetics. *Drug Metab Dispos* 25: 219–27
98 Kuhara T, Hirohata Y, Yamada S, Matsumoto I (1978) Metabolism of sodium dipropylacetate in humans. *Eur. J. Drug Metab. Pharmacokinet.* 3: 171–177
99 Kochen W, Scheffner H (1980) On unsaturated metabolites of the valproic acid (VPA) in serum of epileptic children. In: SI Johannessen, PL Morselli, CE Pippenger, A Richens, D Schmidt, H Meinardi (eds): *Antiepileptic therapy: Advances in drug monitoring*. Raven Press, New York, 111–120
100 Löscher W, Nau H, Siemes H (1988) Penetration of valproate and its active metabolites into cerebrospinal fluid of children with epilepsy. *Epilepsia* 29: 311–6
101 Kondo T, Otani K, Hirano T, Kaneko S, Fukushima Y (1990) The effects of phenytoin and carbamazepine on serum concentrations of mono-unsaturated metabolites of valproic acid. *Br J Clin Pharmacol* 29: 116–9
102 Koch KM, Wilensky AJ, Levy RH (1989) Beta-oxidation of valproic acid. I. Effects of fasting and glucose in humans. *Epilepsia* 30: 782–9
103 Schafer H, Luhrs R, Reith H (1980) Chemistry, Pharmacokinetics and biological activity of some metabolites of valproic acid. In: SI Johannessen, PL Morselli, CE Pippinger, A Richens, D Schmidt, H Meinardi (eds): *Antiepileptic therapy: Advances in drug monitoring*. Raven Press, New York, 103–110
104 Heinemeyer G, Nau H, Hildebrandt AG, Roots I (1985) Oxidation and glucuronidation of valproic acid in male rats-influence of phenobarbital, 3-methylcholanthrene, beta-naphthoflavone and clofibrate. *Biochem Pharmacol* 34: 133–9
105 Abbott FS, Kassam J, Orr JM, Farrell K (1986) The effect of aspirin on valproic acid metabolism. *Clin Pharmacol Ther* 40: 94–100
106 Koch KM, Prickett KS, Rettenmeier AW, Baillie TA, Levy RH (1989) Beta-oxidation of valproate. II. Effects of fasting, glucose, and clofibrate in rats. *Epilepsia* 30: 790–6
107 Mao LF, Millington DS, Schulz H (1992) Formation of a free acyl adenylate during the activation of 2-propylpentanoic acid. Valproyl-AMP: a novel cellular metabolite of valproic acid. *J Biol Chem* 267: 3143–6
108 Becker CM, Harris RA (1983) Influence of valproic acid on hepatic carbohydrate and lipid metabolism. *Arch Biochem Biophys* 223: 381–92
109 Bjorge SM, Baillie TA (1985) Inhibition of medium-chain fatty acid beta-oxidation *in vitro* by valproic acid and its unsaturated metabolite, 2-n-propyl-4-pentenoic acid. *Biochem Biophys Res Commun* 132: 245–52
110 Mortensen PB (1980) Inhibition of fatty acid oxidation by valproate [letter]. *Lancet* 2: 856–7
111 Prickett KS, Baillie TA (1984) Metabolism of valproic acid by hepatic microsomal cytochrome P-450. *Biochem Biophys Res Commun* 122: 1166–73
112 Shirley MA, Hu P, Baillie TA (1993) Stereochemical studies on the beta-oxidation of valproic acid in isolated rat hepatocytes. *Drug Metab Dispos* 21: 580–6
113 Ito M, Ikeda Y, Arnez JG, Finocchiaro G, Tanaka K (1990) The enzymatic basis for the metabolism and inhibitory effects of valproic acid: dehydrogenation of valproyl-CoA by 2-methyl-branched-chain acyl-CoA dehydrogenase. *Biochim Biophys Acta* 1034: 213–8
114 Schulz H (1991) Inhibition of fatty acid oxidation and mitochondrial metabolism of valproic acid *in vitro*. In: RH Levy, JK Penry (eds): *Idiosyncratic reactions to valproate: Clinical risk patterns and mechanisms of toxicity*. Raven Press, New York, 47–52
115 Yao KW, Mao LF, Luo MJ, Schulz H (1994) The relationship between mitochondrial activation and toxicity of some substituted carboxylic acids. *Chem Biol Interact* 90: 225–34
116 Draye JP, Vamecq J (1987) The inhibition by valproic acid of the mitochondrial oxidation of monocarboxylic and omega-hydroxymonocarboxylic acids: possible implications for the metabolism of gamma-aminobutyric acid. *J Biochem (Tokyo)* 102: 235–42
117 Horie S, Suga T (1985) Enhancement of peroxisomal beta-oxidation in the liver of rats and mice treated with valproic acid. *Biochem Pharmacol* 34: 1357–62

118 Veitch K, Draye JP, Van Hoof F (1989) Inhibition of mitochondrial beta-oxidation and peroxisomal stimulation in rodent livers by valproate. *Biochem Soc Trans* 17: 1070–1
119 Ponchaut S, Draye JP, Veitch K, Van Hoof F (1991) Influence of chronic administration of valproate on ultrastructure and enzyme content of peroxisomes in rat liver and kidney. Oxidation of valproate by liver peroxisomes. *Biochem Pharmacol* 41: 1419–28
120 Vamecq J, Vallee L, Fontaine M, Lambert D, Poupaert J, Nuyts JP (1993) CoA esters of valproic acid and related metabolites are oxidized in peroxisomes through a pathway distinct from peroxisomal fatty and bile acyl-CoA beta-oxidation. *FEBS Lett* 322: 95–100
121 Rettenmeier AW, Prickett KS, Gordon WP, Bjorge SM, Chang SL, Levy RH, Baillie TA (1985) Studies on the biotransformation in the perfused rat liver of 2-n-propyl-4-pentenoic acid, a metabolite of the antiepileptic drug valproic acid. Evidence for the formation of chemically reactive intermediates. *Drug Metab Dispos* 13: 81–96
122 Baillie TA (1988) Metabolic activation of valproic acid and drug-mediated hepatotoxicity. Role of the terminal olefin, 2-n-propyl-4-pentenoic acid. *Chem Res Toxicol* 1: 195–9
123 Ortiz de Montellano PR (1989) Cytochrome P-450 catalysis: radical intermediates and dehydrogenation reactions. *Trends Pharmacol Sci* 10: 354–9
124 Kingsley E, Gray P, Tolman KG, Tweedale R (1983) The toxicity of metabolites of sodium valproate in cultured hepatocytes. *J Clin Pharmacol* 23: 178–85
125 Granneman GR, Wang SI, Kesterson JW, Machinist JM (1984) The hepatotoxicity of valproic acid and its metabolites in rats. II. Intermediary and valproic acid metabolism. *Hepatology* 4: 1153–8
126 Porubek DJ, Grillo MP, Baillie TA (1989) The covalent binding to protein of valproic acid and its hepatotoxic metabolite, 2-n-propyl-4-pentenoic acid, in rats and in isolated rat hepatocytes. *Drug Metab Dispos* 17: 123–30
127 Zimmerman HJ, Ishak KG (1982) Valproate-induced hepatic injury: analyses of 23 fatal cases. *Hepatology* 2: 591–7
128 Gerber N, Dickinson RG, Harland RC, Lynn RK, Houghton LD, Antonias JI, Schimschock JC (1979) Reye-like syndrome associated with valproic acid therapy. *J Pediatr* 95: 142–4
129 Nagata K, Liberato DJ, Gillette JR, Sasame HA (1986) An unusual metabolite of testosterone. 17 beta-Hydroxy-4,6-androstadiene-3-one. *Drug Metab Dispos* 14: 559–65
130 Guan X, Fisher MB, Lang DH, Zheng YM, Koop DR, Rettie AE (1998) Cytochrome P450-dependent desaturation of lauric acid: isoform selectivity and mechanism of formation of 11-dodecenoic acid. *Chem Biol Interact* 110: 103–21
131 Porubek DJ, Barnes H, Meier GP, Theodore LJ, Baillie TA (1989) Enantiotopic differentiation during the biotransformation of valproic acid to the hepatotoxic olefin 2-n-propyl-4-pentenoic acid. *Chem Res Toxicol* 2: 35–40
132 Acheampong A, Abbott FS (1985) Synthesis and stereochemical determination of diunsaturated valproic acid analogs including its major diunsaturated metabolite. *J Lipid Res* 26: 1002–8
133 Kochen W, Sprunck HP, Tauscher W, Klemens M (1984) Five doubly unsaturated metabolites of valproic acid in urine and plasma of patients on valproic acid therapy. *J Clin Chem Clin Biochem* 22: 309–17
134 Gopaul SV, Farrell K, Abbott FS (1996) The analysis of N-acetylcysteine (NAC) conjugates of VPA in patients on VPA therapy. *ISSX Proceedings* 7: 131
135 Rettie AE, Rettenmeier AW, Beyer BK, Baillie TA, Juchau MR (1986) Valproate hydroxylation by human fetal tissues and embryotoxicity of metabolites. *Clin Pharmacol Ther* 40: 172–7
136 Kuhara T, Inoue Y, Matsumoto M, Shinka T, Matsumoto I, Kawahara N, Sakura N (1990) Markedly increased omega-oxidation of valproate in fulminant hepatic failure. *Epilepsia* 31: 214–7
137 Matsumoto I, Kuhara T, Inoue Y, Matsumoto M (1984) Effects of valproic acid on amino acid and organic acid metabolism. In: RH Levy, WH Pitlick, M Eichelbaum, J Meijer (eds): *Metabolism of antiepileptic drugs*. Raven Press, New York, 169–175
138 Dickinson RG, Hooper WD, Dunstan PR, Eadie MJ (1989) Urinary excretion of valproate and some metabolites in chronically treated patients. *Ther Drug Monit* 11: 127–33
139 Anderson GD, Acheampong AA, Wilensky AJ, Levy RH (1992) Effect of valproate dose on formation of hepatotoxic metabolites. *Epilepsia* 33: 736–42

140 Brouwer KL, Hall ES, Pollack GM (1993) Protein binding and hepatobiliary distribution of valproic acid and valproate glucuronide in rats. *Biochem Pharmacol* 45: 735–42
141 Dickinson RG, Harland RC, Rodgers RM, Gordon WP, Lynn RK, Gerber N (1978) Dose dependent metabolism and enterohepatic recirculation of valproic acid in the rat. *Proc West Pharmacol Soc* 21: 217–20a
142 Ogiso T, Ito Y, Iwaki M, Yamahata T (1986) Disposition and pharmacokinetics of valproic acid in rats. *Chem Pharm Bull (Tokyo)* 34: 2950–6
143 Watkins JB, Klaassen CD (1982) Effect of inducers and inhibitors of glucuronidation on the biliary excretion and choleretic action of valproic acid in the rat. *J Pharmacol Exp Ther* 220: 305–10
144 Watkins JB, Klaassen CD (1981) Choleretic effect of valproic acid in the rat. *Hepatology* 1: 341–7
145 Howell SR, Hazelton GA, Klaassen CD (1986) Depletion of hepatic UDP-glucuronic acid by drugs that are glucuronidated. *J Pharmacol Exp Ther* 236: 610–4
146 Watkins JB, Gregus Z, Thompson TN, Klaassen CD (1982) Induction studies on the functional heterogeneity of rat liver UDP-glucuronosyltransferases. *Toxicol Appl Pharmacol* 64: 439–46
147 Taburet AM, Aymard P (1983) Valproate glucuronidation by rat liver microsomes. Interaction with parahydroxyphenobarbital. *Biochem Pharmacol* 32: 3859–61
148 Spahn-Langguth H, Benet LZ (1992) Acyl glucuronides revisited: is the glucuronidation process a toxification as well as a detoxification mechanism? *Drug Metab Rev* 24: 5–47
149 Dickinson RG, Hooper WD, Eadie MJ (1984) pH-dependent rearrangement of the biosynthetic ester glucuronide of valproic acid to beta-glucuronidase-resistant forms. *Drug Metab Dispos* 12: 247–52
150 Bailey MJ, Dickinson RG (1996) Chemical and immunochemical comparison of protein adduct formation of four carboxylate drugs in rat liver and plasma. *Chem Res Toxicol* 9: 659–66
151 Williams AM, Worrall S, de Jersey J, Dickinson RG (1992) Studies on the reactivity of acyl glucuronides-III. Glucuronide-derived adducts of valproic acid and plasma protein and anti-adduct antibodies in humans. *Biochem Pharmacol* 43: 745–55
152 Dickinson RG, Kluck RM, Eadie MJ, Hooper WD (1985) Disposition of beta-glucuronidase-resistant "glucuronides" of valproic acid after intrabiliary administration in the rat: intact absorption, fecal excretion and intestinal hydrolysis. *J Pharmacol Exp Ther* 233: 214–21
153 Dickinson RG, Kluck RM, Hooper WD, Patterson M, Chalk JB, Eadie MJ (1985) Rearrangement of valproate glucuronide in a patient with drug-associated hepatobiliary and renal dysfunction. *Epilepsia* 26: 589–93
154 Singh K, Orr JM, Abbott FS (1988) Pharmacokinetics and enterohepatic circulation of 2-n-propyl-4-pentenoic acid in the rat. *Drug Metab Dispos* 16: 848–52
155 Singh K, Orr JM, Abbott FS (1990) Pharmacokinetics and enterohepatic circulation of (*E*)-2-ene valproic acid in the rat. *J Pharmacobiodyn* 13: 622–7
156 Jurima-Romet M, Abbott FS, Tang W, Huang HS, Whitehouse LW (1996) Cytotoxicity of unsaturated metabolites of valproic acid and protection by vitamins C and E in glutathione-depleted rat hepatocytes. *Toxicology* 112: 69–85
157 Zanelli U, Puccini P, Acerbi D, Ventura P, Gervasi PG (1996) Selective depletion of mitochondrial glutathione content by pivalic acid and valproic acid in rat liver: possibility of a common mechanism. *Adv Exp Med Biol* 387: 107–11
158 Gopaul SV, Tang W, Burton R, Kevin F, Abbott FS (1997) Amino acid conjugates of valproic acid in patients on VPA therapy. *ISSX Proceedings* 11: 205
159 Millington DS, Bohan TP, Roe CR, Yergey AL, Liberato DJ (1985) Valproylcarnitine: a novel drug metabolite identified by fast atom bombardment and thermospray liquid chromatography-mass spectrometry. *Clin Chim Acta* 145: 69–76
160 Liberato DJ, Millington SD, Yergey AL (1985) Analysis of acylcarnitines in human metabolic disease by thermospray liquid chromatography/mass spectrometry. In: AL Burlingame, NJ Castagnoli (eds): *Mass spectrometry in the health and life sciences*. Elsevier, Amsterdam, 333–346
161 Bohles H, Richter K, Wagner-Thiessen E, Schafer H (1982) Decreased serum carnitine in valproate induced Reye syndrome. *Eur J Pediatr* 139: 185–6

162 Coulter DL (1984) Carnitine deficiency: a possible mechanism for valproate hepatotoxicity [letter]. *Lancet* 1: 689
163 Laub MC, Paetzke-Brunner I, Jaeger G (1986) Serum carnitine during valproic acid therapy. *Epilepsia* 27: 559–62
164 Lamb MC, Paetzke-Brunner I, Gaeger G (1986) Sodium carnitine during valproic acid therapy. *Epilepsia* 27: 559–562
165 Levy RH, Lin JMH, Acheampong AA, Russell RG (1991) Dose effect of L-carnitine on delta4-VPA-associated hepatotoxicity. In: RH Levy, JK Penry (eds): *Idiosyncratic reactions to valproate: Clinical risk patterns and mechanisms of toxicity*. Raven Press, New York, 25–29
166 Sugimoto T, Araki A, Nishida N, Sakane Y, Woo M, Takeuchi T, Kobayashi Y (1987) Hepatotoxicity in rat following administration of valproic acid: effect of L-carnitine supplementation. *Epilepsia* 28: 373–7
167 Thurston JH, Hauhart RE (1992) Amelioration of adverse effects of valproic acid on ketogenesis and liver coenzyme A metabolism by cotreatment with pantothenate and carnitine in developing mice: possible clinical significance. *Pediatr Res* 31: 419–23
168 Thurston JH, Hauhart RE (1993) Reversal of the adverse chronic effects of the unsaturated derivative of valproic acid–2-n-propyl-4-pentenoic acid-on ketogenesis and liver coenzyme A metabolism by a single injection of pantothenate, carnitine, and acetylcysteine in developing mice. *Pediatr Res* 33: 72–6
169 Taillandier G, Benoit-Guyod JL, Laruelle C, Boucherle A (1977) Investigation in the dipropylacetic acid series, C8 and C9 branched chain ethylenic acids and amides. *Arch Pharm (Weinheim)* 310: 394–403
170 Honack D, Rundfeldt C, Löscher W (1992) Pharmacokinetics, anticonvulsant efficacy, and adverse effects of trans-2-en-valproate after acute and chronic administration in amygdala-kindled rats. *Naunyn Schmiedebergs Arch Pharmacol* 345: 187–96

Milestones in Drug Therapy
Valproate
ed. by W. Löscher

Absorption, distribution, and excretion

Danny D. Shen

Department of Pharmaceutics, University of Washington, Box 357610, H272
Health Sciences Building, Seattle, WA 98195, USA

Owing to its unique fatty acid structure, valproate exhibits a number of distinct pharmacokinetic characteristics in comparison to other aromatic and/or heterocyclic anticonvulsants. The pharmacokinetics of valproate have been the subject of regular reviews over the two decades of valproate's usage as an anticonvulsant [1–4]. This chapter provides a summary of the salient features of the pharmacokinetics of valproate, and additionally, perspectives on emerging areas of investigation, such as the role of cell membrane barrier transport in governing the brain distribution of valproate and the biochemical basis of drug interactions between valproate and other anticonvulsants.

Absorption

The main issue of interest with respect to the oral absorption of valproate relates to the wide variety of available pharmaceutical formulations. Three different chemical forms of valproate are in use: free acid, sodium salt, and divalproex sodium which is a coordinated 1 : 1 complex of sodium valproate and valproic acid. These chemical forms of valproate are available in a variety of commercial dosage forms: syrup (acid or sodium salt), tablets (acid or sodium salt), capsules (acid), sprinkle capsules (divalproex), enteric-coated tablets (sodium salt and divalproex), and controlled-release tablets (divalproex or sodium salt or acid plus sodium salt), and miniretard tablets (sodium salt). An intravenous formulation is also available for rapid loading of valproate.

Davies et al. [3] summarized the biopharmaceutical characteristics of the range of valproate formulations that have been reported in the literature. The gastrointestinal absorption of valproate from all available formulations appears to be complete (>90%) and is not significantly affected by food. Both the onset and rate of absorption, however, depend on the formulation and differ between fasted and fed states [4].

Absorption from the syrup, conventional tablet formulations, or the soft-gelatin capsule of free acid (Depakane) is very rapid and not affected by food; peak plasma concentration is consistently reached in less than 2 h. As expected, a variable time lag (~2–3 h) is observed in the onset of absorption from enteric-coated tablets (Depakote, Epilim EC), which is delayed further in the presence of food [5–8]. Peak plasma concentration after dosing with enteric-coated tablets is usually reached after about 3 to 5 h, although delays longer than 5 h are observed occasionally. A small diurnal variation (i.e. between morning and evening doses) in absorption from the enteric-coated formulations has been reported [9, 10].

Depakote Sprinkle consists of coated divalproex particles in a pull-apart capsule that is particularly suited to pediatric use. The onset of absorption from the sprinkle formulation is faster than divalproex tablet (~1 h), but the rate of absorption is slightly slower; as a result, peak-to-trough difference in valproate plasma concentration is smaller during chronic dosing. To permit once daily dosing, controlled release valproate formulations have been released recently in Europe and the USA [8, 11]. These formulations maintain a 50% fluctuation in plasma valproate concentration over the course of 24 h. Food does not appear to affect the absorption characteristics of the controlled-release formulations [12].

Rectal administration of valproate syrup has been shown to be effective in the treatment of refractory status epilepticus in both adults and children [13, 14]. Plasma valproate concentration time-course after rectal valproate syrup administration is comparable to that observed with the oral capsule [15], which indicates near complete absorption of valproate in the lower bowel.

The absorption of valproate, a reasonably lipophilic molecule, had been thought to involve simple passive diffusion across the gastrointestinal mucosa. Recently, Tsuji and his colleagues demonstrated carrier-mediated transport of valproate across the rabbit intestinal mucosa by both proton-coupled symport and anion-antiport mechanisms [16, 17]. The quantitative role of these putative mucosal transporters in the intestinal permeability of valproate has not been ascertained. The relatively high half-saturation constant (K_M) of these monocarboxylate transporters (in the millimolar range) may explain why capacity-limited oral absorption of valproate is not evident.

Distribution

Tissue distribution of valproate has received considerable attention because of its remarkable organ toxicities (hepatotoxicity and teratogenicity), and surprisingly limited brain distribution. Plasma protein binding governs both the distribution volume and plasma clearance of valproate, and is a major mechanism involved in valproate drug interactions.

Plasma protein binding

Valproate has a relatively small apparent volume of distribution (~0.15–0.2 L/kg); extravascular distribution is limited by its high binding affinity for plasma albumin. A unique feature of valproate binding to plasma proteins is its dependence on concentration over the therapeutic range (350–1040 μM or 50–150 μg/ml); i.e. therapeutic concentrations are within range of valproate's equilibrium dissociation constant for albumin binding. The average plasma free fraction of valproate (slightly below 10%) remains constant up to a plasma concentration of 75 μg/ml, and increases to 15% at 100 μg/ml, 22% at 125 μg/ml and 30% at 150 μg/ml [4, 18]. This concentration-dependent behavior in valproate's plasma protein binding leads to an apparent increase in the clearance of total valproate in plasma at high doses (see later section). It also gives rise to a greater fluctuation in free concentration compared to total concentration in plasma during a steady-state dosing interval. In their investigation, Cramer et al. [18] also noted a high degree of inter-patient variability in the plasma free fraction of valproate.

Diminished plasma valproate protein binding has been demonstrated in special populations (women during late pregnancy, elderly patients) and pathophysiologic states (liver disease, head trauma, end-stage renal diseases) that are commonly characterized by significant hypoalbuminemia [4]. Recently, Dasgupta and McLemore [19] reported elevated free fraction of valproate in sera of patients infected with human immunodeficiency (HIV) virus, which was attributed in part to hypoalbuminemia associated with HIV infection. Reduced plasma protein binding has also been observed in situations where there is elevation in bilirubin (e.g. liver disease and jaundiced neonates), and free fatty acids (e.g. insulin-dependent diabetes milletus), as well as in drug interactions involving plasma protein binding displacement (e.g. with salicylate, non-steroidal anti-inflammatory drugs, anti-HIV drugs) [3, 4, 19, 20].

The clinical significance of a reduced plasma protein binding depends on whether there is a concomitant change in metabolism. In the absence of a change in metabolic enzyme activity, a decrease in valproate binding to plasma proteins leads to an apparent lowering of total plasma valproate concentrations at steady state, but no change in free or unbound plasma valproate concentrations. Assuming that antiepileptic effect is related to free valproate concentration, changes in plasma protein binding alone would not alter clinical response. Therefore, dosage regimens should be maintained despite a fall in total plasma valproate. In situations where concurrent changes in hepatic drug-metabolizing enzymes occur, total plasma concentration of valproate can vary in either direction depending on the relative magnitude of the two competing interacting mechanisms. Dosage adjustments may be needed to compensate for changes in free drug concentration. Measurement of total plasma valproate is not informative; instead, monitoring of free valproate concentration is recommended.

Liver Uptake

Rapid uptake of valproate and its unsaturated metabolites into the liver, the extent of which depends on their plasma protein binding, has been demonstrated in rodents [21]. Moreover, the liver:plasma partitioning ratio decreased with an increase in dose or plasma concentration in the mouse. Brouwer et al. [22] investigated the subcellular distribution of valproate in livers of rats that had been infused with valproate to reach steady-state. They showed that valproate was localized mainly in the cytosolic fraction. The distribution of VPA between hepatic cytosol and serum was highly concentration-dependent. Cytosol:serum concentration ratio decreased as unbound valproate concentration in serum increased, which reflected saturation in binding to cytosolic proteins. Most interestingly, while the *in vivo* metabolic rate of valproate appeared to be a nonlinear function of unbound valproate concentration in serum (i.e. apparent Michaelis-Menten kinetics), it was linearly related to cytosolic concentration. This suggests that dose-dependent metabolic clearance of valproate (see later section) may be related to nonlinearity in the intrahepatic distribution process rather than saturable biotransformation.

Placental uptake

Animal studies have shown that valproate crosses the placenta very rapidly [23, 24]. Studies with blood samples collected from human mothers and their fetuses indicate that partitioning of valproate across the placenta is governed by its respective binding to maternal and fetal serum proteins. Equivalent free concentrations of valproate are observed in maternal and fetal serum at steady-state. Protein binding in fetal serum at term is higher than in the mother (90% versus 80% bound), which results in a fetal:maternal serum concentration ratio of greater than unity [25]. The opposite is observed at early gestation, when binding to fetal serum proteins is weak (50%) relative to maternal serum protein binding (90%) [26]. Elevated plasma free fraction associated with high peak serum concentrations of valproate (>75 μg/ml) may promote fetal accumulation. Hence, it has been suggested that the daily dose be divided or sustained-release formulation be used in pregnant women receiving valproate.

Autoradiography studies have also demonstrated that valproate accumulates in the neuroepithelium of the mouse embryo. The exact mechanism(s) for this fetal accumulation of valproate is not completely understood. Active sequestration mechanisms, or ion trapping at the relatively high intracellular pH during early organogenesis have been suggested [24, 27].

Breast milk

Very low concentrations of valproate are found in human breast milk of nursing mothers, with a reported range of 1–6% of maternal serum concentrations [25, 28–30]. Although the amount of unchanged valproate ingested by the infant is considered negligible, significant transfer of valproate metabolites, some of which may be toxic, remains a possibility. Careful monitoring is warranted in view of the recent case report of thrombocytopenia purpura and anemia in a breast-fed infant whose mother was treated with valproate [29].

Brain distribution

Rapid entry of valproate into the brain was demonstrated in some of the early studies in rodents and the dog [32, 33], which is entirely consistent with the prompt appearance of anticonvulsant effect after a single administration of valproate. The problem associated with the distribution of valproate to the brain is the unexpectedly low steady-state concentrations of valproate in the CNS relative to free drug concentration in plasma. This was first reported some years ago in a study of the cerebrospinal fluid concentration of valproate in monkeys during intravenous infusion [34]. Several later studies demonstrated low concentrations in human brain specimens obtained from surgical patients [4]. The average brain:free serum concentration ratio of valproate is about 0.5, which suggests asymmetric transport of valproate (i.e. efflux rate exceeds influx rate) across the blood-brain barrier [35]. The limited steady-state distribution of valproate into the brain explains the need for high micromolar levels (>350 µM) of valproate in blood to achieve seizure control. This results in a high systemic exposure and certainly contributes to the organ toxicity and systemic side-effects of this branched-chain fatty acid. In a study by Shen et al. [35], the brain:total concentration or brain:free concentration ratios varied widely across patients, which may explain the inter-individual variability in the concentration-effect relationship of valproate.

Recent animal studies have revealed that the bidirectional movement of valproate across the blood-brain barrier is mediated jointly by passive diffusion and carrier transport. The uptake of valproate from blood into the brain is facilitated by a medium- and long-chain fatty acid selective anion exchanger at the brain capillary epithelium, which accounts for two-thirds of the barrier permeability [34]. The mechanism(s) governing the efficient transport of valproate in the reverse direction, i.e. from brain to blood, appears to involve a probenecid-sensitive, active transport system at the brain capillary endothelium [37, 38]. Huai-Yun et al. [39] recently showed that valproate is a potent inhibitor of the ATP-dependent transporter(s) of the multidrug resistance-associated protein (MRP) family in the blood-brain barrier. This raises the possibility that MRPs may serve as the efflux transporter(s) of valproate.

A brain microdialysis study in rabbits suggests that another set of transporters exists at the neural cell membranes that shuttles valproate between the extracellular fluid and intracellular compartments within the brain parenchyma [40]. The putative parenchymal cell transport system is able to concentrate valproate within the cellular compartment, which has important implications in our understanding of the pharmacological mechanisms of valproate (i.e. membrane action versus intracellular mechanisms). Moreover, the efflux component is inhibited by probenecid, which provides another mechanism by which probenecid co-treatment increases the steady-state brain:plasma distribution ratio.

Limitation of brain distribution by avid efflux pumping appears to occur with a number of valproate analogs. Adkison et al. [41] have shown that several of the mono- and di-unsaturated metabolites of valproate also have low brain:free serum concentration ratios (< 1) as revealed by analysis of cortical specimens taken during epilepsy surgery. A full elucidation of the brain transport mechanism of valproate may hold the key to the design of "second generation" alkanoic acid type antiepileptics that have improved brain delivery, reduced systemic burden, and lower risk of organ toxicity.

Elimination

Hepatic metabolism is the principal route of valproate elimination (see Chapter 2.2 on "Chemistry and Biotransformation" for more details). Primary metabolism of valproate involves both Phase I (oxidative) and Phase 2 (conjugative) biotransformation pathways. About 30–40% of the usual clinical dose undergoes glucuronidation. Mitochondrial β-oxidation yielding (Δ^2-VPA and 3-oxo-VPA accounts for another 20–35% of the dose. Cytochrome P450-mediated formation of Δ^4-VPA and oxidation at the ω and ω-1 positions of the side-chain constitute a minor portion of the metabolic clearance.

Intrinsic hepatic clearance (i.e. drug-metabolizing enzyme activities) and plasma protein binding are the main determinants of clearance of an extensively metabolized, low extraction ratio drug such as valproate (i.e. $Cl = f_u \cdot Cl_{int}$). The following is a brief summary of how dose, age, physiologic states and diseases modulate these two variables to jointly effect changes in plasma clearance of valproate.

Dose dependence

Plasma clearance of valproate is dose-dependent. In single dose studies, plasma clearance of valproate shows a progressive increase, as the dose is incremented from 500 mg to as high as 3000 mg [42, 43]. The accelerated clearance at high doses is attributed to the increase in average plasma free fraction. In contrast, clearance expressed in terms of plasma free concentration or intrinsic metabolic clearance shows a downward trend with the increase in dose, which is thought to reflect saturation in β-oxidation based on urinary metabolite studies [44]. The magnitude of change in plasma free fraction exceeded the partial saturation of intrinsic metabolic clearance, which explains the net increase in clearance of total drug in plasma with an increase in dose.

Age

The effects of age on valproate clearance have been reviewed in detail elsewhere [4, 45]. A continuous change in valproate elimination kinetics occurs over the years of childhood development (see Table 1 for a summary of literature data). Clearance of total plasma valproate in neonates is comparable to the average value in adults despite the higher plasma free fractions due to elevated plasma free fatty acids and bilirubin (> 13% vs. < 10%). This reflects a slowed clearance of unbound valproate, which is consistent with immature microsomal drug-metabolizing enzyme activities in the neonates. Prolonged elimination half-life (> 17 h) is observed largely as a result of increases in plasma free fraction and distribution volume. Rapid maturation in microsomal enzyme function occurs over the first 2 years of life. Valproate clearance in infants and pre-school children is nearly twice as high as in adults and elimination half-life is relatively short. From the pre-school years onto adolescence, there is a gradual decline in total drug clearance and an increase in elimination half-life towards adult values.

Table 1. Valproate pharmacokinetic parameters in various age groups[a]

Age group	Subject/ AED therapy	Free fraction (%)	Distribution volume (L/kg)	Clearance (ml/h/kg)	$T_{1/2,\beta}$ Neonates (h)
Neonates (0–2 mo)	None or Polytherapy	~ 13	0.28–0.43	11–18	17–40
Infants (2–16 mo)	Monotherapy	~ 15	0.22–0.32	18–20	8–13
	Polytherapy	–	~ 0.28	~ 36	~ 6
Children (3–16 yr)	Monotherapy	< 12	0.18–0.22	12–14	9–12
	Polytherapy	10–13	0.18–0.30	19–28	7–9
Adults (16–60 yr)	Healthy Volunteers/ Monotherapy	4–10	0.13–0.20	6–9	12–15
	Polytherapy	–	0.14–0.19	15–18	5–9
Elderly (> 60 yr)	Healthy Volunteers	9–11	~ 0.16	7–8	~ 15

[a] Extracted from [4].

Clearance of total valproate in plasma appears to remain stable during the adult years. In elderly patients, however, hypoalbuminemia occurs commonly, which results in elevated valproate free fraction. The average clearance of total valproate in elderly patients is slightly lower than that of young adults. Intrinsic clearance of valproate is surmised to be lower in elderly patients than in young adult subjects, which is consistent with the commonly recognized age-related decline in hepatic drug-metabolizing function, especially oxidative metabolism.

In both children and adults, the average valproate clearance is consistently higher (by 50–100%) in patients receiving polytherapy (i.e. use of multiple anticonvulsants) than in patients receiving valproate alone. This reflects induction of valproate metabolism by several anticonvulsants used in conjunction with valproate, including phenobarbital, phenytoin and carbamazepine (see later section). As a result, high daily dosages per kilogram body weight (>30 mg/kg) along with frequent dosing or the use of sustained release formulations are needed, especially in young children, to maintain fluctuations in plasma valproate concentrations within the therapeutic range.

Pregnancy

Only a small number of studies on valproate clearance in pregnant women have been reported [46, 47]. The available data suggest that valproate clearance begins to increase late in the second trimester, and continues through the mid-portion of the third trimester. Much of the apparent increase in drug clearance can be attributed to the previously mentioned increase in maternal plasma free fraction as a result of elevated free fatty acids and low albumin level [26].

Diseases

Clearance kinetics of valproate have been investigated in liver and renal diseases. Liver diseases lead to both a decrease in plasma protein binding and intrinsic metabolic clearance of valproate; the changes tend to be equal in magnitude resulting in no apparent change in clearance of total drug in plasma [48].

The primary effect of end-stage renal disease appears to be a decrease in plasma protein binding (free fraction >20%) due to displacement by uremic constituents [49, 50]. In a study of one uremic child, Orr et al. [51] observed an apparently higher clearance of total valproate from plasma, but a normal clearance of free valproate. Hence, uremia did not induce a change in intrinsic metabolic clearance of valproate.

Pharmacokinetics of valproate have recently been investigated in head trauma patients as part of a clinical trial of valproate prophylaxis against post-traumatic seizures [52, 53]. Head trauma causes an acute and remarkable drop in plasma albumin, which is reflected in a variable and significant increase in plasma free fraction of valproate. This is accompanied by an increase in intrinsic clearance of valproate reflecting an induction of the oxidative pathways. The mechanism underlying the trauma-induced acceleration in valproate metabolism is not known.

Active Metabolites

A number of the mono-ene and di-ene metabolites of valproate have been shown to possess anticonvulsant activity at a dose potency near that of valproate [54, 55] (see Löscher, this volume, for more details). Early studies focused on the role of the predominant unsaturated metabolite in circulation, Δ^2-VPA, in the pharmacodynamics of valproate. However, the circulating levels of the unsaturated metabolites are typically lower than the parent drug [56]. Moreover, recent studies have shown that in epileptic patients the brain cortical and cerebrospinal fluid concentrations of these pharmacologically active metabolites are much lower than their plasma concentrations and low relative to valproate concentrations at the respective sites [41, 57]. Although sequestration or active concentration of some of the unsaturated species at the critical target sites in the brain has been suggested [58], it is unlikely that the unsaturated metabolites would play a quantitatively significant role in the anticonvulsant pharmacology of valproate. Lastly, Δ^4-VPA and $\Delta^{2,4}$-VPA have been implicated in the pathogenesis of hepatotoxcitiy [56] (see further details in Abbott and Anari, this volume).

Interactions

Drug interactions have been a major issue in the therapeutic management of anticonvulsants over the past two decades as polytherapy became the standard of practice for the treatment of many generalized and partial seizures. Anticonvulsant interactions commonly involve plasma protein displacement and alteration in drug metabolizing enzyme activity. While the pharmacokinetics of plasma protein binding interactions are well understood, the biochemical mechanisms underlying drug interactions have only been elucidated in recent years. Two general types of metabolic interactions are recognized: enzyme induction and competitive metabolic inhibition. The following is a brief overview of the metabolic mechanisms governing the effects of other anticonvulsants on valproate clearance, and the reciprocal effects of valproate on co-administered anticonvulsants.

The effects of other anticonvulsants on valproate metabolism are best reviewed in accordance to the hepatic enzymes responsible for the primary pathways of valproate metabolism. As mentioned, a major fraction of the valproate dose (40%) undergoes glucuronidation. Induction of the UDP glucuronosyltransferases involved in this pathway (UGT1A6, UGT1A8/9, UGT2B7) accounts primarily for the two-fold increase in valproate clearance associated with coadministration of enzyme inducers, such as phenytoin, phenobarbital or carbamazepine [59, 60]. A modest induction of valproate clearance by lamotrigine, possibly involving glucuronidation, has also been observed [61]. Valproate dosage adjustments are often required when these drugs are added to or withdrawn from a patient's regimen. It should be noted that inhibition of valproate glucuronidation by other anticonvulsants has not been reported.

Phenytoin and carbamazepine also induce the cytochrome P450-mediated oxidation of valproate [62–64]. Since these oxidative pathways account for less than 10% of the dose, the impact of induction of those pathways on the total clearance of valproate is relatively minor. However, the increased formation of reactive and toxic metabolites, such as Δ^4-VPA and $\Delta^{2,4}$-VPA, may render an individual susceptible to valproate-induced hepatotoxicity. Polytherapy is recognized to be an important factor associated with severe hepatotoxicity in young children [65].

Beta-oxidation of valproate, which accounts for 20–30% of dose, appears to be involved in the dose-dependent inhibition of valproate metabolism associated with felbamate co-administration [66, 67]. The enzymatic mechanism of this inhibitory interaction is not understood.

Interactions in which valproate is the "precipitant" (i.e. where it affects the plasma levels of other drugs) are related primarily to inhibition of four enzyme systems: UDP glucuronyltransferases, glucosyltransferases, cytochrome P450 2C9, and epoxide hydrolases. Valproate is a potent and broad-spectrum inhibitor of drug glucuronidation; notable examples include lorazepam [68], zidovudine [69] and lamotrigine [61, 70]. In the case of lamotrigine, the effect is pronounced and may require dosage adjustments. In patients who are not concomitantly treated with enzyme inducers (e.g. phenytoin or carbamazepine), the half-life of lamotrigine increases from 30 to 60 h in the presence of valproate, and in patients receiving enzyme inducers, lamotrigine half-life increases from approximately 15 h to 30 h. This interaction is due to inhibition in the UGT1A4-mediated formation of lamotrigine N-glucuronide [71].

Recently, the inhibition spectrum of valproate toward several cytochrome P450 enzymes has been investigated in human liver microsomes; CYP2C9 was the only enzyme inhibited at therapeutic levels of valproate [72]. This is consistent with the study by Sadeque et al. [64], which showed that human CYP2C9 along with CYP2A6 mediate the oxidation of valproate to Δ^4-VPA. Inhibition of CYP2C9 has been proposed to explain the elevations in phenytoin and phenobarbital levels associated with valproate coadministration. Inhibition of the formation of phenobarbital N-glucoside contributes also to the decrease in phenobarbital clearance [73].

Inhibition of the enzyme epoxide hydrolase by valproate was discovered several years ago and provided an explanation for the elevations in plasma levels of the metabolite of carbamazepine, carbamazepine-10,11-epoxide [74–76]. Carbamazepine epoxide contributes to both the anticonvulsant effect and neurotoxicity of carbamzepine. Elevation of the epoxide metabolite may result in carbamzepine-related toxicity [77, 78].

Conclusions

Over the past decade, we have gained a thorough understanding of the metabolism and pharmacokinetics of valproate in the clinical setting. However, there are still gaps in our knowledge concerning the mechanisms governing the entry, retention and clearance of valproate in the brain and organs susceptible to toxicity. Detailed understanding of these cellular disposition processes at the cellular level is important in the search for a valproate substitute that will retain the broad-spectrum anticonvulsant efficacy of valproate and avoid the adverse impacts on the liver and fetal development. Finally, there is still a lack of understanding of the link between valproate pharmacokinetics in the brain structures and the pharmacodynamics of its anticonvulsant and mood stabilizing effects.

References

1 Gugler R, van Unruh GE (1980) Clinical pharmacokinetics of valproic acid. *Clin Pharmacokinet* 5: 67–83
2 Zaccara G, Messori A, Moroni F (1988) Clinical pharmacokinetics of valproic acid – 1988. *Clin Pharmacokinet* 15: 367–389
3 Davis R, Peters DH, McTavish (1994) Valproic acid: a reappraissal of its pharmacological properties and clinical efficacy in epilepsy. *Drugs* 47: 332–372
4 Levy RH, Shen DD (1995) Valproic acid: absorption, distribution, and excretion. In: Levy RH, Mattson RH, Meldrum BS, (eds): *Antiepileptic drugs*. Raven Press, New York: 605–620
5 Levy RH, Cenraud B, Loiseau P, Akbaraly R, Brachet-Liermain A, Guyot M, Gomeni R, Morselli PL (1980) Meal-dependent absorption of enteric-coated sodium valproate. *Epilepsia* 21: 273–280
6 Fischer JH, Barr AN, Paloucek FP, Dorociak JV, Spunt AL (1988) Effect of food on the serum concentration profile of enteric-coated valproic acid. *Neurology* 38: 1319–1322
7 Carrigan PJ, Brinker DR, Cavanaugh JH, Lamm JE, Cloyd JC (1990) Absorption characteristics of a new valproate formulation of divalproex sodium-coated particles in capsules (Depakote Sprinkle). *J Clin Pharmacol* 30: 743–747
8 Roberts D, Easter D, O'Bryan-Tear G (1996) Epilimchrono: a multidose, crossover comparison of two formulations of valproate in healthy volunteers. *Biopharm Drug Dispos* 17: 175–182
9 Loiseau P, Brachet-Liesmain A, Guyot M, Morselli P (1982) Diurnal variations in steady state plasma concentrations of valproic acid in epileptic patients. *Clin Pharm* 7: 544–552
10 Yoshiyama Y, Nakano S, Ogawa N (1989) Chronopharmacokinetics study of valproic acid in man: comparison of oral and rectal administration. *J Clin Pharmacol* 29: 1048–1052
11 Samara E, Granneman R, Achari R, Locke C, Cavanaugh J, Boellner S (1997) Bioavailability of a controlled-release formulation of depakote. *Epilepsia* 38: S102

12 Cavanaugh JH, Granneman R, Lamm J, Linnen P, Chun AHC (1997) Effect of food on the bioavailability of a controlled-release formulation of depakote under multiple-dose conditions. *Epilepsia* 38: S54
13 Vajda FJE, Mihaly GW, Miles JL, Donnan GA, Bladin PF (1978) Rectal administration of sodium valproate in status epilepticus. *Neurology* 28: 897–899
14 Snead OC III, Miles MV (1985) Treatment of status epilepticus in children with rectal sodium valproate. *J Pediatr* 106: 323–325
15 Cloyd JC, Kriel RL (1981) Bioavailability of rectally administered valproic acid syrup. *Neurology* 31: 1348–1352
16 Tamai K, Takanaga H, Maeda H, Yabuuchi H, Sai Y, Suzuki Y, Tsuji A (1997) Intestinal brush-border membrane transport of monocarboxylic acids mediated by proton-coupled transport and anion antiport mechanisms. *J Pharm Pharmacol* 49: 108–112
17 Yabuuchi H, Tamai I, Sai Y, Tsuji A (1998) Possible role of anion exchanger AE2 as the intestinal monocarboxylic acid/anion antiporter. *Pharm Res* 15: 411–416
18 Cramer JA, Mattson RH, Bennett DM, Swick CT (1986) Variable free and total valproic acid concentrations in sole- and multidrug therapy. *Ther Drug Monit* 8: 411–415
19 Dasgupta A, McLemore JL (1998) Elevated free phenytoin and free valproic acid concentrations in sera of patients infected with human immunodeficiency virus. *Ther Drug Monit* 20: 63–67
20 Dasgupta A, Volk A (1996) Displacement of valproic acid and carbamazepine from protein binding in normal and uremic sera by tolmetin, ibuprofen, and naproxen: presence of inhibitor in uremic serum that blocks valproic acid – naproxen interactions. *Ther Drug Monit* 18: 284–287
21 Löscher W, Nau H (1984) Comparative transfer of valproic acid and of an active metabolite into brain and liver: possible pharmacological and toxicological consequences. *Arch Int Pharmacodyn Ther* 270: 192–202
22 Brouwer KLR, Hall ES, Pollack GM (1993) Protein binding and hepatobiliary distribution of valproic acid and valproate glucuronide in rats. *Biochem Pharmacol* 45: 735–742
23 Nau H, Hauck R-S, Ehlers K (1991) Valproic acid-induced neural tube defects in mouse and human: aspects of chirality, alternative drug development, pharmacokinetics and possible mechanisms. *Pharmacol Toxicol* 69: 310–321
24 Cotariu D, Zaidman JL (1991) Developmental toxicity of valproic acid. *Life Sci* 48: 1341–135
25 Nau H, Helge H, Luck W (1984) Valproic acid in the perinatal period: decreased maternal serum protein binding results in fetal accumulation and neonatal displacement of the drug and some metabolites. *J Pediatr* 104: 627–634
26 Nau H, Krauer B (1986) Serum protein binding of valproic acid in fetus-mother pairs throughout pregnancy: correlation with oxytocin administration and albumin and free fatty acid concentrations. *J Clin Pharmacol* 26: 215–221
27 Dencker L, Nau H, D'Argy R (1990) Marked accumulation of valproic acid in embryonic neuroepithelium of the mouse during early organogenesis. *Teratology* 41: 699–706
28 Nau H, Rating D, Koch S, Häuser I, Helge H (1981) Valproic acid and its metabolites: placental transfer, neonatal pharmacokinetics, transfer via mother's milk and clinical status in neonates of epileptic mothers. *J Pharmacol Exper Ther* 219: 768–777
29 von Unruh GE, Froescher W, Hoffmann F, Niesen M (1984) Valproic acid in breast milk: How much is really there? *Ther Drug Monit* 6: 272–276
30 Wisner KL, Perel JM (1998) Serum levels of valproate and cabamazepine in breastfeeding mother-infant pairs. *J Clin Pharmacol* 18: 167–169
31 Stahl MMS, Neiderud J, Vinge E (1997) Thrombocytopenic purpura and anemia in a breastfed infant whose mother was treated with valproic acid. *J Pediatr* 130: 1001–1003
32 Frey H-H, Löscher W (1978) Distribution of valproate across the interface between blood and cerebrospinal fluid. *Neuropharmacology* 17: 637–642
33 Pollack GM, Shen DD (1985) A timed intravenous pentylenetetrazol infusion seizure model for quantitating the anticonvulsant effect of valproic acid in the rat. *J Pharmacol Methods* 13: 135–146
34 Levy RH (1980) CSF and plasma pharmacokinetics: relationship to mechanisms of action as exemplified by valproic acid in monkey. In: J Lockard, A Ward (eds): *Epilepsy: A window to brain mechanisms*. Raven Press, New York, 11: 191–200.
35 Shen DD, Ojemann GA, Rapport RL, Dills RL, Friel PN, Levy RH (1992) Low and variable presence of valproic acid in human brain. *Neurology* 42: 582–585

36 Adkison KD, Shen DD (1996) Uptake of valproic acid into rat brain is mediated by a medium-chain fatty acid transporter. *J Pharmacol Exp Ther* 276: 1189–1200
37 Adkison KDK, Artru AA, Powers KM, Shen DD (1994) Contribution of probenecid-sensitive anion transport processes at the capillary endothelium and choroid plexus to the efficient efflux of valproic acid from the central nervous system. *J Pharmacol Exper Ther* 268: 797–805
38 Naora K, Shen DD (1995) Mechanism of valproic acid uptake by isolated rat brain microvessels. *Epilepsy Res* 22: 97–106
39 Huai-Yun H, Secrest DT, Mark KS, Carney D, Brandquist C, Elmquist WF, Miller DW (1998) Expression of multidrug resistance-associated protein (MRP) in brain microvessel endothelial cells. *Biochem Biophys Res Comm* 243: 816–820
40 Scism JL, Powers KM, Artru AA, Shen DD (1996) The effect of probenecid on extracellular and intracellular compartmentation of valproic acid in the rabbit brain as determined by microdialysis. *Pharm Res* 13: S456
41 Adkison KDK, Ojemann GA, Rapport R, Dills RL, Shen DD (1995) Distribution of unsaturated metabolites of valproate in human and rat brain – pharmacologic relevance? *Epilepsia* 36: 772–782
42 Bowdle TA, Patel IH, Levy RH, Wilensky AJ (1980) Valproic acid dosage and plasma protein binding and clearance. *Clin Pharmacol Ther* 28: 486–492
43 Gomez Bellver MJ, Garcia Sanchez MJ, Alonso Gonzalez AC, Santo Buelga D, Dominquez-Gil A (1993) Plasma protein binding kinetics of valproic acid over a broad dosage range: therapeutic implications. *J Clin Pharmacol Ther* 18: 191–197
44 Granneman GR, Marriott TB, Wang SI, Sennello LT, Hagen NS, Sonders RC (1984) Aspects of the dose-dependent metabolism of valproic acid. In: RH Levy, WH Pitlick, M Eichelbaum, J Meijer (eds): *Metabolism of antiepileptic drugs*. Raven Press, New York, 97–104.
45 Battino D, Estienne M, Avanzini G (1995) Clinical pharmacokinetics of antiepileptic drugs in pediatric patients. Part I: phenobarbital, primidone, valproic acid, ethosuximide and mesuximide. *Clin Pharmacokinet* 29: 257–286
46 Plasse J-C, Revol M, Chabert G, Ducerf F (1979) Neonatal pharmacokinetics of valproic acid. In: D Schaaf, E van der Kleijn (eds): *Progress in clinical pharmacy*. Amsterdam: Elsevier/North-Holland Biomedical Press, 247–252
47 Philbert A, Dam M (1982) The epileptic mother and her child. *Epilepsia* 23: 85–99
48 Klotz U, Rapp T, Müller WA (1978) Disposition of valproic acid in patients with liver disease. *Eur J Clin Pharmacol* 13: 55–60
49 Gugler R, Mueller G (1978) Plasma protein binding of valproic acid in healthy subjects and in patients with renal disease. *Br J Clin Pharmacol* 5: 441–446
50 Brewster D, Muir NC (1980) Valproate plasma protein binding in the uremic condition. *Clin Pharmacol Ther* 27: 76–82
51 Orr JM, Farrell K, Abbott FS, Ferguson S, Godolphin WJ (1983) The effects of peritoneal dialysis on the single dose and steady state pharmacokinetics of valproic acid in a uremic epileptic child. *Eur J Clin Pharmcol* 24: 387–390
52 Anderson GD, Gidal BE, Hendryx RJ, Awan AB, Temkin NR, Wilensky AJ, Winn HR (1994) Decreased plasma protein binding of valproate in patients with acute head trauma. *Br J Clin Pharmacol* 37: 559–562
53 Anderson GD, Awan AB, Adams CA, Temkin NR, Winn, HR (1998) Increases in metabolism of valproate and excretion of 6β-hydroxycortisol in patients with traumatic brain injury. *Br J Clin Pharmacol* 45: 101–105
54 Löscher W, Nau H (1985) Pharmacological evaluation of various metabolites and analogues of valproic acid: anticonvulsant and toxic protencies in mice. *Neuropharmacology* 24: 427–435
55 Abbott FS, Acheampong AA (1988) Quantitative structure-anticonvulsant activity relationships of valproic acid, related carboxylic acids and tetrazoles. *Neuropharmacology* 27: 287–294
56 Baillie TA, Sheffels PR (1995) Valproic acid: chemistry and biotransformation. In: RH Levy, RH Mattson, BS Meldrum (eds): *Antiepileptic drugs*. Raven Press, New York, 589–604
57 Löscher W, Nau H, Siemes H (1988) Penetration of valproate and its active metabolites into cerebrospinal fluid of children with epilepsy. *Epilepsia* 29: 311–316

58 Löscher W, Fisher JE, Nau H, Honack D (1989) Valproic acid in amygdala-kindled rats: alteration in anticonvulsant efficacy, adverse effects and drug and metabolite levels in various brain regions during chronic treatment. *J Pharmacol Exp Ther* 250: 1067–1078
59 Abbott F, Panesar S, Orr J, Burton R, Farrell K (1986b) Effect of carbamazepine on valproic acid metabolism. *Epilepsia* 27: 591
60 Scheyer RD, Mattson RH (1995) Valproic acid: interactions with other drugs. In: RH Levy, RH Mattson, BS Meldrum (eds): *Antiepileptic drugs*. Raven Press, New York, 621–631
61 Anderson GD, Yau MK, Gidal BE, Harris SJ, Levy RH, Lai AA, Wolf KB, Wargin WA, Dren AT (1996) Bidirectional interaction of valproate and lamotrigine in healthy subjects. *Clin Pharmacol Ther* 60: 145–156
62 Levy RH, Rettenmeier AW, Anderson GD, Wilensky AJ, Friel PN, Baillie TA, Acheampong A, Tor J, Guyot M, Loiseau P (1990) Effects of polytherapy with phenytoin, carbamazepine, and stiripentol on formation of 4-ene-valproate, a hepatotoxic metabolite of valproic acid. Clin Pharmacol Ther 48: 225–235
63 Kondo T, Otani K, Hirano T, Kaneko S, Fukushima Y (1990) The effects of phenytoin and carbamazepine on serum concentration of mono-unsaturated metabolites of valproic acid. Br J *Clin Pharmacol* 29: 116–119
64 Sadeque AJM, Fisher MB, Korzekwa KR, Gonzalez FJ, Rettie AE (1997) Human CYP2C9 and CYP2A6 mediate formation of the hepatotoxin 4-ene-valproic acid. *J Pharmacol Exp Ther* 283: 698–703
65 Bryantt AE III, Dreifuss FE (1996) Valproic acid hepatic fatalities. III. US experience since 1986. *Neurology* 46: 465–469
66 Wagner ML, Graves NM, Leppik IE, Remmel RP, Shumaker RC, Ward DL, Perhach JL (1994) The effect of felbamate on valproic acid disposition. *Clin Pharmacol Ther* 56: 494–502
67 Hooper WD, Franklin ME, Glue P, Banfield CR, Radwanski E, McLaughlin DB, McIntyre ME, Dickinson RG, Eadie MJ (1996) Effect of falbamate on valproic acid disposition in healthy volunteers: inhibition of β-oxidation. *Epilepsia* 37: 91–97
68 Samara EE, Granneman RG, Witt GF, Cavanaugh JH (1997) Effect of valproate on the pharmacokinetics and pharmacodynamics of lorazepam. *J Clin Pharmacol* 37: 442–450
69 Lertora J, Rege A, Greenspan D, Akula S (1994) Pharmacokinetic interaction between zidovudine and valproic acid in patients infected with immunodeficiency virus. *Clin Pharmacol Ther* 56: 272–278
70 Yuen AWC, Land G, Weatherley BC, Peck AW (1992) Sodium valproate acutely inhibits lamotrigene metabolism. *Br J Clin Pharmacol* 33: 511–513
71 Green MD, Bishop WP, Tephley TR (1995) Expressed human UGT-1.4 protein catalyzes the formation of quaternary ammonium-linked glucuronides *Drug Metab Dispos* 23: 299–302
72 Hurst S, Labroo R, Carlson S, Mather G, Levy R (1997) *In vitro* inhibition profile of valproic acid for cytochrome P450. International Society for the Study of Xenobiotics. Hilton Head, South Carolina (abstract): 64
73 Bernus I, Dickinson RG, Hooper WD, Eadie MJ (1994) Inhibition of phenobarbitone N-glucosidation by valproate. *Br J Clin Pharmacol* 38: 411–416
74 Kerr BM, Rettie AW, Eddy AC, Loiseau P, Guyot M, Wilensky AJ, Levy RH (1989) Inhibition of human liver microsomal epoxide hydrolase by valproate and valpromide: *in vitro/in vivo* correlation. *Clin Pharmacol Ther* 46: 82–93
75 Pisani F, Caputo M, Fazio A, Oteri G, Russo M, Spina E, Perucca E, Bertilsson L (1990) Interaction of carbamazepine 10,11-epoxide, an active metabolite of carbamazepine, with valproate: a pharmacokinetic study. *Epilepsia* 31: 339–342
76 Robbins DK, Wedlund PJ, Kuhn R, Baumann RJ, Levy RH, Chang S-L (1990) Inhibition of epoxide hydrolase by valproic acid in epileptic patients receiving carbamazepine. *Br J Clin Pharmacol* 29: 759–762
77 Rambeck B, Salke-Treumann A, May T, Boenigk HE (1990) Valproic acid induced carbamazepine 10,11-epoxide toxicity in children and adolescents. *Eur Neurol* 30: 79–83
78 McKee PJ, Blacklaw J, Butler E, Gillham RA, Brodie MJ (1992) Variability and clinical relevance of the interaction between sodium valproate and carbamazepine in epileptic patients. *Epilepsy Res* 11: 193–198

Milestones in Drug Therapy
Valproate
ed. by W. Löscher

Toxicity

Matthias Radatz and Heinz Nau

Department of Food Toxicology, Center of Food Science, School of Veterinary Medicine, Bischofsholer Damm 15, D-30173 Hannover, Germany

Introduction

Valproic acid is widely used in the treatment of epilepsy and as a prophylactic agent for migraine [1]. It is also useful as a mood stabilizer in bipolar disorders [2] or as an adjunct therapeutic for alcohol addiction during withdrawal periods [3]. Other indications for this drug are under investigation like the treatment of intractable hiccups [4] and as an anti-emetic in chemotherapy [5]. The side-effects caused by this important drug will therefore remain to be a subject of discussion and intensive research.

The main purpose of this chapter will be a mechanistical view of the main toxic effects exerted by valproic acid. Hereby, we will discuss animal models and *in vitro* studies in the context of clinically and experimentally derived human data. The principal topics are the hepatic failure, the developmental toxicity and some related drug interactions. A detailed description of the side-effects from a clinical point of view is provided in the chapter by D. Schmidt, this volume.

General considerations about interspecies variations

Experimental models are necessary for the study of the mechanisms of useful and unwanted activities of drugs, and are thus important for the development of more potent, less toxic medications. The extrapolation of experimentally derived data to the human situation is critical. The interspecies variations in the metabolism of a particular drug as well as the interaction with the target site involved in the specific action of this drug have to find consideration in the interpretation of the obtained results [6].

Pharmacokinetic aspects of valproate toxicity

Although the metabolite pattern of valproic acid (VPA) is not very different between laboratory animals and the human, the variations of pharmacokinetic parameters like protein binding, plasma half-life and elimination rate can be considerable [7]. And with regard to valproate's embryotoxic properties, the placental transfer is also an aspect of species comparison. However, in the human as well as in rodents and rhesus monkeys valproic acid and its main metabolites readily cross the placental barrier [8–12].

The mouse is the predominant animal model for the teratogenic effects of valproic acid (see also below: "Developmental toxicity"). Two major points separate rodent teratogenicity from that of the human. The developmental defect affecting the nervous system occurs at a different location of the neural tube (anterior vs. posterior), and the doses needed to induce this defect are about 10-fold greater in the mouse [13]. To account for these differences other animal models had to be developed [14, 15]. A constant-rate infusion via osmotic minipumps using human therapeutic drug concentrations could only produce fetal growth retardation and increased resorption rates but no exencephaly in the mouse [16]. Further studies led to the important findings that the total valproate exposure over time (area under the concentration-time curve) is correlated with embryolethality and fetal weight retardation, whereas peak plasma levels correlate with the prevalence of neural tube defects. Valproate plasma levels have to reach about 230 µg/ml to induce exencephaly in the mouse [17]. The human therapeutic range seldom goes beyond 100 µg/ml, but higher peak serum concentrations up to 180 µg/ml have been reported if single high doses had been administered [18, 19]. Because valproic acid is highly protein-bound in man several factors like pregnancy, stress, other drugs and dietary factors can cause a sudden increase of its unbound fraction leading to higher and possibly harmful exposure of the fetus [20].

Considering the pharmacokinetic species-differences the human seems more susceptible to the teratogenic activity of valproic acid. However, because of the lack of data on human embryonic drug concentrations during the sensitive stage of organogenesis a direct species comparison cannot be made.

Acute and chronic toxicity of valproic acid in experimental animals

Two types of adverse reactions to valproate treatment can be distinguished. The most severe effects are of the idiosyncratic type which include hepatic failure [21], acute pancreatitis [22], hematological disorders [23] and possible induction of systemic lupus erythematosus [24]. More recently, a case of multiorgan system failure has been reported [25]. The developmental toxicity caused by valproic acid only partly belongs to this group because a dose-related teratogenicity has been shown in animal models [17].

Besides these idiosyncrasies several dose-related adverse effects have been described in the human patient. They include gastrointestinal discomfort like nausea and vomiting, neurological signs such as drowsiness, mild ataxia, tremor, and coma, metabolic disturbances and hair changes [26].

Additionally, endocrinological disorders like excessive weight gain, polycystic ovaries and hyperandrogenism are a common complication when treating epileptic women [27]. Cognitive and behavioral effects caused by therapeutic levels of valproic acid are rather mild when compared to other classical antiepileptic drugs [28]. Many of these side-effects have not been observed or were not investigated in animals. Some comments are given as far as information is available in the literature.

Acute toxicity

Most acute overdose of valproate in the human results in no ill effects [29]. Patients have survived doses of up to 36 g [30, 31]. In animals LD50 values were determined for cats, rabbits, rats and mice ranging from 565 to 1700 mg/kg (see Tab. 1). Conine et al. described an inverse relationship of animal size and oral LD50 values [32]. Clinical signs of toxicity include ataxia, sedation, muscular weakness, intermittent head shaking, hypothermia, catalepsy and ptosis in rodents, and sedation, loss of coordination, sleep, tremors and/or emesis in dogs [32–34].

Chronic toxicity

Chronic valproate exposure of experimental animals mainly affects the liver. After feeding rats with 200 mg/kg daily for 3 months, first ultrastructural changes were noted in mitochondria and peroxisomes of the

Table 1. Acute toxicity of valproic acid in experimental animals

Species	LD_{50} [mg/kg]			References
	p.o.	i.p.	i.v.	
Mice	1,700	1,060		[269]
	1,197	838	766	[270]
			750	[271]
		1,104		[272]
		960		[33]
		870	630	[273]
	2,474			[34]
Rats	1,530	790		[269]
	1,519	1,045	964	[270]
				[272]
	1,788			[34]
Rabbits		1,200		[269]
	1,650			[270]
Cats		565		[269]

Modified from [31].

hepatocytes [35]. The greatest damage was seen at 9 and 12 months of treatment, in the form of microvesicular steatosis and cholesterosis. Ponchaut et al. found a 30% reduction in the coupled respiration rates with a decrease of mitochondrial cytochrome aa3 content in hepatocytes from rats fed a diet containing 1% valproic acid for 75 days [36]. Infant mice, injected subcutaneously with a single daily dose of 100–200 mg/kg for 5 days, had reduced plasma β-hydroxybutyrate levels and depleted liver glycogen stores [37]. These effects were similar to the reduced fasting ketonemia detected in epileptic children treated with valproic acid [38].

Valproic acid also influences brain metabolism and development. Brain weights and metabolic flux through the glycolytic pathway and the citric acid cycle were significantly reduced in 2–4 day old mice [37]. Significant reductions of the brain weights were also found in 4 day old rats treated with 75 or 200 mg/kg for 14 day [39]. Chronic valproic acid administration caused ultrastructural changes in the cerebellum of rats [40]. Sobaniec-Lotowska et al. noticed necrosis and swelling of capillary endothelial cells with reduced and occluded capillary lumens. They also found a proliferation of glial cells and structural changes of the blood-brain-barrier associated with neuronal cell damage. Functional deficits such as ataxia after chronic valproate administration have also been described for young rats [39]. The authors thought the reduced cerebellar weights of these animals to be partly responsible. Vorhees reported a behavioral teratogenicity of valproic acid at non-malforming doses [41]. Post-weaning offspring of valproate-treated Sprague-Dawley rats exhibited reduced open-field central activity, lengthened straight channel swimming times and reduced startle responses. It has not yet been determined if there is any relation to the cognitive and behavioral effects in human patients.

Other adverse effects

Pancreatitis as a serious side-effect of valproate therapy in man is rarely seen in experimental animals. The only report known to us comes from Walker et al. who found a dose-dependent increased incidence of atrophic pancreatitis in rats treated with calcium valproate over 1 year [34].

Even less is known about the endocrinological disturbances in animals under valproate treatment. The influence of valproic acid on body weight has been studied by Squadrito et al. in 4-week old genetically obese (Zucker) rats [42]. After a period of 6 weeks the rats had a lower body weight and reduced hyperinsulinemia than the untreated Zucker controls. This is in contrast to the findings in female patients on valproate therapy who develop a metabolic syndrome including weight gain, hyperinsulinemia and hyperandrogenism [43]. In a retrospective study, 50% of the women had a significant weight gain and 64% polycystic ovaries and/or hyperandrogenism [27]. Since these side-effects are a common reason for a discontinuation of the therapy further studies including experimental animals should be encouraged.

Hepatotoxicity

There are two forms of hepatotoxicity caused by valproic acid. The more common fast onset, mild and reversible type, and a rare, idiosyncratic form which can be fatal. After initiation of treatment with valproate up to 44% of patients have elevated liver enzyme activities without clinical symptoms [44]. Because these laboratory anomalies are dose-related and can be reversed by discontinuation of the drug it is likely that enzyme induction is responsible rather than enzyme release through hepatocellular damage [45].

First cases of fatal hepatic failure due to valproic acid therapy were reported by several authors in 1979 [46–48]. The overall incidence for this severe side-effect has been estimated to be between 1:5000–20000 depending on the literature source [45, 49, 50]. Dreifuss et al. found that the primary risk for fatal hepatotoxicity was directly related to the age of the patient and the type of treatment. The patient group with the highest risk is under the age of 2 years and on anticonvulsant polytherapy (1:500) [45]. Additional predisposing factors are mental retardation, intractable epilepsy and other neurological diseases [50]. Furthermore, viral infections [51], starvation, familial metabolic disorders [52], i.e. urea cycle enzyme defects [53] may enhance the individual susceptibility. The idiosyncratic type of hepatotoxicity is most likely to appear within 6 months from the start of therapy [50]. The mechanism of liver toxicity is still not understood. In contrast to the teratogenic activity which is caused by the parent compound, the metabolites, in particular 2-propyl-4-pentenoic-acid (4-ene-VPA), may also be responsible for some of the liver damage.

Valproic acid and its metabolites

Valproic acid is mainly metabolized in the liver and less than 4% are excreted unchanged by the kidneys [54, 55]. The metabolite pattern of laboratory animal species is comparable with the human. The major part of valproate is β-glucuronidated and excreted as such in the urine. The dog seems to be an exception in that the β-oxidation is the predominant pathway. In all species the main plasma metabolites are 2-propyl-2-pentenoic acid (2-ene-VPA) and 2-propyl-3-keto-pentanoic acid (3-keto-VPA) [56] although the ω-oxidation plays an important role in the rhesus monkey [57]. The rat and the mouse have rather similar metabolite distributions compared to man.

Mechanisms of liver toxicity

There is probably not one particular mechanism by which valproic acid exerts its toxic action on hepatic tissue. Its influences on metabolism are manifold and, therefore, toxic effects may occur in various biochemical systems. Several reviews have given detailed information on this subject [21, 58, 59].

Effects on lipid and carnitine metabolism
Becker and Harris found an increase in the medium chain acyl-CoA fraction mainly made up of valproyl-CoA in isolated rat hepatocytes after valproate administration [60]. Coenzyme A, acetyl-CoA and the long chain acyl-CoA fraction were decreased. They suggested that a rapid accumulation of valproyl-CoA and its (CoA-ester) metabolites or a sequestration of coenzyme A caused an inhibition of fatty acid oxidation. Evidence for an increased usage of the alternative pathway of ω-oxidation has been given by Mortensen et al. who found an elevation of C6 to C10 dicarboxylic acids in the urine of patients treated with valproate [61]. Furthermore, an inhibition of ketogenesis has been indicated by a lowered hepatic β-hydroxybutyrate/acetoacetate ratio and β-hydroxybutyrate content in valproate-treated rats and mice [37, 60]. Coude et al. had similar findings with a concomitant reduction of acetyl-CoA levels in rat hepatocytes [62].

Absolute deficiency and high acyl- to free carnitine ratios have been reported after valproate administration in children and mice [63, 64]. Carnitine is responsible for the transport of long chain fatty acids through the inner mitochondrial membrane for β-oxidation and to regulate the ratio of free coenzyme A to acyl-CoA in the mitochondrium [65]. The importance of carnitine in the CoA depletion mechanism has been stressed by Thurston and Hauhart. Concurrent treatment with carnitine and pantothenic acid, the precursor of CoA, could reduce valproate's diminishing effects on β-hydroxybutyrate, free CoA and acetyl-CoA levels in mouse plasma [66]. Coulter found an alleviation of valproate induced hyperammonemia [67].

In a recent study of 41 epileptic patients on valproate mono- and polytherapy total and free carnitine plasma levels were significantly reduced, but there was no significant correlation between carnitine concentrations and valproate doses [68]. Possible causes for the serum carnitine depletion are a decreased renal threshold for free carnitine [69] and the inhibition of the β-oxidation leading to an increased formation of acyl-carnitine esters [70]. A renal leak for carnitine esters has also been suggested since valproyl- and acetyl-carnitine have been detected in urine of patients receiving valproic acid [71].

Effects on carbohydrate metabolism

Valproic acid and some of its metabolites impair gluconeogenesis in rat hepatocytes and renal tubules, possibly via inhibition of pyruvate carboxylase, an acetyl-CoA dependent enzyme at the outset of glucose synthesis [72, 73]. Low plasma glucose levels have been found after valproate treatment in mice, fasted rats and man during liver failure [64, 72, 74].

Valproic acid also interferes with energy metabolism through an inhibition of oxidative phosphorylation [75–77]. The depletion of cytochrome aa3, the final electron carrier of the inner mitochondrial membrane, has been suggested as a possible mechanism [36, 78].

Effects on urea cycle and ammonia

Intramitochondrial enzymes of the urea cycle are also effected by valproate. The first step enzyme to bind ammonia is carbamyl phosphate synthetase which is inhibited either by propionyl-CoA or by a high acyl-CoA/CoA-SH ratio. Valproic acid may raise propionate and propionyl-CoA levels and, as shown above, acyl-CoA/CoA ratios [79, 80]. In rat hepatocytes valproate also markedly decreased concentrations of N-acetylglutamate, an essential activator of carbamyl phosphate synthetase [81]. A consequence of carbamyl phosphate synthetase deficiency is hyperammonemia which has been reported with valproate treatment by many authors [82, 83].

Hyperammonemia has also been detected in rats treated with 250 mg/kg valproic acid three times daily for 7 consecutive days [84]. In a series of studies Löscher et al. examined the effects of valproic acid and the metabolite 2-ene-VPA on rat liver function and morphology. With phenobarbital comedication, plasma ammonia, alanine transferase (ALT) and microvesicular steatosis increased as compared to valproate treatment alone [85]. In contrast, fasting valproate-treated rats for 28 or 40 h would cause an attenuation of these effects, possibly by inducing increased plasma carnitine levels [86]. 2-ene-VPA alone did not cause hepatotoxicity.

A significant relationship between blood ammonia and plasma carnitine values in human patients was found by some authors [87], others found no correlation [88]. Increased protein intake can elevate ammonia levels but these increases can be attenuated by carnitine supplementation [89, 90]. The role of carnitine as a buffer of toxic levels of acyl-CoA metabolites has been emphasized by Stumpf et al. [91].

Patients with deficient carbamyl phosphate synthetase and ornithine transcarbamylase activities are predisposed to hyperammonemic states because ammonia in the form of carbamyl cannot sufficiently enter the urea cycle. Treatment with valproic acid in these patients can increase blood ammonia concentrations to toxic levels [92, 93].

Histopathology

The pathognomonic liver lesion is a widespread microvesicular steatosis, which in case of human valproate toxicity is often accompanied by centrilobular necrosis, and occasional cirrhosis [74]. An explanation of this pathological feature has recently been given by Fromenty and Pessayre [94]. Fatty acids, poorly β-oxidized in the damaged mitochondria, are mainly esterified into triglycerides. These triglycerides possibly emulsified by a rim of non-esterified fatty acids accumulate as small vesicles in the hepatocytes. Similar lesions have been found in rat [95] and mouse livers [96] after valproate treatment and in livers of young rats after 4-ene-VPA administration [97]. Liver histopathology in valproate-treated rats has by no means been consistent. Several studies have found immature rats to be resistant to the steatogenic effects of valproic acid [84, 95, 97]. Other authors reported typical microvesicular steatosis in this age group as well [98, 99]. Liver necrosis which often accompanies steatosis in human liver failure patients can usually not be seen in animal models, except in a single study by Cotariu et al. [100] in young rats.

Toxic effects of valproate metabolites

Siemes et al. investigated the possible toxicity of valproate metabolites in a cohort of 470 epileptic children [101]. The most frequent change of the metabolite pattern was an impaired β-oxidation with a high concentration of 2-ene-VPA and 2,3′-diene-VPA and/or a decreased concentration of its oxidation product 3-keto-VPA. In several other patients valproate degradation was diverted to alternative pathways resulting in increasing amounts of unsaturated metabolites such as 3-ene-, 4-ene, 2,3′-ene-, 4-OH-,5-OH-, 4-keto-VPA and 2-propylglutaric acid. Overall, there was no particular metabolite pattern associated with hepatotoxicity; some children even had a very active β-oxidation with high concentrations of 3-keto-VPA. There was no correlation between elevated 4-ene-VPA levels and liver toxicity.

Special attention has been paid to the microsomal δ-dehydrogenation product 4-ene-VPA for its resemblance with methylene cyclopropylacetic acid, the active toxin of Jamaican vomiting sickness. Both compounds as well as the experimental hepatotoxin 4-pentenoic acid cause similar hepatic lesions which can also be seen with Reye's syndrome [45, 48].

4-ene-VPA is formed in microsomes by desaturation of valproic acid, most likely via cytochrome P450 catalyzation [102]. Several subtypes have shown increased terminal desaturation activities: In rabbit microsomes CYP4B1 [103] and in human microsomes CYP2A6 and CYP2C9 [104].

4-ene-VPA is cytotoxic in cultured rat hepatocytes [105] and causes microvesicular steatosis in rats and mice (see above). It directly inhibits important enzymes of the β-oxidation pathway [97] and, thus, possibly enhances its own generation via alternative valproate metabolism. The hepatotoxic potential of 4-ene-VPA and its metabolites has been attributed to the likely formation of reactive, electrophilic species with subsequent alkylation of cellular proteins and other macromolecules [106]. Its main metabolite 2,4-diene-VPA was found in *in vitro* studies with perfused rat liver [106] and in the blood of rhesus monkeys after 4-ene-VPA administration [107]. The theory of radical formation is supported by the works of Kassahun et al. who detected glutathione (GSH) and N-acetylcysteine conjugates with 4-ene- and 2,4-diene-VPA in urine of humans and rats [108]. Tang et al. also found a depletion of total liver and mitochondrial GSH in Sprague-Dawley rats after 4-ene-VPA treatment [109]. Another research group fed sodium valproate to mice for up to 21 days and found a time dependent reduction of glutathione concentrations in liver and kidney tissue possibly due to an increased lipid peroxidation [110]. Another putative toxic alkylating species after β-oxidation of 4-ene-VPA was proposed by Baillie. This synthetic 3-keto-4-ene-VPA has yet to be recovered *in vivo* [111].

Granneman et al. and other authors found evidence for an unstable 2-(2′-carboxypentanyl) oxirane (4,5-epoxide-VPA) metabolite of 4-ene-VPA [112] which is able to bind covalently to the prosthetic heme group of the cytochrome and thereby destroying the enzyme [113]. Recently, Tang and Abbott have detected GSH and other thiol conjugates of 4,5-epoxide-VPA and 2,4-diene-VPA, respectively, in rat bile samples [114].

Another proposed hepatotoxic mechanism for 2,4-diene-VPA is based on its high affinity to the mitochondrial trifunctional protein. The trifunctional protein catalyzes three consecutive reactions in the oxidation of long chain fatty acids. A gastrin cross-linking assay showed that 2,4-diene-VPA reversibly occupied the trifunctional protein α-subunit possibly inhibiting both its hydratase and dehydrogenase activities [115].

There are structural similarities between valproate metabolites and metabolites formed from the β-oxidation of branched-chain amino acids. Anderson et al. compared the concentrations of straight- and branched-chain fatty acids in the urine of patients of valproate treatment [116]. From the observed pattern of isoleucine and other metabolites they concluded that valproate has the ability to inhibit four acyl-CoA dehydrogenases involved in branched-chain amino acid metabolism. A competition of valproic acid and isoleucine metabolites for β-ketothiolase, the final enzyme, has also been hypothesized by Dickinson et al. [117]. Therefore, an additional risk for the occurrence of metabolite related toxicity is given by genetic deficiencies in these enzymes.

Peroxisomal β-oxidation

β-oxidation by peroxisomes seems to contribute only little to the metabolism of valproic acid and its metabolites. Medium and long-term administration to rats led to proliferation of peroxisomes and an increase in their β-oxidation enzyme activity [118, 119]. However, in rat liver homogenates the oxidation of CoA-esters of valproic acid and 4-ene-VPA would not proceed further than to the β-hydroxyacyl-CoA dehydrogenation step. And the responsible initiation enzyme was apparently distinct from fatty and bile acyl-CoA oxidases [120]. The significance of peroxisomal β-oxidation of valproic acid still remains to be investigated.

Drug interactions

Since comedication especially with other antiepileptic drugs is a risk factor for valproate hepatotoxicity some aspects of the subject will be discussed. The classical antiepileptic drugs primidone, phenobarbital, phenytoin and carbamazepine are potent inducers of microsomal enzymes such as cytochrome P450 [121]. This has two effects on valproate metabolism. Firstly, the reduction of total and free plasma valproate levels between 20 and 50% [122, 123]. And secondly, and probably important for toxicity, an increase of the alternate pathways of ω-oxidation and δ-dehydrogenation. Significant elevations of plasma 4-ene-VPA concentrations up to 100% have been found with concurrent carbamazepine and phenytoin administration [124, 125]. The toxicity of 4-ene-VPA has been described above. Valproic acid itself has enzyme inducing and inhibitory abilities. It inhibits both the p-hydroxylation [126] and the N-glucosidation [127] of phenobarbital and significantly reduces its metabolism and total clearance. Carbamazepine concentrations, on the other hand, are hardly altered by valproate addition. Instead, valproic acid seems to inhibit the enzymes of the epoxide-diol pathway like epoxide hydrolase, which are essential for the elimination of the carbamazepine-10,11-epoxide [128, 129]. Moreover, carbamazepine and its epoxide are both displaced from plasma protein binding sites. Phenytoin plasma levels can vary after valproic acid has displaced it from protein binding sites. A subsequent increase of elimination will be counteracted by metabolic inhibition. Depending on the degree of inhibition plasma levels can be above or below pretreatment levels [130, 131].

In conclusion, when valproic acid is added to ongoing treatment with phenobarbital, carbamazepine or phenytoin the doses of these drugs have to be reduced accordingly to avoid neurotoxicity. Of the newer antiepileptic drugs, lamotrigine has been promising for its good effectiveness against partial and generalized seizures including generalized absence seizures [132]. Valproate was shown to interfere with lamotrigine glucuronidation, more than doubling its plasma half-life. This decreased elimination may eventually predispose a patient to the known life-threatening dermatological side-effects of lamotrigine, including toxic epidermal necrolysis and Stevens-Johnson Syndrome [133].

Conclusions

Idiosyncratic hepatic failure is a very rare but severe side-effect of valproic acid treatment which usually occurs within 90 days from the start of therapy [134]. The predisposing factors of age and antiepileptic polytherapy are indicative of an involvement of the drug's metabolism in this reaction. The high risk for patients under 2 years of age may be explainable by the functional immaturity of the glucuronyl transferase system in neonates [135]. An insufficient glucuronidation, which is usually the major catabolic pathway for valproate, may lead to overloading the β-oxidation pathways. Studies in the guinea pig also showed a reduced metabolic capacity for valproic acid in newborns compared to older animals [136].

On the other hand, there was no clinical or histological evidence of a hypersensitivity reaction [74, 134]. The characteristic microvesicular steatosis of the hepatocytes in most of the patients and the biochemical aberrations, i.e. hyperammonemia, fatty acidemia, dicarboxylic aciduria and depletion of free carnitine, show a striking resemblance to Reye's syndrome and other Reye-like diseases [137]. A key issue of valproate's toxicity is impaired mitochondrial function. Swollen mitochondria and other hepatocellular alterations have consistently been found in liver tissue samples of valproate-treated mice, rats and man [74, 78, 138]. Structural changes of the inner mitochondrial membrane were also detected [139]. Fatty acid oxidation, gluconeogenesis, citric acid cycle and oxidative phosphorylation are all located within these cell organelles. As shown above, valproic acid affects these and other biochemical systems, and it still remains to be proven whether a single effect is responsible for the toxic process. More likely at this point is a combination of several factors.

The sequestration of free coenzyme A by the drug and its metabolites as well as the direct inhibition of mitochondrial enzymes by valproate derivatives are likely equally important. Several *in vitro* and *in vivo* models have favored 4-ene-VPA and its β-oxidation product 2,4-diene-VPA as major causative agents in the pathogenesis of hepatotoxicity. Both substances could impair fatty acid β-oxidation and induce microvesicular steatosis. However, clinical cases in support of this theory are sparse [134, 140]. Metabolic patterns in patients with severe liver failure have actually been quite variable, i.e. increased amounts of 2,3′-diene-VPA or 2-propyl-glutaric acid, the final product of ω-oxidation, were found in the absence of 4-ene-VPA [101, 141].

4-ene-VPA, as the parent compound of several electrophilic metabolites, is also the center of the "reactive intermediate" hypothesis [106]. Membrane lipids are sensitive to peroxidative injury, and one of the cell's protective mechanisms is the reduction of oxygen radicals with glutathione via glutathione peroxidase, a selenium-dependent enzyme. Evidence for such a process was provided by the detection of GSH- and NAC-conjugates of 4-ene-VPA metabolites and the concurrent depletion of mitochondrial and total cellular glutathione content after 4-ene-VPA treatment of rats [109]. Selenium-deficient rats treated with valproic acid have developed enhanced liver toxicity with extensive necrosis [142]. Other investigators found liver and kidney necroses in selenium-deficient, glutathione-depleted rats [143]. Selenium-deficient states with valproate treatment have also been reported in rats after 1 week and in human patients after chronic therapy [144, 145]. A single dose of valproic acid given to rats impaired the glutathione peroxidase-reductase system including the NADPH delivering glucose-6-phosphate reductase, responsible for maintaining glutathione levels. The GSH content decreased in a dose-dependent fashion [146]. The same group of authors detected a dose-related inhibition of red blood cell glutathione reductase in children on valproate therapy [147].

The mitochondria can also be a source of toxic oxygen radicals after uncoupling of oxidative phosphorylation as mentioned for valproic acid [148]. A consequence is the onset of mitochondrial permeability transition which has been suggested as common pathophysiological mechanism in Reye's syndrome and related drug toxicities. The importance of radical scavenger systems for the valproate-induced hepatotoxicity is not yet defined. However, there has been some success with treating these patients with the glutathione precursor N-acetylcysteine [149].

In conclusion, the risk of developing fulminant hepatic failure from treatment with valproic acid is very low for the majority of patients. Nevertheless, intercurrent illnesses, the age of the patient and the type of pharmacotherapy have to be considered when prescribing this drug. A risk-benefit assessment is especially important for patients with ongoing liver, pancreatic or metabolic diseases and for children under 2 years currently treated with enzyme-inducing antiepileptic drugs [150]. In addition, the screening for individual, possibly genetic, enzyme deficiencies especially of the β-oxidation, the urea cycle and the branched-chain amino acid metabolism is helpful in preventing this rare but often fatal disease.

Developmental toxicity

Animal studies suggest that all antiepileptic drugs have some potential for teratogenicity [151]. A recent joint European prospective study has reassessed this risk for babies from epileptic mothers taking antiepileptic medication during pregnancy [152]. It was concluded that valproic acid was consistently associated with an increased risk of major congenital malformations, especially neural tube defects.

The possibility of a teratogenic action of valproic acid in the human was first mentioned by Dickinson et al. in 1979 [9], which prompted a response by several clinicians who noticed equivalent incidences [153, 154]. Robert and Guibaud described an unusual increase of neural tube defects in association with valproate treatment in the Rhône-Alpes region of France between 1979 and 1982, and started a collaboration with other members of the International Clearinghouse for birth defects monitoring systems [155, 156]. Since then, much data has been collected in retrospective [157, 158] and prospective studies [19, 159] and the increased risk for congenital malformations (in particular spina bifida aperta) with valproate exposure during the first trimester of pregnancy has been estimated at about 1–2% [160].

Valproic acid affects many organ systems in the developing child. Besides the central nervous system defects, facial, cardio-vascular, musculosceletal, urogenital and other malformations have been described [161–163]. The "fetal valproate syndrome" (FVS) is characterized by a typical set of craniofacial anomalies, major organic malformations, and developmental delay. It was first documented by DiLiberti et al. [164] and later confirmed and reviewed by other authors [165, 166]. A recent case report suggested a relation between FVS and autism [167].

How valproic acid actually disturbs the development of the growing embryo is still unknown. Several *in vivo* and *in vitro* models have been applied to investigate the teratogenic mechanisms. The malformations seen in man can also be found in laboratory animals [12, 168, 169] (see Tab. 2). Most common for all species are growth retardation and skeletal defects. Anomalies of the developing central nervous system (CNS) have probably received the most attention in these studies because of their severe impact on the newborn child. Most CNS defects following maternal valproate use are severe open defects of the posterior neural tube [170].

Neural tube defects (NTD) have only been reproduced consistently in mice [171], and few reports have documented their occurrence in the hamster [172], the rat [173] and the monkey [174]. In contrast to the human, the predominant NTD in the mouse is exencephaly. This anomaly can be elicited by a single injection of valproic acid on gestational day 8. An experimental model to study spina bifida in the mouse has been developed by Ehlers et al. [14] and has recently been applied to rat embryos. Spina bifida occulta in association with an Arnold-Chiari-like malformation of the brain was observed in 10 day old rat embryos treated, however, with a high dose of 1200 mg/kg [175].

Extrapolation of experimental data to the human is difficult because of different pharmacokinetic parameters in the animal species. In the mouse valproic acid is eliminated much faster ($t_{1/2}$ = 0.7–1.3 h) than in the human ($t_{1/2}$ = 8–20 h), and the doses needed to elicit a teratogenic response are about 10-fold higher in the mouse than in man [176]. For this reason the plasma concentration-time curves (AUC) of a particular drug have served to compare the drug's elimination pattern and with it the total exposure in different species. This can be quite misleading as is the case with valproic acid. The AUC values in the human are rather high in comparison to the AUCs occurring in experimental models which could explain the sensitivity of the human embryo. However, studies in mice which were infused with

Table 2. Neural tube and other defects associated with valproic acid monotherapy

	Human[1]	Monkey[2]	Mouse[3]	Rat[4]	Hamster[5]	Rabbit[6]
Type of NTD	Spina bifida aperta (Spina bifida occulta?, anencephaly)	"Spina bifida-like lesion"	Exencephaly, spina bifida aperta, spina bifida occulta	Spina bifida occulta incl. "Arnold-Chiari-like malformation" (Open brain folds on GD11 or GD11.5)	– (Spina bifida aperta on GD9)	
VPA dose (poss. eliciting NTDs)	10–20 mg/kg p.o., (?)	3 × 170 mg/kg p.o., GD24–26	> 230 mg/kg i.p., GD8	2 × 600 mg/kg s.c., GD10 (3 × 150 mg/kg s.c. GD9 or 2 × 330mg/kg s.c. GD10)	300 mg/kg i.p., GD7:15	–
Other defects	"Fetal Valproate Syndrome", hydrocephalus, microcephalus, cardiovascular, hypospadias, inguinal hernia, skeletal, (lung hypoplasia)	Craniofacial, skeletal, (cardiovascular)	Craniofacial, skeletal, cardiovascular, urogenital	Microencephaly, craniofacial, skeletal, urogenital, cardiovascular, intestinal	Skeletal (150 mg/kg p.o., GD6–18), urogenital	

NTD Neural Tube Defect; GD Gestational Day.

[1] [152, 159, 161, 163, 170].
[2] [12, 174].
[3] [171, 274].
[4] [169, 173, 175, 275, 276].
[5] [172].
[6] [168].

valproate via osmotic minipumps over 24 h showed that peak drug levels and not the AUC were responsible for the embryotoxic effects [17].

Therefore, it remains quite unclear why the comparably low therapeutic plasma levels in the human may cause a teratogenic response. Valproic acid is highly protein-bound in human plasma. But there is a high individual diurnal variability [177], and pregnancy, stress, hormones, dietary factors and other drugs can change the free fraction of valproate and cause a sudden increase to embryotoxic levels [20]. Valproate metabolites, in contrast to the hepatotoxicity, may not play an important role in the teratogenic activity of valproate. Nau and Löscher investigated the teratogenic effects of several analogous compounds of valproic acid in mice [178]. Of the tested metabolites it was only the putative hepatotoxin 4-ene-VPA that induced exencephaly in the embryos; however, 4-ene-VPA levels were found to be exceedingly low in human plasma (approx. 0.1 µg/ml) [179]. Another study with different injection regimens clearly showed that the parent drug and not one of its metabolites was the teratogenic agent [180].

Genetic susceptibility

The explanation for the apparent higher sensitivity of the human embryo to valproate-induced teratogenicity may lie in the genetic susceptibility of the species and the individual. Finnell and coworkers compared the effects of valproic acid and its metabolite 4-ene-VPA on three different mouse strains which were characterized by varying heat-induced exencephaly rates [181, 182]. All three strains showed a common hierarchy of susceptibility to the teratogenic effects of hyperthermia or the chemical compounds. Valproate teratogenicity studies in cultured mouse embryos by Naruse et al. suggested that the basis for the strain differences resides within the embryo rather than the mother [183].

Distinct differences in the morphology of cranial neural tube closure have been observed in several mouse strains. The abnormal mechanism of neural tube closure in the SELH/Bc mouse, involving an absence of the initial closure site 2, has been implicated as a probable cause of this strain's high liability to exencephaly [184]. In contrast, SWV/Bc and ICR/Bc mice differ in the location of closure 2 [185] but show the same sensitivity to valproate- and retinoic acid-induced exencephaly. Based on these studies, Hall et al. proposed a multifactorial threshold model, in which conceptuses with a genetically determined elevated risk for neural tube defects could be easily tipped into a high risk situation by mild teratogens [186].

Recently, first evidence was presented for a strain difference in the transcriptional activity after teratogenic exposure [187]. In neural tube defect-sensitive SWV mice gene expression of folate binding protein (FBP-1) was downregulated whereas FBP-1 and 5,10-methylenetetrahydrofolate reductase (MTHFR) were upregulated in the more resistant LM/Bc mice. It can be concluded that genetic susceptibility is an important factor for the expression of embryotoxic effects.

Structural considerations

The chemical structure of valproic acid seems to be the key factor for its teratogenicity (see Fig. 1). The important structural requirements are an sp^3-hybridized α-carbon atom with a carboxyl group, a hydrogen atom and two alkyl side-chains [188]. Introduction of a double bond between C_2 and C_3 or the substitution of the α-hydrogen atom by a methyl group could abolish the teratogenic activity [178, 189]. The amidation of the carboxyl group and/or the branching of a side chain at C_3 or C_4 will also render the compound less teratogenic [189, 190]. For some unsaturated analogues, i.e. the metabolite 4-ene-VPA a stereospecificity has been shown [191]. The S-enantiomer of 2-n-propyl-4-pentynoic acid (4-yn-VPA) was 1.9 times more potent as a teratogen than the racemate and 7.5 times more potent than its antipode [192]. Later it was proven that the stereoselective dysmorphology was an intrinsic effect independent of pharmacokinetic differences which can be observed across species [193–195].

Based on these findings, Bojic et al. proposed a model describing the nature of a stereospecific embryonic receptor site through which valproic

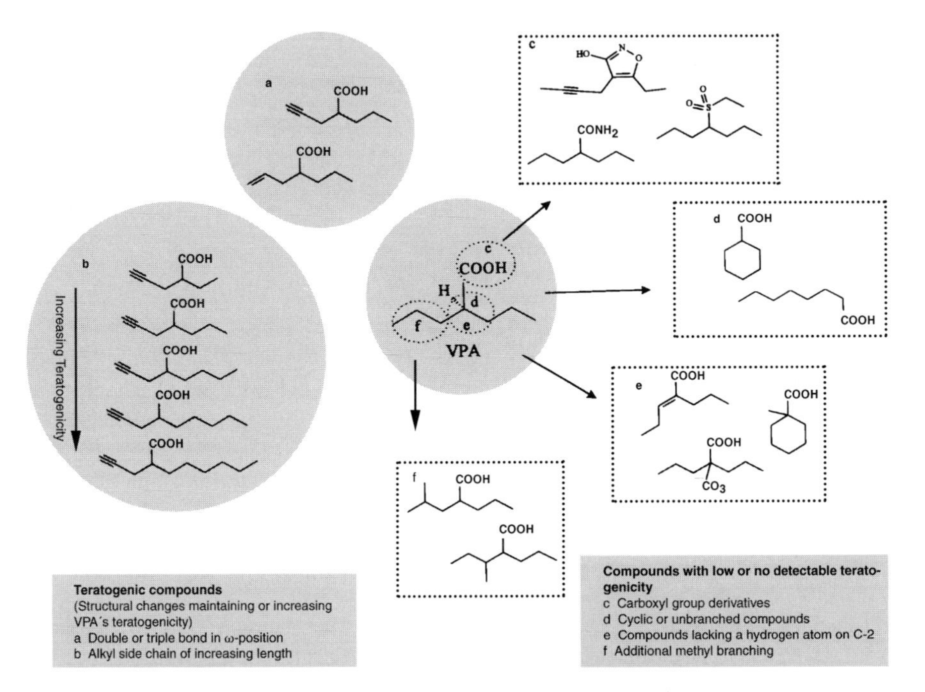

Figure 1. Structure-teratogenicity relationship of VPA and its derivatives.

acid and its analogues mediate their teratogenic action [196]. At this putative receptor, one alkyl side of the valproate molecule is bound to a hydrophobic pocket facilitating an ion and/or H-bonding of the carboxyl group. The hydrophobic bond of the side chain seems to be an important factor responsible for the degree of teratogenicity because an increasing alkyl chain length would lead to a dramatic increase in teratogenicity.

Proposed mechanisms

The neural tube defects are the major congenital malformation after maternal valproate exposure. Of the many proposed mechanisms (see Tab. 3) for the development of these anomalies the folate metabolism has received the most attention. Epidemiological studies showed a reduction of the prevalence of the first occurrence of neural tube defects by periconceptional vitamin supplementation, in particular with folates [197, 198]. In mothers at high risk for the recurrence of neural tube defects because of a previously affected pregnancy, the administration of 400 µg folic acid per day was 72% protective [199].

Table 3. Possible teratogenic mechanisms of valproic acid

Folates
– Alteration of embryonic folate metabolite pattern through inhibition of glutamate formyltransferase [202]
Homocysteine/Methionine
– Deranged one-carbon transfer cycle with altered homocysteine/methionine ratio [207, 277]
– Increased demand for vitamins B_{12} and B_6 [215]
Minerals
– Reduction of embryonic zinc concentrations [235, 237]
Lipid metabolism
– Disturbance of membrane transport processes and/or acetate metabolism [229]
– Increased demand of pantothenic acid, a precursor of Coenzyme A [231]
Alteration of the endogenous retinoid metabolism [243]
Alteration of gene expression
– Transactivation of PPAR δ [261, 262]
– Folate pathway genes [187]
– Pax 3 genes [260]
– AP 1 binding site [265]
Embryonic pH
– Reduction of embryonic intracellular pH [238, 241]
Glutathione depletion
– Reduced embryonic glutathione content [247]
– Reduced plasma glutathione peroxidase activity and selenium concentration [248]

Folates

Folates are required for normal intracellular purine and pyrimidine metabolism, and by transferring one carbon unit, take part in DNA synthesis and cell replication [200]. Whether valproic acid directly affects folate metabolism has been the focus of many investigations. Smith and Carl found decreased plasma folate levels after chronic VPA-treatment in rats [201]. And Wegner and Nau detected a change in the folate metabolite pattern in murine embryos [202]. The levels of 5-formyl-(5-CHO-THF) and 10-formyl-tetrahydrofolates were decreased and those of tetrahydrofolate increased after a single valproate injection on gestational day 9; an indication for the inhibition of the enzyme glutamate formyltransferase. The main non-teratogenic metabolite 2-ene-VPA did not influence the folate metabolism which underlined the structural specificity of this effect.

In concurrence with this study Trotz et al. showed that folinic acid (5-CHO-THF) reduced the VPA-induced exencephaly rate in NMRI mice [203]. Moreover, nutritional folate deficiency enhanced the teratogenic effects [204]. Hansen and coworkers, on the other hand, failed to reproduce the protective effects of folinic acid *in vitro* (CD rats) [205] and *in vivo* (CD-1 mice) [206]. It remains unclear whether strain and species differences in metabolism or genetic susceptibility are responsible for these conflicting results.

The main one-carbon donor for tetrahydrofolate (THF) is the amino acid serine which is converted to glycine by serine-hydroxymethyl-transferase, a pyridoxal-phosphate (vitamin B_6) enzyme. The resulting 5,10-THF is then reduced by 5,10-methylene-THF-reductase (MTHFR) to 5-methyl-THF or directly donates a methyl-group for pyrimidine synthesis. From 5-methyl-THF a methyl-group is then passed on to homocysteine which is in turn converted to methionine by the B_{12}-dependent enzyme methionine synthase. The activated S-adenosyl-methionine is the major donor for numerous biosynthetic methylations, i.e. the synthesis of phosphatidylcholine in biomembranes. It is obvious that deficiencies in the participating enzymes and vitamins may have detrimental effects on multiplying and differentiating cells, i.e. in embryonic development.

Homocysteine and methionine

The close interrelationship between the vitamins folic acid, B_6, B_{12} and the amino acids methionine and homocysteine as a causative factor in the development of NTD has recently been reviewed [207]. Because maternal plasma folate and vitamin B_{12} are independent risk factors for NTD [208] it was suggested that the closely related homocysteine metabolism could play an important role in the pathogenesis of this malformation. In support of this theory, Mills et al. found significantly higher homocysteine levels in NTD mothers and advocated the addition of vitamin B_{12} to a periconceptional supplement [209]. Steegers-Theunissen et al. detected increased homocysteine levels in amniotic fluid of women carrying a fetus with a neural tube defect [210]. They also studied plasma levels of homocysteine and the relevant vitamins after methionine loading in non-pregnant women who had previously given birth to infants with NTD [211]. Both, pre- and post-loading homocysteine levels were significantly increased in these mothers compared to the control group. They concluded that a disorder in the remethylation of homocysteine to methionine was probably due to an acquired or inherited derangement of folate or vitamin B_{12} metabolism. Since vitamin B_6 or folate administration could normalize the homocysteine levels, they suggested that NTD prevention by periconceptional folate administration may effectively correct a mild to moderate homocysteinemia.

Genetic deficiencies of the enzyme MTHFR have also been described as a cause of hyperhomocysteinemia [212, 213]. Homozygosity for the 677T mutation causes decreased plasma folate and increased homocysteine levels and is therefore a risk factor for NTD [214]. This polymorphism has been shown to present 5–18% of different populations [207].

The direct relationship between valproate treatment and folate or homocysteine metabolism in the human has not been established. Animal studies, however, have shown that several vitamins and amino acids can influence VPA-induced effects on the embryo. Supplementation with vitamin B_6 and B_{12} could reduce the rate of exencephaly and other malformations in VPA-treated mice [215]. The protective effect of folinic acid was also enhanced by these vitamins and by the amino acid serine, the main donor for the one-carbon cycle. Methionine, in its activated form is the endpoint of this cycle. Intraperitoneal methionine could significantly reduce the rate of spina bifida occulta in VPA-treated mice [216] and reduced the VPA-related number of resorptions if provided in drinking water of pregnant rats [217]. Methionine application in whole rat embryo culture, on the other hand, did only protect from VPA-induced developmental toxicity when the culture medium was sera from rats consuming methionine [217, 218].

Homocysteine and homocysteine thiolactone induced congenital heart and neural tube defects in avian embryos [219]. The embryotoxic effects of L-homocysteine in cultured rat embryos were stereospecific [220]. 5-methyl-THF, L-serine and L-methionine attenuated this effect, and vitamin B_{12} addition to the culture serum abolished L-homocysteine embryotoxicity [221]. Folate deficiency causes homocysteinemia in rats possibly by the reduction of homocysteine remethylation due to an impaired synthesis of S-adenosyl-methionine [222]. First evidence of a direct influence of valproic acid on the homocysteine-methionine metabolism was presented recently. Homocysteine, S-adenosyl-methionine, and S-adenosyl-homocysteine were significantly increased within 15 min after valproate injection (3 mmol/kg i.p.) of NMRI mice. S-adenosyl-methionine and S-adenosyl-homocysteine returned to baseline levels 15 min later. Plasma methionine and serine concentrations were significantly decreased.

Folate antagonists

To test the beneficial effects of folic acid on the embryonic development several substances have been investigated for their inhibitory action on folate metabolism. The antifolates methotrexate and trimethoprim given in non-teratogenic doses to valproate-treated mice caused an increase in exencephaly rate, embryolethality and fetal weight retardation [223, 224]. The pharmacokinetics of valproic acid were not altered in these experiments. Interestingly, neither antifolate application alone nor nutritional folate deficiency were able to induce neural tube defects in mice. It is, thus, more likely, as shown in previous experiments, that valproic acid disturbs the physiological folate metabolite pattern which in a dose-dependent manner eventually leads to toxic effects in the embryo.

Alcohol

An interference of folate metabolism has also been described for alcohol [225]. Alcohol ingestion disrupts the normal pattern of storage and release of folate by the liver resulting in dramatic fluctuations of serum folate levels. To test the hypothesis that valproic acid teratogenicity is conveyed through a disturbance of folate metabolism, pregnant mice received ethanol prior to valproate injection [226]. And indeed, alcohol administration increased exencephaly and embryolethality rates and the fetal weight reduction in the mouse embryos in comparison to the mice treated with valproate only. In this experiment, however, the plasma-half life of valproic acid was also increased. As a consequence, it is difficult to say whether the increased teratogenicity was due to an impaired elimination of valproic acid or a synergistic effect on folate metabolism. Since alcohol alone has been shown to cause neural tube defects and other malformations in mice [227, 228], a mere additive effect of both agents is also possible.

Effects on fatty acid metabolism

Another biochemical system influenced by valproic acid is the fatty acid metabolism. Besides playing an important role in the pathogenesis of valproate-caused hepatic disease, a mediation of the developmental toxicity through this system has also been contemplated. Clarke and Brown examined the uptake of tritium-labeled acetate in the lipid fraction of cultured rat embryos [229]. After 8 h of valproate exposure [^{3}H]acetate was incorporated into all major neutral lipids, esp. increased in free fatty acids and triacylglycerols. This was considered by the authors as an indication for possible effects of valproic acid on membrane transport processes or acetate metabolism.

Pantothenic acid, another B-vitamin, has been suggested as an important adjunct to valproate therapy of women during the first trimester of pregnancy [230]. Sato and coworkers investigated whether there was an increased demand for pantothenic acid during the time of embryonic development in valproate-treated mice [231]. Pantothenic acid supplementation specifically reduced the prevalence of exencephaly but not of other external malformations usually seen in the offspring. The authors assumed a different protective mechanism of pantothenic acid from that of folinic acid which could also reduce other valproate-induced developmental defects. The supplementation with pantothenic acid, which is an important precursor of coenzyme A, may be beneficial by counteracting the reducing effects of valproic acid on free CoA and acetyl-CoA levels as discussed above.

Zinc metabolism
Zinc deficiency can be teratogenic in the rat, the mouse and the rhesus monkey [232–234]. Two reports stated the reduction of fetal and plasma zinc concentrations with valproate treatment in rats and hamsters [172, 235]. In contrast, Wegner et al. found increased plasma zinc concentrations in mice for 2 and 4 h after zinc injection of valproic acid or 2-ene-VPA, a non-teratogenic metabolite [236]. Other authors reported no changes in tissue zinc concentrations after valproate treatment in rats or mice [206]. Most recently, Bui et al., again, discussed VPA-induced alterations in zinc metabolism as possible cause for developmental toxicity [237]. Pregnant Sprague-Dawley rats were fed a ^{65}Zn-containing diet from gestational day 0 to 12.5. After gavage with valproate on gestational day 11.5 the zinc concentrations significantly increased in maternal liver and decreased in the embryos as compared to controls. Unfortunately, the publication did not include information about the prevalence of congenital defects in these embryos.

Embryonic pH
Another proposed mechanism for valproate teratogenesis is based on the observation that the early mammalian embryo has an alkaline pH in comparison to the maternal plasma pH. Weak acids like valproic acid would thus accumulate in embryonic tissue via ion "trapping" and subsequently exert their toxic effects [238].

Support for this theory comes from research showing that right-sided forelimb postaxial ectrodactyly occurs after CO_2-inhalation and acetazolamide treatment [239, 240], both known to reduce intracellular pH. Scott et al. found pH reductions of up to 0.3 units in the limb buds of VPA-treated C57Bl/6 mice [241]. The frequency and severity of the limb malformations were exacerbated by coadministration of amiloride. This potassium-sparing diuretic acts as an inhibitor of the Na^+/H^+-exchange. Hydrogen ions from dissociated valproic acid would therefore remain in the cell for a longer period of time increasing total embryonic exposure to a lowered pH.

Effects on retinoids
Vitamin A and its metabolite all-*trans*-retinoic acid are crucial for many cellular processes in embryonic development. At the same time, retinol and

some of its derivatives also possess teratogenic activities [242]. Antiepileptic drugs including valproic acid can influence plasma retinoid levels. It is not known how this will affect retinoid tissue levels. Nevertheless, the alteration of the endogenous retinoid metabolism may be significant in regard to valproic acid teratogenicity [243].

Reactive intermediates
Valproic acid derived electrophilic species as a cause for hepatotoxicity are described above. The formation of free radicals as a common mechanism for teratogenicity, especially of anticonvulsants, has been the subject of discussion for quite some time [244]. In the case of phenytoin, evidence exists for the bioactivation to an electrophilic arene oxide intermediate by CYP450 and to a free radical intermediate by embryonic peroxidase [245].

In the latter reaction phenytoin possibly serves as a reducing cosubstrate in the pathway of prostaglandin H and hydroxy-eicosatetranoic acid synthesis. In a study with pregnant CD-1 mice, the incidence of phenytoin-induced cleft palate was significantly reduced by pretreatment with an irreversible inhibitor of prostaglandin H synthase and lipooxygenase [246].

Glutathione (GSH) plays an important role in the detoxification of reactive intermediates. Harris et al. examined the glutathione status and the incidence of neural tube defects in cultured rat embryos after administration of valproic acid and two other teratogens [247]. The embryonic glutathione content significantly decreased with valproic acid and when the cysteine (a GSH precursor) prodrug 2-oxo-thiazolidone-4-carboxylate was added to the culture medium the embryos showed a significant decrease in abnormal neurulation.

Several enzymes such as glutathione peroxidase, superoxide dismutase and catalase are involved in the cytoprotective mechanisms preventing injury by oxygen radicals. Valproic acid therapy reduced glutathione peroxidase enzyme activity in children with intractable seizures [248]. Two of the children showed low glutathione peroxidase and selenium plasma levels. Their clinical state improved with discontinuation of the anticonvulsant and selenium substitution. Based on these observations and the association of valproic acid with the development of neural tube defects, Graf et al. examined several cytoprotective enzyme activities in children with myelomeningocele [249]. The findings of significant deficiencies of glutathione peroxidase and glutathione reductase activities in these children and their healthy parents suggested a possible role of this deficiency for the causation of neural tube defects in general.

Common mechanisms in teratogenicity and hepatotoxicity?

A clear structural dependency has been shown for the teratogenic activity of valproic acid and its derivatives (see Fig. 1). If the same holds true for the liver toxicity has not been established. In both instances unspecific mechanisms involving reactive intermediates could participate to some extent. Liver damage is likely due to the sequestration of CoA-SH with subsequent depletion of acetyl-CoA. In contrast, a significant acetyl-CoA reduction after valproate treatment could not be detected in the embryo [250].The enzymes responsible for activating medium chain fatty acids (valproic acid $\rightarrow$ valproyl-CoA) are apparently not active during early organogenesis. At present there is no indication that hepatotoxicity and teratogenicity of valproic acid have a common mechanism.

Cellular changes in VPA-induced teratogenicity

Valproic acid predominantly accumulates in the neuroepithelium of mouse embryos during early organogenesis [251]. The radioactivity after high doses of ^{14}C-labeled valproic acid (400 mg/kg) remained relatively constant over a 3-h period when plasma valproate levels had already decreased by half. The retention time was longer than in other parts of the embryo, and the neuroepithelium showed a higher uptake of valproic acid than maternal blood, amniotic fluid or other embryonic tissues. Compared to the low dose treatment in this experiment (1 mg/kg) approximately 2000 times more valproic acid accumulated in the nervous tissue, possibly indicating the saturation of metabolic capacities of the maternal organism. Histologically the neuroepithelium exhibits marked disorganization with loss of the normal palisade structure and cellular polarity [252]. Therefore, valproic acid seems to interfere with cellular adhesion and stability at the apical region of the neuroepithelium. Failure of the opposing neural folds to connect rather than proper formation of the folds is the presumed cellular mechanism behind this defect [253].

Valproic acid also reduces protein and DNA content in cultured mouse and rat embryos as an indicator of general growth retardation [247, 254]. At the same time, a reduction in size and vascularization of the yolk sac can be noticed. This may explain the occurrence of abnormally curved embryos, since the growth and expansion of the yolk sac and amnion are necessary for normal rotation [254]. Furthermore, extensive delays in turning have been postulated as one reason for failure of neural tube closure [255].

Studies of valproic acid on human chondrogenesis *in vitro* have provided another aspect for the explanation of teratogenic effects in particular for the digital anomalies [256]. Cultured human chondrocytes showed decreased mitotic activity and a modification of the extracellular matrix after therapeutic doses of valproic acid. The ratio of the collagen types changed, and there were indications for a disturbed proteoglycan synthesis. In another report matrix synthesis and limb bud development were also decreased

[254]. Proteoglycans are present in mesodermal matrix of the cranial region, where they may be involved in the expansion of the mesodermal extracellular spaces during bulging of the cranial neural folds. They have also been observed at the site of developing basement membranes and within the lateral neuroepithelium of the cranial neural folds, where they may be involved in the emigration of the neural crest [253, 257]. The complex involvement of extracellular matrix and basement membranes in the neurulation process classify these structures as another possible target for the teratogenicity of valproic acid.

Molecular aspects of teratogenicity

Several genes are thought to play an important role in regulating mammalian neurulation. The expression of some of these genes (see below) has been studied in murine embryos during the time of neural tube closure. After exposure to teratogenic doses of valproic acid, Wlodarczyk et al. [258] found a significant elevation of the transcription factor genes EMX-1, EMX-2, *c-fos*, *c-jun* and *creb* and the cell cycle genes p53 and bcl-2. The latter two genes are expressed in changing ratios during neural tube closure regulating the cell cycle during this event [259]. p53 is a multifunctional protein capable of inducing apoptosis and growth arrest. The bcl-2 product, on the other hand, is believed to be a potent inhibitor of p53 and, thus, to regulate cell proliferation and differentiation. The relative overexpression of bcl-2 supported the hypothesis that altered neuroepithelial cell proliferation rates rather than increased apoptosis is the underlying mechanism of disturbed neural tube development [258].

Investigations in the murine mutant Splotch, which is a known model for neural tube defects, also demonstrated the involvement of these two and several other genes at the crucial time points of neural tube closure [259]. The adhesion molecules N-cam and N-cad are significant for neural tube formation and in particular crest cell migration. This occurs presumably by reducing adhesiveness between cells following diminished expression of these genes. In the study of Splotch mice N-cam reduction came later than in wild-type controls, and N-cad was even overexpressed, probably impairing neural tube closure. The function of Pax-3 during neural tube closure is not known but it has been expressed prior to neural tube closure in several parts of the neural tissue. Pax-3 was overexpressed, in the Splotch homozygous embryos. Pax-3, Hoxb-4 and Emx-2 have also been observed in mouse and rat embryos to examine the possible influences of valproate exposure on the location of gene expression [260]. Although Pax-3 expression was spatially abnormal, this was coincident with the induced structural defects. The domains of expression for these genes were the same in the two species, and the treatment with valproic acid caused no shifts in boundaries of expression.

Although many of the involved genes during neural tube development are known there are still many questions regarding the influence of valproic acid on gene expression. Recent results from our laboratory suggest a relationship between the transactivation of another set of transcription factors (PPARs, peroxisome proliferator-activated receptors) *in vitro* and induction of exencephaly by valproic acid in mice [261]. PPARs belong to the steroid-thyroid hormone-retinoid receptor superfamily, and several members of this group have been shown to be crucially involved in embryogenesis. Specifically, studies of structure-activity relationships with a number of valproic acid analogues have shown a highly significant relationship between transactivation of PPAR δ *in vitro* and induction of exencephaly *in vivo*. It is presently not known if valproic acid and analogous compounds directly or indirectly transactivate PPAR δ, and which target genes regulated by this receptor might be involved in normal and abnormal development. It should be noted, however, that PPAR δ is present during the time of neural tube closure [262].

Conclusions

At present there is no single substance for which the teratogenic mechanism is fully understood. Two major types of interactions between xenobiotics and embryonic target tissue can be distinguished. The unspecific, irreversible and cumulative cell damage by a reactive intermediary metabolite and the specific, receptor-mediated, action towards one type of molecule usually by the parent compound [263]. Valproic acid seems to exert its toxic action in both ways.

The hallmark of the experimental valproate-induced teratogenicity is its high structural specificity which is pointing towards a stereoselective interaction [196, 264]. It has indeed been shown that valproic acid interacts with PPAR δ [261], with the AP-1 binding site [265], with protein kinase C [266], with β-adrenergic receptors and G-Protein [267], and with Pax-3 [260]. The studies on valproate-treated mouse embryos suggest the involvement of multiple genes being responsible for the induction of neural tube defects [258].

The role of folate binding protein and 5,10-methylenetetrahydrofolate reductase is being explored [187]. Several clinical as well as experimental studies emphasize the importance of vitamin and homocysteine levels in the development of neural tube defects. Folates have been shown to protect from the occurrence/recurrence of neural tube defects, and valproic acid treatment can influence folate metabolite patterns [202]. Delayed delivery of one-carbon units for DNA-synthesis will disturb the sensitive balance of genetic expression during embryonic development.

Another suggestion for a specific cellular VPA-binding site involved in the teratogenic activity comes from Scott et al. [241]. A VPA-induced decrease of intracellular pH may disturb metabolic processes and cellular structures.

An individual sensitivity to the toxic effects of valproic acid has been proven in animal models [186, 268]. This has not been established for the human, but is not unlikely since genetic deficiencies have been associated with the occurrence of neural tube defects or at least a higher risk for such a malformation [214, 249].

A putative metabolite or reactive intermediate which could be responsible for the teratogenic action has yet to be found. The only proven teratogenic metabolite 4-ene-VPA occurs in too low levels to be solely responsible. Nevertheless, valproic acid accumulates in the neuroepithelium [251], and the highly proliferative nature of these cells during early organogenesis may make them more sensitive to oxidative injury. Additionally, the low levels of anti-oxidative enzymes at this age may increase the susceptibility of the embryo to this kind of damage [245]. Patients with a deficient anti-oxidant defense system may therefore be at higher risk for developing VPA-related toxicity.

In summary, valproic acid teratogenicity is dose-related and is probably restricted to the first trimester of pregnancy. The teratogenic mechanism may be a combination of an unspecific, irreversible toxicity and a specific receptor-mediated action.

References

1 Rothrock JF (1997) Clinical studies of valproate for migraine prophylaxis. *Cephalalgia* 17: 81–83

2 Deltito JA, Levitan J, Damore J, Hajal F, Zambenedetti M (1998) Naturalistic experience with the use of divalproex sodium on an in-patient unit for adolescent psychiatric patients. *Acta Psychiatr Scand* 97: 236–240

3 Hammer BA, Brady KT (1996) Valproate treatment of alcohol withdrawal and mania [letter]. *Am J Psychiatry* 153: 1232

4 Friedman NL (1996) Hiccups: a treatment review. *Pharmacotherapy* 16: 986–995

5 Aiken TC, Collin RC (1995) A possible anti-emetic role for sodium valproate in cytotoxic chemotherapy. *Br J Haematol* 89: 903–904

6 Eason CT, Bonner FW, Parke DV (1990) The importance of pharmacokinetic and receptor studies in drug safety evaluation. *Regul Toxicol Pharmacol* 11: 288–307

7 Nau H (1986) Species differences in pharmacokinetics and drug teratogenesis [published erratum appears in Environ Health Perspect 1988 Apr; 77: following 156]. *Environ Health Perspect* 70: 113–129

8 Nau H, Rating D, Koch S, Hauser I, Helge H (1981) Valproic acid and its metabolites: placental transfer, neonatal pharmacokinetics, transfer via mother's milk and clinical status in neonates of epileptic mothers. *J Pharmacol Exp Ther* 219: 768–777

9 Dickinson RG, Harland RC, Lynn RK, Smith WB, Gerber N (1979) Transmission of valproic acid (Depakene) across the placenta: half-life of the drug in mother and baby. *J Pediatr* 94: 832–835

10 Kondo T, Otani K, Hirano T, Kaneko S (1987) Placental transfer and neonatal elimination of mono-unsaturated metabolites of valproic acid [letter]. *Br J Clin Pharmacol* 24: 401–403

11 Nau H (1986) Transfer of valproic acid and its main active unsaturated metabolite to the gestational tissue: correlation with neural tube defect formation in the mouse. *Teratology* 33: 21–27

12 Hendrickx AG, Nau H, Binkerd P, Rowland JM, Rowland JR, Cukierski MJ, Cukierski MA (1988) Valproic acid developmental toxicity and pharmacokinetics in the rhesus monkey: an interspecies comparison. *Teratology* 38: 329–345
13 Nau H, Hendrickx AG (1987) Valproic acid teratogenesis. *ISI Atlas of Science: Pharmacology* 52–56
14 Ehlers K, Sturje H, Merker HJ, Nau H (1992) Valproic acid-induced spina bifida: a mouse model. *Teratology* 45: 145–154
15 Nau H, Spielmann H (1983) Embryotoxicity testing of valproic acid [letter]. *Lancet* 1: 763–764
16 Nau H, Zierer R, Spielmann H, Neubert D, Gansau C (1981) A new model for embryotoxicity testing: teratogenicity and pharmacokinetics of valproic acid following constant-rate administration in the mouse using human therapeutic drug and metabolite concentrations. *Life Sci* 29: 2803–2814
17 Nau H (1985) Teratogenic valproic acid concentrations: infusion by implanted minipumps vs conventional injection regimen in the mouse. *Toxicol Appl Pharmacol* 80: 243–250
18 Rowan AJ, Overweg J, Meijer JWA (1981) Monodose therapy with valproic acid: 24 h telemetric EEG and serum level studies. In: M Dam, L Gram, JK Penry (eds): *Advances in Epileptology*. Raven Press, New York, 533–539
19 Jäger-Roman E, Deichl A, Jakob S, Hartmann AM, Koch S, Rating D, Steldinger R, Nau H, Helge H (1986) Fetal growth, major malformations, and minor anomalies in infants born to women receiving valproic acid. *J Pediatr* 108: 997–1004
20 Nau H, Krauer B (1986) Serum protein binding of valproic acid in fetus-mother pairs throughout pregnancy: Correlation with oxytocin administration and albumin and free fatty acid concentrations. *J Clin Pharmacol* 26: 215–221
21 Eadie MJ, Hooper WD, Dickinson RG (1988) Valproate-associated hepatotoxicity and its biochemical mechanisms. *Med Toxicol Adverse Drug Exp* 3: 85–106
22 Wilmink T, Frick TW (1996) Drug-induced pancreatitis. *Drug Saf.* 14: 406–423
23 Hauser E, Seidl R, Freilinger M, Male C, Herkner K (1996) Hematologic manifestations and impaired liver synthetic function during valproate monotherapy. *Brain Dev* 18: 105–109
24 Gigli GL, Scalise A, Pauri F, Silvestri G, Diomedi M, Placidi F, Grazia PM, Masala C (1996) Valproate-induced systemic lupus erythematosus in a patient with partial trisomy of chromosome 9 and epilepsy. *Epilepsia* 37: 587–588
25 Pinkston R, Walker LA (1997) Multiorgan system failure caused by valproic acid toxicity. *Am J Emerg Med* 15: 504–506
26 Schmidt D (1984) Adverse effects of valproate. *Epilepsia 25 Suppl* 1: S44–S49
27 Isojarvi JI, Laatikainen TJ, Knip M, Pakarinen AJ, Juntunen KT, Myllyla VV (1996) Obesity and endocrine disorders in women taking valproate for epilepsy. *Ann Neurol* 39: 579–584
28 Devinsky O (1995) Cognitive and behavioral effects of antiepileptic drugs. *Epilepsia 36 Suppl 2:* S46–S65
29 Dreifuss FE (1995) Valproic acid: Toxicity. In: RH Levy, R-H Mattson, BS Meldrum (eds): *Antiepileptic Drugs*. Raven Press, New York, 641–648
30 Lakhani M, McMurdo ME (1986) Survival after severe self poisoning with sodium valproate. *Postgrad Med J* 62: 409–410
31 Löscher W (1985) Valproic acid. In: HH Frey, D Janz (eds): *Handbook of experimental pharmacology, Vol. 74*. Springer-Verlag, Berlin, 507–536
32 Conine DL, Majors KR, Lehrer S, Becker BA (1976) Acute toxicity of sodium 2-propylpentanoate, a compound whose toxicity decreases as animal size increases. *Toxicol Appl Pharmacol* 37: 144
33 Löscher W (1980) A comparative study of the pharmacology of inhibitors of GABA-metabolism. *Naunyn Schmiedebergs Arch Pharmacol* 315*:* 119–128
34 Walker RM, Smith GS, Barsoum NJ, Macallum GE (1990) Preclinical toxicology of the anticonvulsant calcium valproate. *Toxicology* 63*:* 137–155
35 Sobaniec-Lotowska ME (1997) Effects of long-term administration of the antiepileptic drug-sodium valproate upon the ultrastructure of hepatocytes in rats. *Exp Toxicol Pathol* 49: 225–232
36 Ponchaut S, Vanhoof F, Veitch K (1992) Cytochrome aa3 depletion is the cause of the deficient mitochondrial respiration induced by chronic valproate administration. *Biochem Pharmacol* 43: 644–647

37 Thurston JH, Hauhart RE, Schulz DW, Naccarato EF, Dodson WE, Carroll JE (1981) Chronic valproate administration produces hepatic dysfunction and may delay brain maturation in infant mice. *Neurology* 31: 1063–1069
38 Thurston JH, Carroll JE, Dodson WE, Hauhart RE, Tasch V (1983) Chronic valproate administration reduces fasting ketonemia in children. *Neurology* 33: 1348–1350
39 Diaz J, Shields WD (1981) Effects of dipropylacetate on brain development. *Ann Neurol* 10: 465–468
40 Sobaniec-Lotowska ME, Sobaniec W (1996) Morphological features of encephalopathy after chronic administration of the antiepileptic drug valproate to rats. A transmission electron microscopic study of capillaries in the cerebellar cortex. *Exp Toxicol Pathol* 48: 65–75
41 Vorhees CV (1987) Behavioral teratogenicity of valproic acid: selective effects on behavior after prenatal exposure to rats. *Psychopharmacology (Berl.)* 92: 173–179
42 Squadrito F, Sturniolo R, Arcadi F, Arcoraci V, Caputi AP (1988) Evidence that a GABA-ergic mechanism influences the development of obesity in obese Zucker rats. *Pharmacol Res Commun* 20: 1087–1088
43 Isojarvi JI, Rattya J, Myllyla VV, Knip M, Koivunen R, Pakarinen AJ, Tekay A, Tapanainen JS (1998) Valproate, lamotrigine, and insulin-mediated risks in women with epilepsy. *Ann Neurol* 43: 446–451
44 Sussman NM, McLain LWJ (1979) A direct hepatotoxic effect of valproic acid. *JAMA* 242: 1173–1174
45 Dreifuss FE, Langer DH (1987) Hepatic considerations in the use of antiepileptic drugs. *Epilepsia 28 Suppl* 2: S23–S29
46 Suchy FJ, Balistreri WF, Buchino JJ, Sondheimer JM, Bates SR, Kearns GL, Stull JD, Bove KE (1979) Acute hepatic failure associated with the use of sodium valproate. *N Engl J Med* 300: 962–966
47 Bowdle TA, Patel IH, Wilensky AJ, Comfort C (1979) Hepatic failure from valproic acid [letter]. *N Engl J Med* 301: 435–436
48 Gerber N, Dickinson RG, Harland RC, Lynn RK, Houghton LD, Antonias JI, Schimschock JC (1979) Reye-like syndrome associated with valproic acid therapy. *J Pediatr* 95: 142–144
49 Scheffner D (1986) Fatal liver failure in children on valproate [letter]. *Lancet* 2: 511
50 Jeavons PM (1984) Non-dose-related side-effects of valproate. *Epilepsia 25 Suppl* 1: S50–S55
51 Gram L (1983) Hepatotoxicity of valproate: Reflections on the pathogenesis and proposal for an international collaborative registration. In: J Oxley, D Janz (eds): *Chronic Toxicity of Antiepileptic Drugs*. Raven Press, New York, 69–78
52 Appleton RE, Farrell K, Applegarth DA, Dimmick JE, Wong LT, Davidson AG (1990) The high incidence of valproate hepatotoxicity in infants may relate to familial metabolic defects. *Can J Neurol Sci* 17: 145–148
53 Hjelm M, de Silva LV, Seakins JW, Oberholzer VG, Rolles CJ (1986) Evidence of inherited urea cycle defect in a case of fatal valproate toxicity. *Br Med J (Clin Res Ed)* 292: 23–24
54 Gugler R, Schell A, Eichelbaum M, Froscher W, Schulz HU (1977) Disposition of valproic acid in man. *Eur J Clin Pharmacol* 12: 125–132
55 Levy RH, Shen DD (1995) Valproic acid: Absorption, distribution, and excretion. In: RH Levy, R-H Mattson, BS Meldrum (eds): *Antiepileptic Drugs*. Raven Press, New York, 605–619
56 Nau H, Löscher W (1984) Valproic acid and metabolites: Pharmacological and toxicological studies. *Epilepsia 25 Suppl* 1: S14–S22
57 Rettenmeier AW, Gordon WP, Prickett KS, Levy RH, Lockard JS, Thummel KE, Baillie TA (1986) Metabolic fate of valproic acid in the rhesus monkey. Formation of a toxic metabolite, 2-n-propyl-4-pentenoic acid. *Drug Metab Dispos* 14: 443–453
58 Powell-Jackson PR, Tredger JM, Williams R (1984) Hepatotoxicity to sodium valproate: a review. *Gut* 25: 673–681
59 Cotariu D, Zaidman JL (1988) Valproic acid and the liver. *Clin Chem 34:* 890–897
60 Becker CM, Harris RA (1983) Influence of valproic acid on hepatic carbohydrate and lipid metabolism. *Arch Biochem Biophys* 223: 381–392
61 Mortensen PB, Gregersen N, Kolvraa S, Christensen E (1980) The occurrence of C6-C10-dicarboxylic acids in urine from patients and rats treated with dipropylacetate. *Biochem Med* 24: 153–161

62 Coude FX, Grimber G, Pelet A, Benoit Y (1983) Action of the antiepileptic drug, valproic acid, on fatty acid oxidation in isolated rat hepatocytes. *Biochem Biophys Res Commun* 115: 730–736
63 Kossak BD, Schmidt-Sommerfeld E, Schoeller DA, Rinaldo P, Penn D, Tonsgard JH (1993) Impaired fatty acid oxidation in children on valproic acid and the effect of L-carnitine. *Neurology* 43: 2362–2368
64 Thurston JH, Carroll JE, Hauhart RE, Schiro JA (1985) A single therapeutic dose of valproate affects liver carbohydrate, fat, adenylate, amino acid, coenzyme A, and carnitine metabolism in infant mice: Possible clinical significance. *Life Sci* 36: 1643–1651
65 Coulter DL (1991) Carnitine, valproate, and toxicity. *J Child Neurol* 6: 7–14
66 Thurston JH, Hauhart RE (1992) Amelioration of adverse effects of valproic acid on ketogenesis and liver coenzyme A metabolism by cotreatment with pantothenate and carnitine in developing mice: Possible clinical significance. *Pediatr Res* 31: 419–423
67 Coulter DL (1995) Carnitine deficiency in epilepsy: Risk factors and treatment. *J Child Neurol 10 Suppl* 2: S32–S39
68 Chung S, Choi J, Hyun T, Rha Y, Bae C (1997) Alterations in the carnitine metabolism in epileptic children treated with valproic acid. *J Korean Med Sci* 12: 553–558
69 Matsuda I, Ohtani Y, Ninomiya N (1986) Renal handling of carnitine in children with carnitine deficiency and hyperammonemia associated with valproate therapy. *J Pediatr* 109: 131–134
70 Laub MC, Paetzke-Brunner I, Jaeger G (1986) Serum carnitine during valproic acid therapy. *Epilepsia* 27: 559–562
71 Millington DS, Bohan TP, Roe CR, Yergey AL, Liberato DJ (1985) Valproylcarnitine: a novel drug metabolite identified by fast atom bombardment and thermospray liquid chromatography-mass spectrometry. *Clin Chim Acta* 145: 69–76
72 Turnbull DM, Bone AJ, Bartlett K, Koundakjian PP, Sherratt HS (1983) The effects of valproate on intermediary metabolism in isolated rat hepatocytes and intact rats. *Biochem Pharmacol* 32: 1887–1892
73 Rogiers V, Vandenberghe Y, Vercruysse A (1985) Inhibition of gluconeogenesis by sodium valproate and its metabolites in isolated rat hepatocytes. *Xenobiotica* 15: 759–765
74 Zimmerman HJ, Ishak KG (1982) Valproate-induced hepatic injury: Analyses of 23 fatal cases. *Hepatology* 2: 591–597
75 Silva MF, Ruiter JP, Illst L, Jakobs C, Duran M, de Almeida IT, Wanders RJ (1997) Valproate inhibits the mitochondrial pyruvate-driven oxidative phosphorylation *in vitro*. *J Inherit Metab Dis* 20: 397–400
76 Rumbach L, Warter JM, Rendon A, Marescaux C, Micheletti G, Waksman A (1983) Inhibition of oxidative phosphorylation in hepatic and cerebral mitochondria of sodium valproate-treated rats. *J Neurol Sci* 61: 417–423
77 Jimenez-Rodriguezvila M, Caro-Paton A, Duenas-Laita A, Conde M, Coca MC, Martin-Lorente JL, Velasco A, Maranon A (1985) Histological, ultrastructural and mitochondrial oxidative phosphorylation studies in liver of rats chronically treated with oral valproic acid. *J Hepatol* 1: 453–465
78 Hayasaka K, Takahashi I, Kobayashi Y, Iinuma K, Narisawa K, Tada K (1986) Effects of valproate on biogenesis and function of liver mitochondria. *Neurology* 36: 351–356
79 Coulter DL, Allen RJ (1980) Secondary hyperammonaemia: A possible mechanism for valproate encephalopathy [letter]. *Lancet* 1: 1310–1311
80 Gruskay JA, Rosenberg CE (1979) Inhibition of hepatic mitochondrial carbamyl phosphate synthetase (CPS-1) by acylCoA-esters. *Pediatric Research* 13: 475
81 Coude FX, Grimber G, Parvy P, Rabier D, Petit F (1983) Inhibition of ureagenesis by valproate in rat hepatocytes. Role of N-acetylglutamate and acetyl-CoA. *Biochem J* 216: 233–236
82 Coulter DL, Allen RJ (1981) Hyperammonemia with valproic acid therapy. *J Pediatr* 99: 317–319
83 Hjelm M, Oberholzer V, Seakins J, Thomas S, Kay JD (1986) Valproate-induced inhibition of urea synthesis and hyperammonaemia in healthy subjects [letter]. *Lancet* 2: 859
84 Löscher W, Wahnschaffe U, Honack D, Wittfoht W, Nau H (1992) Effects of valproate and E-2-en-valproate on functional and morphological parameters of rat liver. I. Biochemical, histopathological and pharmacokinetic studies. *Epilepsy Res* 13: 187–198

85 Löscher W, Nau H, Wahnschaffe U, Honack D, Rundfeldt C, Wittfoht W, Bojic U (1993) Effects of valproate and E-2-en-valproate on functional and morphological parameters of rat liver. II. Influence of phenobarbital comedication. *Epilepsy Res* 15: 113–131
86 Löscher W, Wahnschaffe U, Honack D, Drews E, Nau H (1993) Effects of valproate and E-2-en-valproate on functional and morphological parameters of rat liver. III. Influence of fasting. *Epilepsy Res* 16: 183–194
87 Ohtani Y, Endo F, Matsuda I (1982) Carnitine deficiency and hyperammonemia associated with valproic acid therapy. *J Pediatr* 101: 782–785
88 Thom H, Carter PE, Cole GF, Stevenson KL (1991) Ammonia and carnitine concentrations in children treated with sodium valproate compared with other anticonvulsant drugs. *Dev Med Child Neurol* 33: 795–802
89 Laub MC (1986) Nutritional influence on serum ammonia in young patients receiving sodium valproate. *Epilepsia* 27: 55–59
90 Gidal BE, Inglese CM, Meyer JF, Pitterle ME, Antonopolous J, Rust RS (1997) Diet- and valproate-induced transient hyperammonemia: Effect of L-carnitine. *Pediatr Neurol* 16: 301–305
91 Stumpf DA, Parker WDJ, Angelini C (1985) Carnitine deficiency, organic acidemias, and Reye's syndrome. *Neurology* 35: 1041–1045
92 Horiuchi M, Imamura Y, Nakamura N, Maruyama I, Saheki T (1993) Carbamoylphosphate synthetase deficiency in an adult: Deterioration due to administration of valproic acid. *J Inherit Metab Dis* 16: 39–45
93 Oechsner M, Steen C, Sturenburg HJ, Kohlschutter A (1998) Hyperammonaemic encephalopathy after initiation of valproate therapy in unrecognised ornithine transcarbamylase deficiency. *J Neurol Neurosurg Psychiatry* 64: 680–682
94 Fromenty B, Pessayre D (1997) Impaired mitochondrial function in microvesicular steatosis. Effects of drugs, ethanol, hormones and cytokines. *J Hepatol 26 Suppl* 2: 43–53
95 Lewis JH, Zimmerman HJ, Garrett CT, Rosenberg E (1982) Valproate-induced hepatic steatogenesis in rats. *Hepatology* 2: 870–873
96 Nau H, Merker HJ, Brendel K, Häuser I, Gansau C, Wittfoht W (1984) Disposition, embryotoxicity and hepatotoxicity of valproic acid in the mouse as related to man. In: RH Levy (eds): *Metabolism of Antiepileptic Drugs*. Raven Press, New York, 85–96
97 Kesterson JW, Granneman GR, Machinist JM (1984) The hepatotoxicity of valproic acid and its metabolites in rats. I. Toxicologic, biochemical and histopathologic studies. *Hepatology* 4: 1143–1152
98 Stephens, J. Pathogenetic hypotheses regarding hepatotoxicity of VPA: Beta-oxydative pathway, acyl coA, carnitine, urea cycle, organic acidemia, place of hyperammonemia. 1991. Nijmegen. Workshop on Recent Developments on Valproate and its Metabolites.
99 Levy RH, Lin JMH, Acheampong AA, Russel RG (1992) Dose effect of L-carnitine on delta^4-VPA-associated hepatotoxicity. In: RH Levy, JK Penry (eds): *Idiosyncratic Reactions to Valproate: Clinical Risk Patterns and Mechanisms of Toxicity*. Raven Press, New York, 25–29
100 Cotariu D, Reif R, Zaidman JL, Evans S (1987) Biochemical and morphological changes induced by sodium valproate in rat liver. *Pharmacol Toxicol* 60: 235–236
101 Siemes H, Nau H, Schultze K, Wittfoht W, Drews E, Penzien J, Seidel U (1993) Valproate (VPA) metabolites in various clinical conditions of probable VPA-associated hepatotoxicity. *Epilepsia* 34: 332–346
102 Rettie AE, Rettenmeier AW, Howald WN, Baillie TA (1987) Cytochrome P-450-catalyzed formation of delta 4-VPA, a toxic metabolite of valproic acid. *Science* 235: 890–893
103 Rettie AE, Sheffels PR, Korzekwa KR, Gonzalez FJ, Philpot RM, Baillie TA (1995) CYP4 isozyme specificity and the relationship between omega-hydroxylation and terminal desaturation of valproic acid. *Biochemistry* 34: 7889–7895
104 Sadeque AJM, Fisher MB, Korzekwa KR, Gonzalez FJ, Rettie AE (1997) Human CYP2C9 and CYP2A6 mediate formation of the hepatotoxin 4-ene-valproic acid. *J Pharmacol Exp Ther* 283: 698–703
105 Kingsley E, Gray P, Tolman KG, Tweedale R (1983) The toxicity of metabolites of sodium valproate in cultured hepatocytes. *J Clin Pharmacol* 23: 178–185

106 Rettenmeier AW, Prickett KS, Gordon WP, Bjorge SM, Chang SL, Levy RH, Baillie TA (1985) Studies on the biotransformation in the perfused rat liver of 2-n-propyl-4-pentenoic acid, a metabolite of the antiepileptic drug valproic acid. Evidence for the formation of chemically reactive intermediates. *Drug Metab Dispos* 13: 81–96
107 Rettenmeier AW, Gordon WP, Prickett KS, Levy RH, Baillie TA (1986) Biotransformation and pharmacokinetics in the rhesus monkey of 2-n-propyl-4-pentenoic acid, a toxic metabolite of valproic acid. *Drug Metab Dispos* 14: 454–464
108 Kassahun K, Farrell K, Abbott F (1991) Identification and characterization of the glutathione and N-acetylcysteine conjugates of (E)-2-propyl-2,4-pentadienoic acid, a toxic metabolite of valproic acid, in rats and humans. *Drug Metab Dispos* 19: 525–535
109 Tang W, Borel AG, Fujimiya T, Abbott FS (1995) Fluorinated analogues as mechanistic probes in valproic acid hepatotoxicity: Hepatic microvesicular steatosis and glutathione status. *Chem Res Toxicol* 8: 671–682
110 Raza M, Al-Bekairi AM, Ageel AM, Qureshi S (1997) Biochemical basis of sodium valproate hepatotoxicity and renal tubular disorder: Time dependence of peroxidative injury. *Pharmacol Res* 35: 153–157
111 Baillie TA (1988) Metabolic activation of valproic acid and drug-mediated hepatotoxicity. Role of the terminal olefin, 2-n-propyl-4-pentenoic acid. *Chem Res Toxicol* 1: 195–199
112 Granneman GR, Wang SI, Machinist JM, Kesterson JW (1984) Aspects of the metabolism of valproic acid. *Xenobiotica* 14: 375–387
113 Prickett KS, Baillie TA (1986) Metabolism of unsaturated derivatives of valproic acid in rat liver microsomes and destruction of cytochrome P-450. *Drug Metab Dispos* 14: 221–229
114 Tang W, Abbott FS (1996) Characterization of thiol-conjugated metabolites of 2-propyl-4-pentenoic acid (4-ene VPA), a toxic metabolite of valproic acid, by electrospray tandem mass spectrometry. *J Mass Spectrom* 31: 926–936
115 Baldwin GS, Abbott FS, Nau H (1996) Binding of a valproate metabolite to the trifunctional protein of fatty acid oxidation. *FEBS Lett* 384: 58–60
116 Anderson GD, Acheampong AA, Levy RH (1994) Interaction between valproate and branched-chain amino acid metabolism. *Neurology* 44: 742–744
117 Dickinson RG, Bassett ML, Searle J, Tyrer JH, Eadie MJ (1985) Valproate hepatotoxicity: A review and report of two instances in adults. *Clin Exp Neurol* 21: 79–91
118 Ponchaut S, Draye JP, Veitch K, van Hoof F (1991) Influence of chronic administration of valproate on ultrastructure and enzyme content of peroxisomes in rat liver and kidney. Oxidation of valproate by liver peroxisomes. *Biochem Pharmacol* 41: 1419–1428
119 Horie S, Suga T (1985) Enhancement of peroxisomal beta-oxidation in the liver of rats and mice treated with valproic acid. *Biochem Pharmacol* 34: 1357–1362
120 Vamecq J, Vallee L, Fontaine M, Lambert D, Poupaert J, Nuyts JP (1993) CoA esters of valproic acid and related metabolites are oxidized in peroxisomes through a pathway distinct from peroxisomal fatty and bile acyl-CoA beta-oxidation. *FEBS Lett* 322: 95–100
121 Riva R, Albani F, Contin M, Baruzzi A (1996) Pharmacokinetic interactions between antiepileptic drugs. Clinical considerations. *Clin Pharmacokinet* 31: 470–493
122 Sackellares JC, Sato S, Dreifuss FE, Penry JK (1981) Reduction of steady-state valproate levels by other antiepileptic drugs. *Epilepsia* 22: 437–441
123 Pisani F, Fazio A, Artesi C, Oteri G, Pisani B, Romano F, Perucca E, Di Perri R (1987) An epidemiological study of the clinical impact of pharmacokinetic anticonvulsant drug interactions based on serum drug level analysis. *Ital J Neurol Sci* 8: 135–141
124 Levy RH, Rettenmeier AW, Anderson GD, Wilensky AJ, Friel PN, Baillie TA, Acheampong A, Tor J, Guyot M, Loiseau P (1990) Effects of polytherapy with phenytoin, carbamazepine, and stiripentol on formation of 4-ene-valproate, a hepatotoxic metabolite of valproic acid. *Clin Pharmacol Ther* 48: 225–235
125 Kondo T, Otani K, Hirano T, Kaneko S, Fukushima Y (1990) The effects of phenytoin and carbamazepine on serum concentrations of mono-unsaturated metabolites of valproic acid. *Br J Clin Pharmacol* 29: 116–119
126 Kapetanovic IM, Kupferberg HJ, Porter RJ, Theodore W, Schulman E, Penry JK (1981) Mechanism of valproate-phenobarbital interaction in epileptic patients. *Clin Pharmacol Ther* 29: 480–486

127 Bernus I, Dickinson RG, Hooper WD, Eadie MJ (1994) Inhibition of phenobarbitone N-glucosidation by valproate. *Br J Clin Pharmacol* 38: 411–416
128 Svinarov DA, Pippenger CE (1995) Valproic acid-carbamazepine interaction: Is valproic acid a selective inhibitor of epoxide hydrolase? *Ther Drug Monit* 17: 217–220
129 Bernus I, Dickinson RG, Hooper WD, Eadie MJ (1997) The mechanism of the carbamazepine-valproate interaction in humans. *Br J Clin Pharmacol* 44: 21–27
130 Perucca E, Hebdige S, Frigo GM, Gatti G, Lecchini S, Crema A (1980) Interaction between phenytoin and valproic acid: Plasma protein binding and metabolic effects. *Clin Pharmacol Ther* 28: 779–789
131 Bruni J, Gallo JM, Lee CS, Perchalski RJ, Wilder BJ (1980) Interactions of valproic acid with phenytoin. *Neurology* 30: 1233–1236
132 Coulter DA (1997) Antiepileptic drug cellular mechanisms of action: Where does lamotrigine fit in? *J Child Neurol 12 Suppl 1:* S2–S9
133 Page RL, O'Neil MG, Yarbrough DR, Conradi S (1998) Fatal toxic epidermal necrolysis related to lamotrigine administration. *Pharmacotherapy* 18: 392–398
134 Scheffner D, König S, Rauterberg-Ruland I, Kochen W, Hofmann WJ, Unkelbach S (1988) Fatal liver failure in 16 children with valproate therapy. *Epilepsia* 29: 530–542
135 Lieh-Lai MW, Sarnaik AP, Newton JF, Miceli JN, Fleischmann LE, Hook JB, Kauffman RE (1984) Metabolism and pharmacokinetics of acetaminophen in a severely poisoned young child. *J Pediatr* 105: 125–128
136 Yu HY, Shen YZ, Sugiyama Y, Hanano M (1987) Dose-dependent pharmacokinetics of valproate in guinea pigs of different ages. *Epilepsia* 28: 680–687
137 Osterloh J, Cunningham W, Dixon A, Combest D (1989) Biochemical relationships between Reye's and Reye's-like metabolic and toxicological syndromes. *Med Toxicol Adverse Drug Exp* 4: 272–294
138 Graf R, Gossrau R, Merker HJ, Schwabe R, Stahlmann R, Nau H (1985) Enzyme cytochemistry combined with electron microscopy, pharmacokinetics, and clinical chemistry for the evaluation of the effects of steady-state valproic acid concentrations on the mouse. *Histochemistry* 83: 347–358
139 Rumbach L, Mutet C, Cremel G, Marescaux CA, Micheletti G, Warter JM, Waksman A (1986) Effects of sodium valproate on mitochondrial membranes: electron paramagnetic resonance and transmembrane protein movement studies. *Mol Pharmacol* 30: 270–273
140 Kochen W, Schneider A, Ritz A (1983) Abnormal metabolism of valproic acid in fatal hepatic failure. *Eur J Pediatr* 141: 30–35
141 Kuhara T, Inoue Y, Matsumoto M, Shinka T, Matsumoto I, Kawahara N, Sakura N (1990) Markedly increased omega-oxidation of valproate in fulminant hepatic failure. *Epilepsia* 31: 214–217
142 Hurd RW, Perchalski RJ, Wilder BJ, McDowell LR, Wilkinson NS (1991) Enhanced hepatotoxicity of valproate in the selenium-deficient rat. *Epilepsia 32 Suppl* 3: 9
143 Burk RF, Hill KE, Awad JA, Morrow JD, Lyons PR (1995) Liver and kidney necrosis in selenium-deficient rats depleted of glutathione. *Lab Invest* 72: 723–730
144 Hurd RW, Van RHA, Wilder BJ, Karas B, Maenhaut W, De Reu L (1984) Selenium, zinc, and copper changes with valproic acid: Possible relation to drug side effects. *Neurology* 34: 1393–1395
145 Siemes H, Nau H, Seidel U, Gramm HJ (1992) [Irreversible valproate-associated liver failure]. Irreversibles valproatassoziiertes Leberversagen. *Monatsschr Kinderheilkd.* 140: 869–875
146 Cotariu D, Evans S, Zaidman JL, Marcus O (1990) Early changes in hepatic redox homeostasis following treatment with a single dose of valproic acid. *Biochem Pharmacol* 40: 589–593
147 Cotariu D, Evans S, Lahat E, Theitler J, Bistritzer T, Zaidman JL (1992) Inhibition of human red blood cell glutathione reductase by valproic acid. *Biochem Pharmacol* 43: 425–429
148 Lemasters JJ, Nieminen AL (1997) Mitochondrial oxygen radical formation during reductive and oxidative stress to intact hepatocytes. *Biosci Rep* 17: 281–291
149 Farrell K, Abbott FS, Junker AK, Waddell JS, Pippenger CE (1989) Successful treatment of valproate hepatotoxicity with N-acetylcysteine. *Epilepsia* 30: 700
150 König SA, Elger CE, Vassella D, Schmidt D, Bergmann A, Boenigk HE, Despland PA, Genton P, Krämer G, Löscher W et al (1998) Empfehlungen zu Blutuntersuchungen und

klinischer Überwachung zur Früherkennung des Valproat-assoziierten Lebersersagens. *Schweizerische Ärztezeitung* 79: 580–585
151 Cramer J, Mattson R-H (1995) Phenobarbital: Toxicity. In: RH Levy, R-H Mattson, BS Meldrum (eds): *Antiepileptic Drugs*. Raven Press, New York, 409–420
152 Samren EB, van Duijn CM, Koch S, Hiilesmaa VK, Klepel H, Bardy AH, Mannagetta GB, Deichl AW, Gaily E, Granstrom ML et al (1997) Maternal use of antiepileptic drugs and the risk of major congenital malformations: A joint European prospective study of human teratogenesis associated with maternal epilepsy. *Epilepsia* 38: 981–990
153 Dalens B, Raynaud EJ, Gaulme J (1980) Teratogenicity of valproic acid [letter]. *J Pediatr* 97: 332–333
154 Gomez MR (1981) Possible teratogenicity of valproic acid [letter]. *J Pediatr* 98: 508–509
155 Robert E, Guibaud P (1982) Maternal valproic acid and congenital neural tube defects [letter]. *Lancet* 2: 937
156 Bjerkedal T, Czeizel A, Goujard J, Kallen B, Mastroiacova P, Nevin N, Oakley GJ, Robert E (1982) Valproic acid and spina bifida [letter]. *Lancet* 2: 1096
157 Robert E, Rosa F (1983) Valproate and birth defects [letter]. *Lancet 2: 1142*
158 Lindhout D, Meinardi H (1984) Spina bifida and in-utero exposure to valproate [letter]. *Lancet* 2: 396
159 Lindhout D, Schmidt D (1986) In-utero exposure to valproate and neural tube defects [letter]. *Lancet* 1: 1392–1393
160 Omtzigt JG, Los FJ, Grobbee DE, Pijpers L, Jahoda MG, Brandenburg H, Stewart PA, Gaillard HL, Sachs ES, Wladimiroff JW (1992) The risk of spina bifida aperta after first-trimester exposure to valproate in a prenatal cohort. *Neurology* 42: 119–125
161 Huot C, Gauthier M, Lebel M, Larbrisseau A (1987) Congenital malformations associated with maternal use of valproic acid. *Can J Neurol Sci* 14: 290–293
162 Sharony R, Garber A, Viskochil D, Schreck R, Platt LD, Ward R, Buehler BA, Graham JMJ (1993) Preaxial ray reduction defects as part of valproic acid embryofetopathy. *Prenat Diagn* 13: 909–918
163 Janas MS, Arroe M, Hansen SH, Graem N (1998) Lung hypoplasia – a possible teratogenic effect of valproate. Case report. *APMIS* 106: 300–304
164 DiLiberti JH, Farndon PA, Dennis NR, Curry CJ (1984) The fetal valproate syndrome. *Am J Med Genet* 19: 473–481
165 Ardinger HH, Atkin JF, Blackston RD, Elsas LJ, Clarren SK, Livingstone S, Flannery DB, Pellock JM, Harrod MJ, Lammer EJ (1988) Verification of the fetal valproate syndrome phenotype. *Am J Med Genet* 29: 171–185
166 Clayton-Smith J, Donnai D (1995) Fetal valproate syndrome. *J Med Genet* 32: 724–727
167 Williams PG, Hersh JH (1997) A male with fetal valproate syndrome and autism. *Dev Med Child Neurol* 39: 632–634
168 Petrere JA, Anderson JA, Sakowski R, Fitzgerald JE, de la Iglesia FA (1986) Teratogenesis of calcium valproate in rabbits. *Teratology* 34: 263–269
169 Menegola E, Broccia ML, Nau H, Prati M, Ricolfi R, Giavini E (1996) Teratogenic effects of sodium valproate in mice and rats at midgestation and at term. *Teratog Carcinog Mutagen* 16: 97–108
170 Lindhout D, Omtzigt JG, Cornel MC (1992) Spectrum of neural-tube defects in 34 infants prenatally exposed to antiepileptic drugs. *Neurology* 42: 111–118
171 Kao J, Brown NA, Schmid B, Goulding EH, Fabro S (1981) Teratogenicity of valproic acid: *in vivo* and *in vitro* investigations. *Teratog Carcinog Mutagen* 1: 367–382
172 Moffa AM, White JA, Mackay EG, Frias JL (1984) Valproic acid, zinc and open neural tubes in 9 day-old hamster embryos. *Teratology* 29, 47A
173 Klug S, Lewandowski C, Zappel F, Merker HJ, Nau H, Neubert D (1990) Effects of valproic acid, some of its metabolites and analogues on prenatal development of rats *in vitro* and comparison with effects *in vivo*. *Arch Toxicol* 64: 545–553
174 Michejda M, McCollough D (1987) New animal model for the study of neural tube defects. *Z Kinderchir 42 Suppl* 1:32–5: 32–35
175 Briner W, Lieske R (1995) Arnold-Chiari-like malformation associated with a valproate model of spina bifida in the rat. *Teratology* 52: 306–311
176 Nau H, Hauck RS, Ehlers K (1991) Valproic acid-induced neural tube defects in mouse and human: aspects of chirality, alternative drug development, pharmacokinetics and possible mechanisms. *Pharmacol Toxicol* 69: 310–321

177 Battino D, Estienne M, Avanzini G (1995) Clinical pharmacokinetics of antiepileptic drugs in paediatric patients. Part I: Phenobarbital, primidone, valproic acid, ethosuximide and mesuximide. *Clin Pharmacokinet* 29: 257–286
178 Nau H, Löscher W (1986) Pharmacologic evaluation of various metabolites and analogs of valproic acid: Teratogenic potencies in mice. *Fundam Appl Toxicol* 6: 669–676
179 Omtzigt JG, Nau H, Los FJ, Pijpers L, Lindhout D (1992) The disposition of valproate and its metabolites in the late first trimester and early second trimester of pregnancy in maternal serum, urine, and amniotic fluid: Effect of dose, co-medication, and the presence of spina bifida. *Eur J Clin Pharmacol* 43: 381–388
180 Nau H (1986) Valproic acid teratogenicity in mice after various administration and phenobarbital-pretreatment regimens: The parent drug and not one of the metabolites assayed is implicated as teratogen. *Fundam Appl Toxicol* 6: 662–668
181 Finnell RH, Moon SP, Abbott LC, Golden JA, Chernoff GF (1986) Strain differences in heat-induced neural tube defects in mice. *Teratology* 33: 247–252
182 Finnell RH, Bennett GD, Karras SB, Mohl VK (1988) Common hierarchies of susceptibility to the induction of neural tube defects in mouse embryos by valproic acid and its 4-propyl-4-pentenoic acid metabolite. *Teratology* 38: 313–320
183 Naruse I, Collins MD, Scott WJJ (1988) Strain differences in the teratogenicity induced by sodium valproate in cultured mouse embryos. *Teratology* 38: 87–96
184 Gunn TM, Juriloff DM, Harris MJ (1995) Genetically determined absence of an initiation site of cranial neural tube closure is causally related to exencephaly in SELH/Bc mouse embryos. *Teratology* 52: 101–108
185 Juriloff DM, Harris MJ, Tom C, MacDonald KB (1991) Normal mouse strains differ in the site of initiation of closure of the cranial neural tube. *Teratology* 44: 225–233
186 Hall JL, Harris MJ, Juriloff DM (1997) Effect of multifactorial genetic liability to exencephaly on the teratogenic effect of valproic acid in mice. *Teratology* 55: 306–313
187 Finnell RH, Wlodarczyk BC, Craig JC, Piedrahita JA, Bennett GD (1997) Strain-dependent alterations in the expression of folate pathway genes following teratogenic exposure to valproic acid in a mouse model. *Am J Med Genet* 70: 303–311
188 Hauck RS, Nau H (1989) [Structural bases of the teratogenic effects of the antiepileptic valproic acid. 2-n-propyl-4-pentenic acid, the first structural analogue with significantly higher teratogenic action than VPA]. Zu den strukturellen Grundlagen der teratogenen Wirkung des Antiepileptikums Valproinsaure (VPA). 2-n-Propyl-4-pentinsaure, das erste Strukturanalogon mit signifikant höherer teratogener Aktivitat als VPA. *Naturwissenschaften* 76: 528–529
189 Bojic U, Elmazar MM, Hauck RS, Nau H (1996) Further branching of valproate-related carboxylic acids reduces the teratogenic activity, but not the anticonvulsant effect. *Chem Res Toxicol* 9: 866–870
190 Radatz M, Ehlers K, Yagen B, Bialer M, Nau H (1998) Valnoctamide, valpromide and valnoctic acid are much less teratogenic in mice than valproic acid. *Epilepsy Res* 30: 41–48
191 Hauck RS, Nau H (1989) Asymmetric synthesis and enantioselective teratogenicity of 2-n-propyl-4-pentenoic acid (4-en-VPA), an active metabolite of the anticonvulsant drug, valproic acid. *Toxicol Lett* 41–48
192 Hauck RS, Nau H (1992) The enantiomers of the valproic acid analogue 2-n-propyl-4-pentynoic acid (4-yn-VPA): Asymmetric synthesis and highly stereoselective teratogenicity in mice. *Pharm Res* 9: 850–855
193 Hauck RS, Elmazar MM, Plum C, Nau H (1992) The enantioselective teratogenicity of 2-n-propyl-4-pentynoic acid (4-yn-VPA) is due to stereoselective intrinsic activity and not differences in pharmacokinetics. *Toxicol Lett* 60: 145–153
194 Andrews JE, Ebron MM, Bojic U, Nau H, Kavlock RJ (1995) Validation of an *in vitro* teratology system using chiral substances: Stereoselective teratogenicity of 4-yn-valproic acid in cultured mouse embryos. *Toxicol Appl Pharmacol* 132: 310–316
195 Andrews JE, Ebron MM, Bojic U, Nau H, Kavlock RJ (1997) Stereoselective dysmorphogenicity of the enantiomers of the valproic acid analogue 2-N-propyl-4-pentynoic acid (4-yn-VPA): Cross-species evaluation in whole embryo culture. *Teratology* 55: 314–318
196 Bojic U, Ehlers K, Ellerbeck U, Bacon CL, O'Driscoll E, O'Connell C, Berezin V, Kawa A, Lepekhin E, Bock E et al. (1998) Studies on the teratogen pharmacophore of valpoic acid analogues: Evidence of interactions at a hydrophobic centre. *Eur J Pharmacol* 354:289–299

197 Milunsky A, Jick H, Jick SS, Bruell CL, MacLaughlin DS, Rothman KJ, Willett W (1989) Multivitamin/folic acid supplementation in early pregnancy reduces the prevalence of neural tube defects. *JAMA* 262: 2847–2852
198 Czeizel AE, Dudas I (1992) Prevention of the first occurrence of neural-tube defects by periconceptional vitamin supplementation. *N Engl J Med* 327: 1832–1835
199 MRC Vitamin Study Research Group (1991) Prevention of neural tube defects: Results of the Medical Research Council Vitamin Study. *Lancet* 338: 131–137
200 Steinberg SE (1984) Mechanisms of folate homeostasis. *Am J Physiol* 246: G319–G324
201 Smith DB, Carl GF (1982) Interactions between folates and carbamazepine or valproate in the rat. *Neurology* 32: 965–969
202 Wegner C, Nau H (1992) Alteration of embryonic folate metabolism by valproic acid during organogenesis: Implications for mechanism of teratogenesis. *Neurology* 42: 17–24
203 Trotz M, Wegner C, Nau H (1987) Valproic acid-induced neural tube defects: Reduction by folinic acid in the mouse. *Life Sci* 41: 103–110
204 Wegner C, Trotz MN (1987) Folate supplementation and deficiency in experimental valproate-induced teratogenesis. In: H Nau, WJ Scott (eds): CRC Press, Inc., Boca Raton, Florida, 3–12
205 Hansen DK, Grafton TF (1991) Lack of attenuation of valproic acid-induced effects by folinic acid in rat embryos *in vitro*. *Teratology* 43: 575–582
206 Hansen DK, Grafton TF, Dial SL, Gehring TA, Siitonen PH (1995) Effect of supplemental folic acid on valproic acid-induced embryotoxicity and tissue zinc levels *in vivo*. *Teratology* 52: 277–285
207 Fowler B (1997) Disorders of homocysteine metabolism. *J Inherit Metab Dis* 20: 270–285
208 Kirke PN, Molloy AM, Daly LE, Burke H, Weir DG, Scott JM (1993) Maternal plasma folate and vitamin B_{12} are independent risk factors for neural tube defects. *Q J Med* 86: 703–708
209 Mills JL, McPartlin JM, Kirke PN, Lee YJ, Conley MR, Weir DG, Scott JM (1995) Homocysteine metabolism in pregnancies complicated by neural-tube defects. *Lancet* 345: 149–151
210 Steegers-Theunissen RP, Boers GH, Blom HJ, Nijhuis JG, Thomas CM, Borm GF, Eskes TK (1995) Neural tube defects and elevated homocysteine levels in amniotic fluid. *Am J Obstet Gynecol 172:* 1436–1441
211 Steegers-Theunissen RP, Boers GH, Trijbels FJ, Finkelstein JD, Blom HJ, Thomas CM, Borm GF, Wouters MG, Eskes TK (1994) Maternal hyperhomocysteinemia: A risk factor for neural-tube defects? *Metabolism* 43: 1475–1480
212 Kang SS, Wong PW, Bock HG, Horwitz A, Grix A (1991) Intermediate hyperhomocysteinemia resulting from compound heterozygosity of methylenetetrahydrofolate reductase mutations. *Am J Hum Genet* 48: 546–551
213 Harmon DL, Woodside JV, Yarnell JW, McMaster D, Young IS, McCrum EE, Gey KF, Whitehead AS, Evans AE (1996) The common "thermolabile" variant of methylene tetrahydrofolate reductase is a major determinant of mild hyperhomocysteinaemia. *Q J Med 89:* 571–577
214 Van der Put NM, Thomas CM, Eskes TK, Trijbels FJ, Steegers-Theunissen RP, Mariman EC, De Graaf-Hess A, Smeitink JA, Blom HJ (1997) Altered folate and vitamin B_{12} metabolism in families with spina bifida offspring. *Q J Med* 90: 505–510
215 Elmazar MM, Thiel R, Nau H (1992) Effect of supplementation with folinic acid, vitamin B_6, and vitamin B_{12} on valproic acid-induced teratogenesis in mice. *Fundam Appl Toxicol* 18: 389–394
216 Ehlers K, Elmazar MM, Nau H (1996) Methionine reduces the valproic acid-induced spina bifida rate in mice without altering valproic acid kinetics. *J Nutr* 126: 67–75
217 Nosel PG, Klein NW (1992) Methionine decreases the embryotoxicity of sodium valproate in the rat: *in vivo* and *in vitro* observations. *Teratology* 46: 499–507
218 Hansen DK, Dial SL, Grafton TF (1995) Lack of attenuation of valproic acid-induced embryotoxicity by compounds involved in one-carbon transfer reactions. *Toxic in Vitro* 9: 615–621
219 Rosenquist TH, Ratashak SA, Selhub J (1996) Homocysteine induces congenital defects of the heart and neural tube: Effect of folic acid. *Proc Natl Acad Sci USA* 93: 15227–15232

220 Van Aerts LA, Klaasboer HH, Postma NS, Pertijs JC, Copius-Peereboom JH, Eskes TK, Noordhoek J (1993) Stereospecific *in vitro* embryotoxicity of L-homocysteine in pre- and post-implantation rodent embryos. *Toxic in Vitro* 7: 743–749
221 van Aerts LA, Blom HJ, Deabreu RA, Trijbels FJ, Eskes TK, JH CP-S, Noordhoek J (1994) Prevention of neural tube defects by and toxicity of L-homocysteine in cultured postimplantation rat embryos. *Teratology* 50: 348–360
222 Miller JW, Nadeau MR, Smith J, Smith D, Selhub J (1994) Folate-deficiency-induced homocysteinaemia in rats: Disruption of S-adenosylmethionine's coordinate regulation of homocysteine metabolism. *Biochem J* 298: 415–419
223 Elmazar MM, Nau H (1992) Methotrexate increases valproic acid-induced developmental toxicity, in particular neural tube defects in mice. *Teratog Carcinog Mutagen* 12: 203–210
224 Elmazar MM, Nau H (1993) Trimethoprim potentiates valproic acid-induced neural tube defects (NTDs) in mice. *Reprod Toxicol* 7: 249–254
225 Hillman RS, Steinberg SE (1982) The effects of alcohol on folate metabolism. *Annu Rev Med* 33: 345–354
226 Elmazar MM, Nau H (1995) Ethanol potentiates valproic acid-induced neural tube defects (NTDs) in mice due to toxicokinetic interactions. *Reprod Toxicol* 9: 427–433
227 Chernoff GF (1977) The fetal alcohol syndrome in mice: An animal model. *Teratology* 15: 223–229
228 Sulik KK, Johnston MC, Webb MA (1981) Fetal alcohol syndrome: embryogenesis in a mouse model. *Science* 214: 936–938
229 Clarke DO, Brown NA (1987) Valproic acid teratogenesis and embryonic lipid metabolism. *Arch Toxicol Suppl* 11: 143–147
230 Thurston JH, Hauhart RE (1993) Vitamins to prevent neural-tube defects [letter; comment]. *N Engl J Med 328:* 1641–1642
231 Sato M, Shirota M, Nagao T (1995) Pantothenic acid decreases valproic acid-induced neural tube defects in mice (I). *Teratology* 52: 143–148
232 Rogers JM, Keen CL, Hurley LS (1985) Zinc deficiency in pregnant Long-Evans hooded rats: Teratogenicity and tissue trace elements. *Teratology* 31: 89–100
233 Dreosti IE, Buckley RA, Record IR (1986) The teratogenic effect of zinc deficiency and accompanying feeding patterns in mice. *Nutrition Research* 6: 159–166
234 Leek JC, Vogler JB, Gershwin ME, Golub MS, Hurley LS, Hendrickx AG (1984) Studies of marginal zinc deprivation in rhesus monkeys. V. Fetal and infant skeletal effects. *Am J Clin Nutr* 40: 1203–1212
235 Vormann J, Höllriegel V, Merker HJ, Günther T (1986) Effect of valproate on zinc metabolism in fetal and maternal rats fed normal and zinc-deficient diets. *Biol Trace Elem Res* 10: 25–35
236 Wegner C, Drews E, Nau H (1990) Zinc concentrations in mouse embryo and maternal plasma. Effect of valproic acid and nonteratogenic metabolite. *Biol Trace Elem Res* 25: 211–217
237 Bui LM, Taubeneck MW, Commisso JF, Uriu-Hare JY, Faber WD, Keen CL (1998) Altered zinc metabolism contributes to the developmental toxicity of 2-ethylhexanoic acid, 2-ethylhexanol and valproic acid. *Toxicology* 126: 9–21
238 Nau H, Scott-WJ J (1986) Weak acids may act as teratogens by accumulating in the basic milieu of the early mammalian embryo. *Nature* 323: 276–278
239 Weaver TE, Scott WJJ (1984) Acetazolamide teratogenesis: Association of maternal respiratory acidosis and ectrodactyly in C57BL/6J mice. *Teratology* 30: 187–193
240 Layton WMJ, Hallesy DW (1965) Forelimb deformity in rats: Association with acetazolamide. *Science* 150: 79
241 Scott WJJ, Schreiner CM, Nau H, Vorhees CV, Beliles RP, Colvin J, McCandless D (1997) Valproate-induced limb malformations in mice associated with reduction of intracellular pH. *Reprod Toxicol* 11: 483–493
242 Nau H, Chahoud I, Dencker L, Lammer EJ, Scott WJ (1994) Teratogenicity of vitamin A and retinoids. In: R Blomhoff (eds): *Vitamin A in Health and Disease.* Marcel Dekker, New York, 615–663
243 Nau H, Tzimas G, Mondry M, Plum C, Spohr HL (1995) Antiepileptic drugs alter endogenous retinoid concentrations: a possible mechanism of teratogenesis of anticonvulsant therapy. *Life Sci* 57: 53–60

244 Ashby R, Davis L, Dewhurst BB, Espinal R, Penn RN, Upshall DG (1976) Aspects of the teratology of cyclophosphamide (NSC-26271). *Cancer Treat Rep* 60: 477–482
245 Winn LM, Wells PG (1995) Free radical-mediated mechanisms of anti-convulsant teratogenicity. *European Journal of Neurology 2 Suppl* 4: 5–29
246 Yu WK, Wells PG (1995) Evidence for lipoxygenase-catalyzed bioactivation of phenytoin to a teratogenic reactive intermediate: *in vitro* studies using linoleic acid-dependent soybean lipoxygenase, and *in vivo* studies using pregnant CD-1 mice. *Toxicol Appl Pharmacol* 131: 1–12
247 Harris C, Stark KL, Juchau MR (1988) Glutathione status and the incidence of neural tube defects elicited by direct acting teratogens *in vitro*. *Teratology* 37: 577–590
248 Weber GF, Maertens P, Meng XZ, Pippenger CE (1991) Glutathione peroxidase deficiency and childhood seizures. *Lancet 337:* 1443–1444
249 Graf WD, Pippenger CE, Shurtleff DB (1995) Erythrocyte antioxidant enzyme activities in children with myelomeningocele. *Dev Med Child Neurol* 37: 900–905
250 Brown NA, Farmer PB, Coakley M (1985) Valproic acid teratogenicity: Demonstration that the biochemical mechanism differs from that of valproate hepatotoxicity. *Biochem Soc.Trans.* 13: 75–77
251 Dencker L, Nau H, D'Argy R (1990) Marked accumulation of valproic acid in embryonic neuroepithelium of the mouse during early organogenesis. *Teratology* 41: 699–706
252 Turner S, Sucheston ME, De Philip RM, Paulson RB (1990) Teratogenic effects on the neuroepithelium of the CD-1 mouse embryo exposed in utero to sodium valproate. *Teratology* 41: 421–442
253 Copp AJ, Brook FA, Estibeiro JP, Shum AS, Cockroft DL (1990) The embryonic development of mammalian neural tube defects. *Prog Neurobiol* 35: 363–403
254 Bruckner A, Lee YJ, O'Shea KS, Henneberry RC (1983) Teratogenic effects of valproic acid and diphenylhydantoin on mouse embryos in culture. *Teratology* 27: 29–42
255 Cole WA, Trasler DG (1980) Gene-teratogen interaction in insulin-induced mouse exencephaly. *Teratology* 125–139
256 Aulthouse AL, Hitt DC (1994) The teratogenic effects of valproic acid in human chondrogenesis *in vitro*. *Teratology* 49: 208–217
257 Morriss-Kay G, Tuckett F (1989) Immunohistochemical localisation of chondroitin sulphate proteoglycans and the effects of chondroitinase ABC in 9- to 11-day rat embryos. *Development* 106: 787–798
258 Wlodarczyk BC, Craig JC, Bennett GD, Calvin JA, Finnell RH (1996) Valproic acid-induced changes in gene expression during neurulation in a mouse model. *Teratology* 54: 284–297
259 Bennett GD, An J, Craig JC, Gefrides LA, Calvin JA, Finnell RH (1998) Neurulation abnormalities secondary to altered gene expression in neural tube defect susceptible Splotch embryos. *Teratology* 57: 17–29
260 Williams JA, Mann FM, Brown NA (1997) Gene expression domains as markers in developmental toxicity studies using mammalian embryo culture. *Int J Dev Biol* 41: 359–364
261 Lampen A, Göttlicher M, Siehler S, Ellerbeck U, Nau H (1998) A new molecular bioassay based on cell differentiation *in vitro* for the estimation of the teratogenic potency of valproic acid derivatives. Paris, "Chemical Safety For The 21st Century", International Union of Toxicology
262 Kliewer SA, Forman BM, Blumberg B, Ong ES, Borgmeyer U, Mangelsdorf DJ, Umesono K, Evans RM (1994) Differential expression and activation of a family of murine peroxisome proliferator-activated receptors. *Proc Natl Acad Sci USA* 91: 7355–7359
263 Wells PG, Winn LM (1996) Biochemical toxicology of chemical teratogenesis. *Crit Rev Biochem Mol Biol* 31: 1–40
264 Nau H (1994) Valproic acid-induced neural tube defects. *Ciba Found Symp* 181: 144–152
265 Chen G, Yuan P, Hawver DB, Potter WZ, Manji HK (1997) Increase in AP-1 transcription factor DNA binding activity by valproic. *Neuropsychopharmacology* 16: 238–245
266 O'Brian E, Regan CM (1998) Protein Kinase C inhibitors arrest the C6 glioma cell cycle at a mid-G1 phase restriction point: Implications for the antiproliferative action of valproate. *Toxicology in Vitro* 12: 9–14

267 Chen G, Manji HK, Wright CB, Hawver DB, Potter WZ (1996) Effects of valproic acid on beta-adrenergic receptors, G-protein, and adenylyl cyclase in rat C6 glioma cells. *Neuropsychopharmacology* 15: 271–280
268 Finnell RH (1991) Genetic differences in susceptibility to anticonvulsant drug-induced developmental defects. *Pharmacol Toxicol* 69: 223–227
269 Swinyard EA (1964) The pharmacology of dipropylacetic acid sodium with special emphasis on its effects on the central nervous system. University of Utah, College of Pharmacy. Salt Lake City, Utah, pp 1–25
270 Shuto K, Nishigaki T (1970) The pharmacological studies on sodium dipropylacetate anticonvulsant activities and general pharmacological actions (in Japanese). *Pharmacometrics* 4: 937–949
271 Frey HH, Löscher W (1976) Di-n-propylacetic acid–profile of anticonvulsant activity in mice. *Arzneimittelforschung.* 26: 299–301
272 Kupferberg HJ (1980) Sodium valproate. In: GH Glaser, JK Penry, DM Woodbury (eds): *Antiepileptic drugs: Mechanism of action*. Raven Press, New York, 643–654
273 Löscher W, Nau H (1985) Pharmacological evaluation of various metabolites and analogues of valproic acid. Anticonvulsant and toxic potencies in mice. *Neuropharmacology* 24: 427–435
274 Paulson RB, Sucheston ME, Hayes TG, Weiss HS (1989) Effects of sodium valproate and oxygen on the craniofacial skeletal pattern in the CD-1 mouse embryo. *J Craniofac Genet Dev Biol* 9: 339–348
275 Vorhees CV (1987) Teratogenicity and developmental toxicity of valproic acid in rats. *Teratology* 35: 195–202
276 Binkerd PE, Rowland JM, Nau H, Hendrickx AG (1988) Evaluation of valproic acid (VPA) developmental toxicity and pharmacokinetics in Sprague-Dawley rats. *Fundam Appl Toxicol* 11: 485–493
277 Hishida R, Nau H (1998) VPA-induced neural tube defects in mice. I. Altered metabolism of sulfur amino acids and glutathione. *Teratog Carcinog Mutagen* 18: 49–61

Clinical use of valproate

Milestones in Drug Therapy
Valproate
ed. by W. Löscher

Valproate in the treatment of epilepsies in children

N. Sarisjulis[1] and Olivier Dulac[2]

[1] Hospital de Niños Sor Maria Ludovica, Calle 14 e/65 y 66, CP 1900 La Plata, Argentina
[2] Hôpital Saint Vincent de Paul, INSERM U29, Université René Descartes, 82 Avenue Denfert-Rochereau, F-75674 Paris Cedex, France

The treatment of epilepsy with valproate in childhood must take into account the wide range of clinical presentations of epilepsy, including specific responses to the drug, and pharmacological and toxicological particularities in this age range.

Particularities of epilepsies in infancy and childhood

Aetiology

In childhood as in adulthood, epilepsy may result from *cerebral lesions*, particularly of prenatal occurrence, abnormalities of the program and development of the brain, or of peri- or postnatal occurrence, of clastic, vascular, traumatic, infectious, or toxic origin (including endogenous due to inborn errors of metabolism). Tumours are rare, causing less than 1% of epilepsies in the first decade [1]. In fact, less than one-quarter of childhood epilepsies are due to some cause identifiable by history, clinical examination or neuroradiological investigation.

A *genetic predisposition* can be demonstrated or suspected in nearly half the cases, and it can be either the only causative factor without any evidence of anatomic brain damage, or produce by itself specific brain lesions such as dysplasia or vascular malformations [2].

The third factor which is specific to childhood is due to *maturation phenomena*, since there are age ranges that are at particular risk for the occurrence of certain types of epilepsy, with spontaneous tendency for seizures to become less severe and frequent in the long term [3].

These three groups of etiologic factors account for the epilepsy either independently or in combination, resulting in a wide range of clinical expressions of disease. In particular, the expression of epilepsy may vary for a given aetiology, according to other factors – genetic or maturational predisposition.

Clinical expression

Epilepsy in childhood may produce repeat seizures or episodes of status epilepticus like later in life. However, in contrast with adulthood, progressive motor, sensorial or cognitive deterioration may be the main or exclusive expression of the disease in a patient with rare or mild seizures from the onset. This contrast is usually due to major paroxysmal activity recorded on EEG, affecting both hemispheres, although it can predominate in some part of the brain. In such conditions, called *epileptic encephalopathies*, deterioration mainly involves the function supported by the most affected area. These patients may or may not exhibit seizures that are usually brief and not responsible for the deterioration.

Seizures may be focal or generalised, and there is no *type of seizures* encountered in adulthood that is not, at least occasionally, observed in childhood. In addition to those observed in adulthood, there are seizures specific to infancy and childhood, such as spasms in clusters, and prolonged epileptic conditions like long lasting unilateral clonic seizures followed by hemiplegia – the hemiconvulsion-hemiplegia (HH) syndrome [4] – and myoclonic status in which the patient exhibits continuous and erratic jerks of the face and extremities with drowsiness.

The combination of one or several seizure types with a particular interictal condition, beginning at a specific age, and particular interictal and ictal EEG tracings define *epileptic syndromes*. Proper diagnosis of a given epileptic syndrome is of the utmost importance in terms of prognosis, therapeutic decisions and etiologic research. Some syndromes (i.e. absence epilepsy) are homogeneous enough for a single aetiology to be suspected. In other syndromes (i.e. infantile spasms), the various characteristics, including seizure types and interictal EEG may vary, according particularly to aetiology. Therapeutic decisions greatly depend on the syndromic diagnosis since some drugs may specifically improve and others specifically worsen some epileptic syndromes [5]. Preventing the occurrence of worsening must be a constant preoccupation of the clinician.

Epileptic syndromes are often difficult to identify at onset, and the combination of the specific characteristics must be met in order for the syndromic diagnosis to be reached with sufficient reliability. During the first months or years of the epilepsy, as long as these specific features that allow syndromic diagnosis are not yet combined, therapeutic decisions are difficult to take, and avoiding the risk of worsening is again a major factor of decision.

Formulation and tolerability of valproate in infancy and childhood

Formulation

The most classical formulations of valproate are poorly adapted to children: the solution has a very unpleasant taste and many infants refuse to take the drug, even when it is mixed into food. Some children even refuse to take a given food when valproate has been added to the same type of food days earlier. Sprinkles that would solve the problem of taste became recently available in Europe and the US. Regarding tablets, the absorption is rapid and the half-life quite short: pharmacokinetic characteristics would require the drug to be given three times a day, which is most impractical for patients that go to school for the whole day. Control release formulation has been marketed for school-age children, and this improves persistence of therapeutic blood levels throughout the day [6].

Side-effects

From the *cognitive* point of view, valproate is one of the antiepileptic drugs best tolerated in childhood. *Haematological* side-effects have no particularities. *Loss of hair* is no more severe than in adulthood although a few cases of alopecia have been reported [7]. *Increased appetite* with major weight gain mainly affects adolescent females [8], but it may also occur in childhood.

Hepatic failure affects mainly infants aged less than 2 years suffering from intractable epilepsy and receiving combined therapy [9]. Early symptoms are recurrence of seizures not explained by identifiable intercurrent events, somnolence and vomiting. Decreased values of prothrombine and increase of transaminases are the main biological findings [10, 11]. Prompt withdrawal of valproate treatment often permits reversibility of hepatic failure [12]. In contrast, systematic monitoring of transaminases appears to be useless since hepatic failure may occur within a few days of normal findings, and moderate increase of transaminases is frequent and has no prognostic value.

Abdominal pain may result from *gastritis*, particularly with soluble formulation. Acute *pancreatitis* may be difficult to identify in children with severe mental retardation who have difficulties to express themselves. Lipase dosage should be added to the classical dosage of amylase whose half-life is too short and negative findings are therefore meaningless [13].

Metabolic interactions

Clinically significant metabolic interactions are encountered in infancy and childhood. They mainly concern drugs that are metabolized in the liver since valproate reduces the kinetics of drug catabolism. *Ethosuximide* blood levels are increased in combination with valproate with a high risk of somnolence [14].

Carbamazepine levels are increased with increase of epoxide which reduces tolerability and produces somnolence and ataxia, partly because of increased levels of carbamezepine epoxide [15].

Clonazepam in combination with valproate has been reported to produce somnolence, even at doses supposed to be moderate [16].

Values of *phenytoin* unlinked to proteins are not modified by valproate, although the total blood level of the drug is decreased [17]. Therefore, phenytoin dose should not be altered when introducing valproate.

The combination of *lamotrigine* and valproate has been shown to be particularly effective in some types of epilepsy. However, it may be difficult to manage because of a higher risk of skin rash. In practice, lamotrigine should be introduced very slowly, over an 8-week period.

Interaction between epilepsy and drug treatment

Several clinical patterns may occur as the first manifestation of epilepsy. Patients may exhibit *convulsions* at any age, but particularly in the newborn period or in infancy. They may have *spasms* in infancy, and *loss of consciousness* or *drop attacks* in childhood. *Partial seizures* occur at any age. *Deterioration* of cognitive or motor functions may be the major sign of epilepsy in neonate, infancy or childhood. Syndromic diagnosis may be difficult at onset.

Various epileptic conditions

The pattern of onset of the epilepsy varies from patient to patient, but it is possible to consider a few major patterns: infants presenting with convulsions, infants with cognitive deterioration, infants and children presenting with partial seizures, children presenting with typical or atypical absences, patients presenting with cognitive deterioration and/or drop attacks.

Convulsions and epilepsies in neonates and infancy
When faced with convulsions in the newborn or in infancy, it is most difficult to determine whether it is due to epilepsy or an occasional event, and whether an eventual epilepsy is focal or generalised. Indeed, some patients with generalised epilepsy may exhibit focal seizures affecting different areas of the body over a period of several days or weeks. Only the observation of repeat seizures is thus likely to distinguish these two conditions. The consequences of diagnosis are major since antiepileptic drugs when inappropriately chosen may worsen the condition, particular-

ly vigabatrin and carbamazepine in generalised epilepsy with myoclonus and/or absences.

Occasional seizures: Convulsions due to brain damage usually require intravenous administration of antiepileptic drugs. To date, there is no evidence that the intravenous formulation of valproate has any efficacy for the treatment of neonatal convulsions. In addition, chronic antiepileptic treatment is not required following neonatal acute brain injury, provided the EEG no longer shows paroxysmal activity [18]. Therefore, there is no indication for either intravenous or oral administration of valproate in this context.

Benign neonatal and infantile idiopathic or familial convulsions: They occur as clusters of brief episodes of focal or secondarily generalised convulsions, each one lasting one to a few minutes and being repeated several times a day during a period of 2 or 3 days. Familial cases are dominantly inherited, and causative genes have been identified for neonatal cases [19–21]. Like in occasional convulsions, there is no indication for intravenous administration of valproate. However, there is a sizeable risk of recurrence when patients receive no chronic treatment following the acute period. Therefore, valproate which is well tolerated in these patients without any evidence of metabolic encephalopathy or intractability, is indicated for periods of 6 to 12 months.

Epileptic encephalopathies with suppression bursts: This condition is characterised by the combination of focal and generalised seizures with bursts of polyspikes alternating with lack of activity on EEG, this alternation being called suppression-bursts. Cases with myoclonus and lack of identifiable cause (except for rare cases with inborn error of metabolism) are called *neonatal myoclonic encephalopathy*, and cases with tonic spasms and an identifiable, clastic or malformative cause are called *Ohtahara syndrome*. However, this classical nomenclature remains a matter of dispute, and this distinction is certainly most difficult to perform during the first weeks of the disease, apart from cases with a cause identifiable on radiology [22, 23]. For therapeutic purposes, it is reasonable at first to consider the disorder as being non progressive. Valproate is useful for treating first seizures, before thorough diagnosis is made. However, as soon as this type of encephalopathy is recognised, vigabatrin, eventually combined with carbamazepine, seems to be more likely to be effective (unpublished personal data). In addition, since inborn errors of metabolism may cause this condition, it is wise to avoid the administration of valproate in order to reduce the risk of hepatic failure.

Febrile convulsions: Its definition concerns seizures *provoked by* fever, without evidence of central nervous system (CNS) infection. These are the most frequent cause of convulsions in children because, according to the different regions in the world, 2–7% of children present a least one febrile convulsion before the age of 5 years.

Febrile convulsions (FC) are an age-related condition, occurring between 6 months and 5 years, with a peak of frequency in the middle of the second year of life; it has a genetic contribution, 30% of cases having a familial history of FCs. The genetic transmission seems to be multifactorial or dominant with low penetrance. Causes of fever are various, being due to some viral infection in 90% of the cases [24].

The main difficulty, after excluding CNS infection, is the differential diagnosis between simple FC and the first seizure of chronic epilepsy; the second difficulty is to identify risk factors for prolonged seizures, in case of relapse.

Age of appearance of seizures and their clinical characteristics are determinant. In 80% of cases, the convulsion is considered simple: it is single, short, isolated, generalised tonic or, sometimes, followed by some clonic jerks. In the other cases it is considered complex: it is unilateral, prolonged, followed by focal motor deficit, repeated in the same day, or it has occurred in the first year of life.

The risk to develop a severe epilepsy mainly concerns complex FCs, especially when they appear in the first year of life [25].

The risk of status epilepticus decreases after the first year of age and is very low after the second year. On the other hand, generalised FCs lasting more than 30 min and occurring in the second year of life may produce no sequel. On the contrary, in the first year of life, FCs that result in status epilepticus and hemiconvulsion-hemiplegia (HH) syndrome represent the major risk of motor and mental sequel and severe epilepsy.

The two main types of severe epilepsy following FC are: *severe myoclonic epilepsy in infants* and *symptomatic partial epilepsy*. In *severe myoclonic epilepsy in infants*, the first seizure occurs always in the first year of life and, in symptomatic partial epilepsy, the first FC occurs often at this age in the context of fever, and is often unilateral and prolonged. In both instances severity of epilepsy can be reduced by adequate prophylactic treatment. On the contrary, some idiopathic generalised and partial benign epilepsies sometimes follow simple FCs. In these cases, a prophylactic treatment following FC is not indicated in the absence of risk factors for later epilepsy.

Moreover, results of interictal EEG are often paradoxical and disappointing. After an FC in the first year of life, the EEG is most frequently normal although at this age the risk to develop severe epilepsy is the most important. In the second or third year of life, EEG can show focal spikes or generalised spike-waves, although the only significant risk is the potential development of benign epilepsy [26].

Discussions about treatment of FC remain open. Several studies have demonstrated that adequate medication can reduce the recurrence risk for FC. On the basis of benign status of FC in most of the cases, the question is: when to give a prophylactic treatment for FC, and which drug to use for effective continuous or intermittent treatment?

Regarding the first question, according to several studies [27, 28], prophylactic treatment should be indicated when the risk of recurrence is high (70–90%), thus:

- Age less than 15 months at first FC.
- History of FC or chronic epilepsy in first degree relatives.
- Complex FC.

Prophylactic treatment could reduce the risk of recurrence in this group to reach that of children with a low recurrence risk (10–15%) [28].

Intermittent prevention with valproate, administered during febrile episodes, can reduce the recurrence risk. This has been suggested in a study comparing efficacy and side-effects between diazepam and valproate that showed no significant difference in efficacy [29].

Therefore, for intermittent prophylaxis, the first line drug is diazepam, administered by intrarectal route. This route can be utilised by parents and by non specialised operators, without loss of time and with good efficacy.

Continuous prophylaxis is no longer a standard recommendation, except for highly selected cases and for a limited period of time, usually until the age of 2 years in seizure-free patients. In these cases, the administration of phenobarbital can negatively affect cognitive functions [30, 31], and behaviour with hyperactivity, irritability, aggressiveness, and inversion of sleep pattern. Valproate is preferred because of efficacy and moderate side-effects. Moreover, the rare cases of fatal hepatotoxicity reported concerned patients with seizure disorders different from FC.

Severe myoclonic epilepsy in infants: In this syndrome, seizures appear during the first year of life, in infants without any personal antecedent; they are usually clonic, either unilateral or generalised, occurring sometimes as status epilepticus, with temporary postictal hemiplegia [32].

The first seizures are febrile in two thirds of the cases, therefore being considered as pure FCs since the neurological examination, psychomotor development and EEG are usually normal at that moment [33]. Criteria such as early age (under 1 year old), occurrence with moderate fever (<39°C) or postictal motor asymmetry tend to justify prophylactic drug treatment. Unfortunately, seizures soon recur, both febrile and non-febrile, and episodes of status epilepticus may occur that contribute to the development of brain lesions, thus the poor prognosis of this syndrome [34].

Both segmentary and massive myoclonus may appear later, during the second year of life, together with EEG anomalies consisting of discharges of generalised spike and waves, activated by sleep and photic stimulation. Conversely, clonic seizures become milder, whereas myoclonic status may last several days. The neurologic condition deteriorates progressively, and ataxia, language retardation, and hyperkinesia appear together with severe psychomotor retardation.

Valproate has proved to be effective, mainly when administered together with a benzodiazepine. It diminishes frequency, and above all, duration of non-febrile convulsive seizures [32]. Among benzodiazepines, clobazam is the first choice since it causes fewer side effects (e.g. somnolence) than clonazepam [35, 36].

Partial epilepsy in infancy and childhood
All ages and all the areas of the brain may be affected. The most frequently encountered symptoms in infancy consist of arrest of activity, jerks of the limbs or upper eyelids, oculoclonia, erythrosis, mastication automatisms, and lateral deviation of the head and eyes. It is the area of the brain that experiences the discharge that determines the clinical expression of the seizures. However, the latter claim mainly concerns patients after the age of 1 year. Before this age, only occipital, rolandic and temporal seizures produce specific clinical expression, consisting of oculoclonic or limb jerks and mastication respectively.

Among partial epilepsies, *symptomatic* cases are defined by brain lesions disclosed on neuroradiology. The place for valproate in the treatment of symptomatic partial epilepsy is quite restricted, except for patients who experience secondary generalisation of the seizures. *Idiopathic* cases are defined by the lack of brain lesion, favourable outcome and link to maturation phenomena. This group comprises benign infantile convulsions [37], benign rolandic epilepsy and benign occipital epilepsy, which both respond favourably to various compounds [38], of which valproate is preferred because of lack of significant side-effects and of risk of worsening. In fact, valproate is often considered the first line drug for this group [39–41]. There are cases of familial nocturnal frontal epilepsy that do not seem to respond to valproate as well as they do to carbamazepine [42].

Cryptogenic partial epilepsy designs cases with neither evidence of brain damage, or of clinical and EEG characteristics of idiopathic epilepsy. Early diagnosis of cryptogenic partial epilepsy in infancy is particularly difficult during the first months of the disorder, before evidence of a single type of recurrent partial seizures has been collected, and evolution to generalised epilepsy with infantile spasms remains possible until the second year of life. During this period, valproate or vigabatrin are preferable to other medications in order to prevent the occurrence of drug-induced spasms, mainly carbamazepine [5]. Thus, the place for valproate is precisely to handle this difficult period during which diagnostic work-up is performed, trying to identify either the signs of a brain lesion, or those of idiopathic epilepsy [43].

Mental deterioration in infancy
West Syndrome: The incidence of this syndrome often called infantile spasms is approximately 1 for 4000 to 6000 births, it usually appears be-

tween 3 months and 1 year of age, with a peak frequency at 5 months. However, cases with onset up to 4 years of age have been reported.

West Syndrome (WS) is typically defined by a triad consisting of spasms in clusters as the main or exclusive seizure type, diffuse, asynchronous paroxysmal EEG abnormalities called ”hypsarrythmia” and psychomotor impairment. Each of these three components may apparently be missing at onset.

Infantile spasm are a heterogeneous group; improvement in neuroradiological technology has reduced the rate of cryptogenic cases to less than 30% of those for which there is evidence of a symptomatic cause. The four main causes are represented by malformations, neurocutaneos syndromes, ante- or perinatal anoxic-ischemia and non-labelled congenital encephalopathies. A small number of patients present idiopathic spasm, meaning that there is neither direct evidence or indirect signs of brain damage (5–10%) [44]; lack of indirect signs of brain damage comprises: normal development prior to the appearance of spasms and lack of significant psychomotor regression, no other seizure type, clinically symmetric spasms that from the EEG point of view are so-called “independent”, meaning that hypsarrhythmia recurs between spasms of a cluster [45].

This distinction regarding aetiology is of great clinical relevance, since the choice of anti-epileptic drugs as well as evolution depend on etiology.

Before 1981, occasional reports described cessation of spasms using valproate [46]. Later on, more systematic studies with valproate as an initial monotherapy were carried out. Combining the results of two studies [47, 48], 37 patients were administered 20–60 mg/kg/d; seizures stopped in 27%, and another 27% had over 50% decrease in seizure frequency; however, two patients presented an increase in the number of seizures. Blood levels of VPA remained between 40 and 80 μg/ml, with excellent tolerance, apart from occasional lethargy or hypotonia. Another study showed no relapse 12 to 18 months after cessation of treatment, in cases in which total initial control was achieved with 30–40 mg/kg/d [49].

In a prospective study, including 22 patients diagnosed with West Syndrome, of which 18 were symptomatic and four cryptogenic [50], valproate doses were steadily increased. After 4 weeks monotherapy, spasms were controlled in 11 patients with doses ranging from 40 to 100 mg/kg (mean 74 mg/kg) and a mean serum level of 113 μg/ml (range 46–177 μg/ml); there were seven relapses. For the four patients with cryptogenic West Syndrome, spasms were controlled within 4 weeks, with slight or no psychomotor impairment. Hypotonia was observed in most patients, and in spite of the high doses of valproate, no case of hepatitis occurred.

The greatest restriction to the use of valproate in West Syndrome is its potential to cause side-effects. Hepatic toxicity is estimated to affect between 1/10000 and 1/50000 patients receiving this drug [51]; the risk increases with polytherapy, and when the age is under 2 years; therefore, children affected by West Syndrome theoretically have a high toxicity risk of hepatic toxicity, approximately 1/5000 [9].

Since vigabatrin has demonstrated major effectiveness in infantile spasm due to tuberous sclerosis and other causes, including cryptogenic case [52], valproate is no longer the drug of first choice for the treatment of infantile spasms. The use is restricted to cases resistant to vigabatrin and steroids, particularly those combining other types of seizures, especially generalised tonic or tonic clonic seizures. The combination of valproate with lamotrigine seems to be particularly promising for children who, following infantile spasms, develop tonic seizures [53].

When spasms are combined with partial seizures, the most adequate decision is to administer vigabatrin [53].

Non progressive myoclonic encephalopathy: This epilepsy is seen in infants with severe prenatal encephalopathy, which from the very first months of life is related to axial hypotonia, abnormal polymorphous movements, and severe psychomotor retardation. The characteristics described for this type of epilepsy are also observed in Angelman Syndrome.

At the beginning, seizures consist of erratic myoclonus or prolonged clonic seizures that may suggest tremor. The EEG with polygraphy allows to identify the epileptic nature of the abnormal movements, since each jerk occurs at the same time as a slow spike-wave; frequently, myoclonic jerks are continuous, but asynchronous upon the different muscles in a given individual, therefore it is more difficult to see their relation to the EEG discharges.

Myoclonus is present very early in the disorder, but it is often difficult to determine the precise age of onset. Children are usually unable to walk, due to severe ataxia and hyperkinesia.

When the epileptic nature of the condition is identified, these episodes of status are refractory to all treatments. EEG discharges and rhythmic myoclonus, which are continuous both during wake and sleep, are not controlled by benzodiazepines, or only temporarily. ACTH has also been tried without success. In fact, only valproate combined with ethosuximide seems to produce some kind of improvement in the clinical picture [54]. Piracetam as co-medication sometimes produces good results. When the condition is at least partially controlled by treatment, the clinical picture improves notoriously.

Inborn errors of metabolism (*IEM*). This group of diseases comprises a great number of different conditions with variable clinical expression and specific therapeutic options. In most instances, seizures are a rare event that occurs late in the evolution of the disease, but some diseases produce epileptic seizures as the first clinical expression. Myoclonus is particularly frequent in this context in children, but rarely in infants whose seizures due to IEM rarely consist of myoclonus or spasms, but more often of convulsive seizures. Some infants with IEM exhibit a whole variety of different kinds of seizure types, including spasms, myoclonus and both focal and generalised convulsive seizures.

Since the cause of the disorder affects the brain diffusely, and therefore the epilepsy is a generalised phenomenon, valproate is usually the first line drug. However, because of metabolic interferences, this drug is likely to trigger some kind of worsening of the neurological condition, and the administration of valproate in this context needs therefore to be restricted to patients for which metabolic interactions can be ruled out. In practice, several diseases should prevent the administration of valproate:

- Suspicion of Alpers disease, which is characterised by epilepsia partialis continua and often produces hepatic failure. Many cases of hepatic failure occurring in conjunction with the administration of valproate may have resulted from this disease [55].
- Abnormalities affecting metabolism of the urea cycle, since hyperammoniemia is an identified complication of valproate therapy [56].
- Mitochondrial disorders are likely to affect beta oxydation, therefore to trigger hepatic failure [57].

Absences

Childhood and juvenile absence epilepsy (CAE, JAE): Is probably the most benign form of generalised idiopathic epilepsy (GIE). It prevails in females (78%) and there are familial antecedents of GIE in 15 to 40% of the cases [58]. Seizures only consist of absences, which may be accompanied by eyelid clonias in over 80% of the cases. Hypotonic, tonic, autonomic including enuretic, or single automatic components are rare, and there are neither limb myoclonus or sudden falls. Absences are brief, with sudden onset and end; very frequently, their occurrence is favoured by emotions, awakening and hyperpnea, a manoeuvre easy to perform in the consulting room. The EEG recording of a seizure is necessary to confirm the diagnosis: it shows generalised and regular 3 Hz spike-wave discharges with sudden onset and end; there is a sub-group of these patients that is sensitive to photic stimulation.

Absences disappear in 90% of the cases with treatment and long-term prognosis is excellent. The later appearance of generalised tonic clonic seizures, between 10 and 15 years of age, is more frequent in boys; in this case, absences start after 8 years of life and photo-sensitivity is more frequent, and this group more often resists treatment. This chronic evolution may explain the difficulties in learning and in the socialization of these children. Thus, early treatment with valproate and/or ethosuximide yield more favourable prognoses [59].

The choice between these two drugs as first medication is difficult and habits vary from country to country, valproate being used in Europe, whereas ethosuximide is preferred in the USA, because of concerns with hepatic failure, although the risk is clearly particularly low in those patients which are older than the high risk age range, and who are pharmacosensitive and on monotherapy. In addition, valproate is active against generalised tonic-clonic seizures in contrast with ethosuximide. However, the risk of GTC is particularly low with CAE.

If valproate is not efficient, ethosuximide or lamotrigine are used in monotherapy or even in combination with valproate if necessary, since both drugs exhibit a beneficial synergism.

Myoclonic Absences: This type of absence lasts longer than pure absences, from 10 to 60 s and is accompanied by upper limb myoclonus. First seizures begin at a mean age of 7 years (range 2–12); in approximately half the cases, there is pre-existing mental retardation [60].

Unconsciousness, with sudden beginning and end, is not complete, and the child usually seems to mask motor phenomena by a voluntary gesture. Myoclonic shakes mainly affect shoulders, arms and legs, more than the face; they are rhythmic, sometimes asymmetrical. Absences may generally be triggered by hyperpnea. The EEG shows, as in simple absences, rhythmic, bilateral, symmetric and synchronous 3 Hz spike-waves, which are rarely sensitive to photic stimulation.

Other types of epileptic seizures, particularly tonic-clonic, are quite rare, and tonic seizures do not occur during the absence period.

Regarding treatment, contrary to pure absences, myoclonic absences do not produce a favourable response to monotherapy, either valproate or ethosuximide; their association is slightly more efficient [13]. A new molecule, Lamotrigine, has proved efficacions and deserves to be tried, mainly when the first two have failed [61].

Evolution sometimes leads to disappearance of seizures, whereas in most cases, seizures persist over 10 years, yet in some other cases, it evolves to Lennox-Gastaut syndrome. Pre-existing mental retardation is clearly a major risk factor for intractability.

Lennox-Gastaut Syndrome (LGS) begins early, between 2 and 8 years of age; it starts even earlier when the syndrome is symptomatic, that is to say, when it appears in patients with prior mental retardation, including after West Syndrome and/or a subsequent to partial epilepsy.

In patients without evidence of previous brain injury (Cryptogenic Lennox-Gastaut Syndrome), onset is usually later, 4 to 8 years of age; it rarely begins after the age of 10.

The typical triad of this syndrome is [62]:

- epileptic seizures, mainly axial tonic seizures, atonic seizures and atypical absences.

- EEG abnormalities consisting of diffuse slow spike and wave discharges and bursts of fast rhythms at around 10 Hz during sleep.
- Slow mental development combined with personality disorders.

Tonic seizures are frequent and sometimes necessary for the diagnosis; they are particularly frequent during sleep, sometimes discrete or limited to brief upper eye deviation, EEG shows recruiting rhythm during the tonic seizure, whatever its clinical expression.

Atypical absences have a less sudden beginning and end than typical absence seizures, and consist of slow head ante-flexion (atonic absences), or, on the contrary, a retropulsion (retropulsive absence); generalised spike-waves on EEG are slow and irregular.

Drop attacks may result from tonic seizures, atonic absences or, more rarely from massive myoclonus, the latter being accompanied by a poly-spike discharge on the EEG. Certain patients present, likewise, partial seizures, sometimes of multifocal origin. Absences, tonic or mixed episodes of status may last from several hours to several days; on the contrary, episodes of generalised tonic status are rare.

Interictal EEG shows bursts of slow spike-waves, sometimes focal abnormalities and an increasingly slow basic activity. In slow sleep, diffuse bursts of polyspike discharges are observed during the periods in which seizures are frequent.

Cognitive disturbances are dominated by an increasingly slow motion, emotional instability, and attention impairment. They appear soon and may lead to insanity; sometimes these disturbances are more elective, and more marked in the periods in which seizures are more frequent.

Treatment is difficult; considering the polymorphism of the seizure types, it is usually impossible to avoid polytherapy. Valproate together with ethosuximide are used to control atypical absences, taking into account that valproate possesses moderate effectiveness for this type of absences.

To date, several studies [63, 64] have shown that lamotrigine exhibits major efficacy with cessation of seizures in one-third of the patients with a 2-year follow up [65]. Likewise, felbamate has proved to have significant effectiveness on seizures of LGS [66]. These have currently become the main alternatives.

When facing the need of polytherapy, interactions between valproate and other drugs should be taken into account as well as, reducing lamotrigine or ethosuximide doses, thus obtaining an optimal efficiency and minimal adverse effects [67].

Continuous spike waves in slow sleep (CSWS) are characterised by mental deterioration combined with bilateral, synchronous spike-wave complexes affecting over 85% of slow wave sleep time and disappearing in REM sleep [68]. During waking there are generalised spike-wave discharges and sometimes focal spikes. The EEG aspect may appear in children that have earlier presented normal or moderately slower acquisitions, or with a focal neurologic deficit. Diverse types of neuropsychologic disturbances correlate according to the topography of predominant disturbance in the EEG.

(A) Speech disturbance combined with temporal predominance of paroxysmal activity defines the Landau-Kleffner syndrome that appears between 2 and 8 years of age as a deficit of oral language comprehension. Some patients are hyperkinetic or frankly psychotic. The EEG typically shows temporal lobe predominance of paroxysmal disturbances, mainly on tracings performed early in the evolution. Epileptic seizures are rare or absent in approximately 1/3 of the patients, and in the other cases, they disappear altogether between 10 and 15 years of age. Evolution of language disturbances varies greatly. They are more severe in patients with early onset, those without remission and those that last the longest.

If by means of medication the paroxysmal activity of sleep can be controlled, improvement in oral communication occurs within a few months.

(B) In a second group absences have an atonic component, and polygraphy shows that atonia is contemporaneous with the slow wave of the spike wave discharge recorded on EEG, the negative myoclonus.

(C) The third group appears as a Frontal Syndrome, with disturbances in behaviour and programming, accompanied by frontal predominance of EEG paroxysmal disturbances.

(D) In a fourth group oral dyspraxia is associated with rolandic predominance of paroxysmal anomalies.

(E) Finally, in a fifth group there are no clinical phenomena in the interval between seizures. Thus, some patients with apparently the same kind of EEG anomalies suffer no neuropsychologic impairment. However, these patients need careful follow-up since cognitive deterioration may be insidious and overlooked for months before it becomes evident. It is clearly during this period that appropriate treatment is most likely to prevent severe sequelae.

Motor seizures can precede continuous spike-wave activity in slow sleep; they mainly consist of partial seizures appearing at a mean age of 4 to 5 years, and the EEG shows frontal, frontal-temporal and central-temporal focal spikes, sometimes associated with generalised spike-waves. In no case have tonic seizures been observed.

This epilepsy evolves favourably with seizures disappearing between 10 and 15 years of age, and the period of typical EEG pattern disappears at a mean age of 10; thereafter, there is a steady recovery of the neuropsychologic disturbances, which most often fails to reach complete recovery [69].

Early identification of this type of epilepsy is important since various antiepileptic drugs, including carbamazepine, and occasionally valproate are likely to determine the appearance of continuous spike-wave complexes during sleep; thus, when these drugs are indicated for partial epilepsy with abundant interictal paroxysmal discharges, a control EEG with sleep recording should be carried out early in the follow-up, in order to verify that there is no major increase of paroxysmal activity. This measure could contribute to prevent neuropsychologic impairment. Once this syndrome has been diagnosed, optimal treatment consists of benzodiazepines, and in the resistant cases, prolonged steroid therapy for several months, usually 1 or 2 years, is currently considered as the treatment of choice [43].

Drop attacks and/or cognitive deterioration

Late onset infantile spasms, beginning after the age of 1 year [70], LGS, CSWS and Myoclonic-astatic Epilepsy may all produce drop attacks and cognitive deterioration. The distinction between these syndromes may be difficult at onset and may need sleep EEG recording, and eventually ictal video EEG. This is most important in view of therapeutic decisions since some drugs may worsen the spike wave activity; others worsen or even trigger myoclonus.

Myoclonic-astatic Epilepsy (Doose Syndrome): Doose defined "myoclonic-astatic petit mal" as affecting a group of patients that begin between 7 months and 7 years of age with massive myoclonic, atonic and myoclonic-atonic seizures, frequently associated with atypical absences and generalised tonic clonic seizures. Patients beginning in the first year of life are clearly affected by "severe" or "benign" myoclonic epilepsies of infancy, two subgroups of Myoclonic-astatic Epilepsy identified by Dravet et al. [32, 71, 72]. In the following, we include only the patients beginning later in life, usually after the age of 2 years.

This group has a variable outcome. For most patients, the total period of frequent seizures lasts from 6 months to 1 year, followed by a period with few generalised convulsive seizures lasting a few months, until total cessation, and they no longer have seizures after 1 or 2 years. Some patients even suffer an episode of status epilepticus lasting for several hours before seizures stop altogether. The EEG of these children shows brief bursts of spike-wave discharges at 3 HZ. Unfortunately, even in this condition described as favourable, children frequently present sequelae consisting of language retardation and attention deficit.

Other children develop myoclonic status that persists for several days or even weeks, with drowsiness and erratic jerks of the extremities of the limbs and face, combined with vibrating tonic seizures that occur mainly at the end of night sleep, and cognitive functions severely deteriorate. Subsequently, the children are left with severe mental impairment and persistent nocturnal tonic seizures. These children with unfavourable outcome have irregular slow spike and wave discharges late in the evolution [73].

Valproate in combination with ethosuximide and/or benzodiazepines have long been the most likely effective drugs for this condition [74]. The addition of lamotrigine to valproate has produced remarkable effects in a number of cases, achieving favourable outcome [75].

On the contrary, it is important to point out that some drugs, particularly carbamazepine and GABAergic compounds are likely to worsen the picture.

Syndromic diagnosis not feasible at the beginning

Before the proper syndromic diagnosis can be made, it is necessary to prevent the occurrence of convulsive seizures by administering a compound that is unlikely to worsen the condition of the patient and to affect the EEG picture, thus delaying the syndromic diagnosis, whether caused by focal or generalised epilepsy. Valproate is the only drug that meets these requirements. However, before the proper etiologic diagnosis can be performed, it is most important to exclude any risk for hepatic functions, especially in infants beginning epilepsy, the age range at highest risk for hepatic failure from toxic origin [9]. Indeed, efficacy of valproate has been shown to be similar to that of carbamazepine in newly diagnosed patients, including for those with focal epilepsy [39].

Practical aspects of treatment with valproate

The medication should be introduced progressively, over a period of 1 or 2 weeks, in order to prevent side-effects triggered by too rapid increase in blood levels, particularly drowsiness with diffuse slow wave activity, nausea, vomiting, tremor and hallucinations [76]. For the same reason, any change in drug dose or formulation should also be progressive in order to avoid too rapid modification of blood level.

An initial dose of 10 to 15 mg/kg is advised, with an increase of 5 mg/kg every other day, to a total dose of 20 to 30 mg/kg, according to age, the youngest patients needing the highest dose according to weight. It is only in rare instances, particularly in infants with severe epilepsy, that doses up to 40 mg/kg may be helpful.

Although there are theoretical arguments recommending to administer the drug three times a day, the experience shows that twice is often sufficient. In adolescents receiving the control release formulation, it is even usually possible to restrict the number of intakes to once per day, at night time [16].

Although it is classical to advise blood level measures soon after starting treatment, the data collected are rarely if at all useful, and it is our feeling that this tradition should be dispensed with [77], particularly since it is inappropriate to increase the dose on the basis of low blood level values if the patient is seizure-free, and since for valproate the efficacy/blood level relations still remain to be established [78].

Regarding systematic measures of transaminases, again there is no evidence that they contribute by any means to prevent hepatic failure. On the other hand, effective and well tolerated valproate treatment is too often inappropriately stopped based on moderate increases of transaminases that are very frequent in the course of valproate treatment. It seems more appropriate to inform the parents of the rare but clinically relevant risk that exists during the first 6 months of treatment, the initial manifestations, particularly recurrence of seizures, drowsiness and vomiting, and of what to do in case the latter occurs. Presenting written information is best, to be given in addition to the drug package insert.

Valproate is no doubt the first line drug for idiopathic generalised epilepsy (IGE), although some may consider ethosuximide for childhood absence epilepsy. For atypical absences, valproate is also first line. For idiopathic partial epilepsy, it is also first line, although some may prefer carbamazepine, but the risk of side-effects, including rare cases of worsening with appearance of CSWS should be considered. For patients who fall or suffer generalised convulsions combined with other type of seizures, be it due to Lennox-Gastaut or Doose syndromes, or cases not classifiable from the syndromic point of view, valproate is useful as first line, although lamotrigine offers further benefit, especially when added to valproate [79]. In CSWS, valproate has no indication, and in rare cases it may even worsen the condition, thus benzodiazepines are clearly first line. For infants with convulsive seizures, unless there is evidence of a focal lesion, valproate is the drug of first choice before the type of epilepsy has been identified. For

Table 1. Therapeutic algorithm for first and second line drugs according to the various epilepsy syndromes

Syndrome	Worsening Likely	First line	Second line
West	Carbamazepine	Vigabatrin Steroids	Valproate
Severe myoclonic epilepsy of infancy	Phenobarbital, Carbamazepine, Vigabatrin, Lamotrigine	Valproate	Benzodiazepines
Benign myoclonic epilepsy of infancy	Carbamazepine	Valproate	Ethosuximide
Infantile absences	Carbamazepine, Phenobarbital, Phenytoin, Vigabatrin	Valproate Ethosuximide	Lamotrigine
Lennox-Gastaut	Carbamazepine	Valproate Lamotrigine	Ethosuximide Felbamate
Myoclonic-Astatic	Carbamazepine, Vigabatrin	Valproate Lamotrigine	Ethosuximide Benzodiazepines
Continuous spike waves during slow sleep and Landau-Kleffner	Carbamazepine, Phenobarbital, Valproate, Phenytoin	Benzodiazepines	Steroids
Febrile convulsions		Diazepam (Intermittent)	Valproate (Continuous)

patients with infantile spasms, valproate is no longer a first line drug. However, it may be indicated in addition to more specific compounds in case of occurrence of additional generalised convulsive seizures or intractability by vagabatrin and steroids.

When given in addition to other therapies, potential metabolic interactions must be taken in account to determine the appropriate dose.

Cessation of treatment should be considered when the patient is seizure-free for over 1 year with absence or benign partial epilepsy, and over 5 years for symptomatic cases. Two seizure-free years are usually considered reasonable for the other cases, except for juvenile myoclonic epilepsy that needs prolonged therapy.

Conclusion

Valproate is the drug of choice for the great majority of epilepsies beginning in infancy and childhood. This is due to both a wide spectrum of efficacy and excellent tolerability, and to the low risk of worsening. However, there are special conditions in which more specific indications need to be applied, as soon as the proper diagnosis can be made. In practice, there is very little place for biological monitoring, including blood level measures, and most aspects of follow-up should be covered clinically.

References

1 Aicardi J, Praud E, Bancaud J, Mises J, Chevrie JJ (1970) Epilepsies cliniquement primitives et tumeurs cérébrales chez l'enfant. *Arch Fr Pediatr* 27: 1041–1055
2 Beck-Mannagetta G, Anderson E, Doose H, Janz D (eds) (1989) *Genetics of the epilepsies.* Springer-Verlag, Berlin
3 Dulac O (1995) Epileptic syndromes in infancy and childhood: recent advances. *Epilepsia* 36: S51–S57
4 Gastaut H, Vigouroux M, Trevisan C et al (1957) Le syndrome hémiconvulsions-hémiplégie-épilepsie (syndrome H.H.E.). *Rev Neurol* 97: 37–52
5 Perucca E, Gram L, Avanzini G, Dulac O (1998) Antiepileptic drugs as a cause of worsening seizures. *Epilepsia* 39: 5–17
6 Klotz U (1982) Bioavailability of a slow release preparation of valproic acid under steady state conditions. *Int J Pharmacol ther Toxicol* 20: 24–26
7 Jeavons PM (1984) Non-dose-related side effects of valproate. *Epilepsia* 25 (suppl 1): S50–S55.
8 Dinesen H, Gram L, Anderson T, Dam M (1984). Weight gain during treatment with valproate. *Acta Neurolog Scand*; 70: 65–69
9 Scheffner D, König S, Rauterberg-Ruland I, Kochen W, Hofmann WJ, Unkelbach S (1988) Fatal liver failure in 16 children with valproate therapy. *Epilepsia* 29: 530–542
10 Hellström-Westas L, Blennow G, Lindroth M, Rosén I, Svenningson N (1995) Low risk of seizure recurrence after early withdraw of antiepileptic treatment in the neonatal period. *Arch Dis Child* 72 (suppl 2): F97-F101.
11 Sussman NM, McLain LW. (1979). A direct hepatotoxic effect of valproic acid. *JAMA* 242: 1173–1174
12 Collettir B, Trainer TD, Karwisz BR. (1986). Reversible valproate fulminant hepatic failure. *J Pediatr Gastroenterol Nutr* 5: 990–994
13 Asconapé JJ, Penry JK, Dreifuss FE et al. (1992). Valproate-associated pancreatitis. *Epilepsia* 34: 117–123

14 Pisani F, Narbona MC, Trunfio C, Fazio A, La Rosa G, Oteri G, Di Perri R (1984) Valproic acid-ethosuximide interaction: a pharmacokinetic study. *Epilepsia* 25: 229–233
15 Ramsay RE, Mc Manus D, Guterman A, Briggle T, Vasques D, et al (1990) Carbamazepine metabolism in humans: effect of concurrent anticonvulsant therapy. *The Drug Monit* 12: 235–241
16 Jeavons PM, Clark JE, Maheswari MC (1977) Treatment of generalised epilepsies of childhood with sodium valproate (epilim). *Dev Med Child Neurol* 19: 9–25
17 Lai ML, Huang JD (1993) Dual effect of valproic acid on the pharmacokinetic of phenytoin *Biopharm Drug Dispos* 14: 365–370
18 Dulac O (1998). Convulsions et epilepsies du nouveau-né et du nourrisson In: Ponsot G, Arthuis M, Pinsard N, Dulac O, Mancini J (eds): *Neurologie pediatrique, deuxieme édition.* Flammarion, Paris, 367–406
19 Vigevano F, Fusco L, Di Capua M et al (1994) Benign infantile familial convulsions: Clinical and genetics aspects. *Pediatric Neurology* 11: 94
20 Malafosse A, Beck C, Bellet H, Di Capua M et al (1994) Benign infantile familial convulsions are not an allelic form of the benign familial neonatal convulsions gene. *Ann Neurol* 35: 479–482
21 Leppert M, Anderson VE, Quateblaum T, Stauffer D et al (1989) Benign familial neonatal convulsions linked to genetic markers on chromosome 20. *Nature* 337 (6208): 647–648
22 Othahara S, Ohtsuka Y, Yamatogi Y (1987) The early infantile epileptic encephalopaty with suppression-bursts: Developmental aspects. *Brain Dev* 9: 371–376
23 Schlumberger E, Dulac O, Pluin P (1992) Syndrome (s) d'epilepsie néonatale avec "suppression-burst": approche nosologique. In: Roger J, Bureau M, Dravet Ch, Dreifuss FE, Perret A, Wolf P (eds): *Epileptic syndromes in infancy, childhood and adolescence.* John Libbey, London-Paris: 35–42
24 Nelson K, Ellenberg J (eds) (1981) *Febrile seizures.* Raven Press, New York.
25 Dalla Bernardina B, Colamaria V, Capovilla G, Bondavalli S, Bureau M (1982) Epilepsie myoclonique grave de la première année. *Rev Electroencephalogr Neurophysiol Clin* 12: 21–25
26 Aicardi J, Chevrie JJ (1973) The significance of electroencephalographic paroxysms in children less than three years of age. *Epilepsia* 14: 47–55
27 Martin-Fernandez JJ, Molto-Jorda JM, Villaverde R, Salmeron P, Prieto-Muñoz I, Fernandez-Barreiro A. (1996) Risk factors in recurrent febrile seizures. *Rev Neurol* 24 (136): 1520–1524
28 Knudsen FU (1996) Febrile seizures- Treatment and outcome. *Brain Dev* 18: 438–449
29 Daugbjerg P, Brems M, Mai J, Ankerhus J, Knudsen FU (1990) Intermittent prophilaxis in febrile convulsions: diazepam or valproic acid ? *Acta Neurol Scand* 82: 17–20
30 Camfield CS, Chaplin S, Doyle AB, Shapiro SH, Cunning C, Camfield PR (1979). Side effects of phenobarbital in toddlers: Behavioral and cognitive aspects. *J Pediatr* 95: 361–365
31 Byrne JM, Camfield PR, Clark-Touesnard M, Hondas BJ (1987) Effects of phenobarbital on early intellectual and behavioral development: A concordance twin case study. *J Clin-Exp Neuropsychol* 9: 393–398
32 Dravet Ch, Bureau M, Guerrini R, Giraud N, Roger J (1992) Severe myoclonic epilepsy in infants. In: Roger J, Bureau M, Dravet Ch, Dreifuss FE, Perret A, Wolf P (eds): *Epileptic syndromes in infancy, childhood and adolescence.* John Libbey, London & Paris, 75–88
33 Yakoub M, Dulac O, Jambaque I et al (1992) Early diagnosis of severe myoclonic epilepsy in infancy. *Brain Dev* 14: 299–303
34 Chiron C, Dulac O (1994) Les différents syndromes épileptiques chez le nourrisson. In: Pons G, Dulac O, Ben-Ari Y (eds): *Les médicaments antiépileptiques chez l'enfant.* Springer-Verlag, Paris, 47–55
35 Chapman AG, Horton RW, Meldrum BS (1979) Anticonvulsant Action of a 1,5-benzodiazepine, clobazam, in epilepsy. *Epilepsia* 19: 293–299
36 Schmidt D, Rohde M, Wolf P, Roeder-Wanner U (1986) Tolerance to the antiepileptic effect of clobazam. In: Koella WP, Frey HH, Froscher W, Meinardi H, (eds): *Tolerance to beneficial and adverse effects of antiepileptic drug.* Raven Press, New York, 109–115
37 Watanabe K (1996) Benign partial epilepsies. In: Wallace S (ed): *Epilepsy in children.* Chapman and Hill, London, 293–313

38 Degen R, Dreifuss FE (1992) Localised and generalised epilepsies of early childhood. *Epilepsy Res*. Suppl 6
39 Verity CM, Hosking G, Easter DJ (1995) A multicentre trial of sodium valproate and carbamazepine in paediatric epilepsy. The Paediatric EPITEG Collaborative Group. *Dev Med Child Neurol* 37 (2): 97–108
40 Chaigne D, Dulac O (1989) Carbamazepine versus valproate in partial epilepsies of childhood. *Advances in epileptology* 17: 198–200
41 Bureau M, Dravet Ch, Genton P, Roger J, Weber M (eds) (1993) *Les épilepsies de l'enfant*. Doc médicale Labaz, Paris
42 Berkovic S, Scheffer I (1997) Epilepsies with single gene inheritance. *Brain Dev* 19 (1): 13–18.
43 Dulac O (1998) Epilepsies et convulsions de l'enfant. *Encyclop Med Cirurg*. Elsevier: Paris. Pédiatrie 4-091-A-10, 26 p
44 Vigevano F, Fusco L, Cusmai R, Claps D, Ricci S, Milani L (1993) The idiopatic form of West syndrome. *Epilepsia* 34: 743–747
45 Dulac O, Plovin P, Jambaqué I (1993) Predicting favorable outcome in idiopathic West syndrome. *Epilepsia* 34: 747–756
46 Simon D, Penry JK (1975) Sodium de-N-propylacetate in the treatment of epilepsy. *Epilepsia* 16: 549–573
47 Bachman D (1982) Use of valproic acid in treatment of infantile spasm. *Arch Neurol* 39: 49–52
48 Pavone L, Incorpora G, La Rosa M, Li Volti S, Mollica F (1981) Treatment of infantile-spasm with sodium dipropylacetic acid. *Dev Med Child Neurol* 23: 454–461
49 Dulac O, Steru D, Rey E, Perret A, Arthuis M (1986) Sodium valproate monotherapy in childhood epilepsy. *Brain Dev* 8: 47–52
50 Siemes H, Short HL, Michael TH, Nau H (1988) Therapy of infantile spasm with valproate: results of a prospective study. *Epilepsia* 29: 553–560
51 Dreifuss FE, Santilli N (1986) Valproic acid fatalities: Analysis of US cases. *Neurology* 36 (suppl 1)
52 Chiron C, Dumas C, Jambaqué I, Mumford J, Dulac OP (1996) Ramdomized trial comparing vigabatrin and hydrocortisone in infantile spasm due to tuberous sclerosis. *Epilepsy Research* 26: 389–395
53 Dulac O, Chugani HT, Dalla Bernardina B (eds) (1994) *Infantile spasm and West syndrome*. WB Saunders, London
54 Dalla Bernardina B, Fontana E, Sgro V, Colamaria V, Elia M (1992) Myoclonic epilepsy (myoclonic status) in non-progressive encephalopathies In: Roger J, Bureau M, Dravet Ch, Dreifuss FE, Perret A, Wolf P (eds): *Epileptic syndromes in infancy, childhood and adolescence*. John Libbey, London-Paris, 89–96
55 Narkewicz MR, Sokol RJ, Beckwith B et al (1991) Liver involvement in Alper's disease. *J Pediatr* 119: 260–267
56 Zaret BS, Becker RR, Marini MM, Wagle W, Pasarelli C (1982) Sodium valproate induced hyperammonemia without clinical hepatic dysfunction. *Neurology* 32: 206–208
57 Dreifuss FE (1995) Valproic acid toxicity, In: RH Levy, RH Mattson, BS Meldrum. (eds): *Antiepileptic drugs, fourth edition*, Raven Press, New York, 641–648
58 Loiseau P (1992) Childhood absence epilepsy In: Roger J, Bureau M, Dravet Ch, Perret A, Wolf P (eds): *Epileptic syndromes in infancy, childhood and adolescence*. John Libbey, London-Paris, 135–150
59 Olsson I, Campenhausen G (1993) Social adjustement in young adults with absence epilepsies. *Epilepsia* 34: 846–851
60 Tassinari C, Bureau M, Thomas P (1992) Epilepsy with myoclonic absences. In: Roger J, Bureau M, Dravet Ch, Perret A, Wolf P (eds): *Epileptic syndromes in infancy, childhood and adolescence*. John Libbey, London-Paris, 151–161
61 Schlumberger E, Chavez F, Palacios L, Rey E, Pajot N, Dulac O (1994) Lamotrigine in treatment of 120 children with epilepsy. *Epilepsia* 35: 359–367
62 Beaumanoir A, Dravet Ch (1992) The Lennox-Gastaut syndrome. In: J Roger, M Bureau, Ch Dravet, A Perret, P Wolf (eds): *Epileptic syndromes in infancy, childhood and adolescence*. John Libbey, London-Paris, 115–132
63 Oller LFV, Russi A, Oller Daurella L (1991) Lamotrigine in Lennox-Gastaut syndrome. *Epilepsia* 32 (suppl 1): 58

64 Besag FMC, Wallace SJ, Dulac O, Alving J, Spencer SC, Hosking G (1995) Lamotrigine for the treatment of epilepsy in childhood. *The Journal of Pediatrics* 6: 991–997
65 Dulac O, N'Guyen T (1993) The Lennox-Gastaut syndrome. *Epilepsia* 34 (suppl 7): 7–17
66 Felbamate study group in Lennox-Gastaut syndrome (1993) Efficacy of felbamate in childhood epileptic encephalopaty (Lennox-Gastaut syndrome). *N Engl J Med* 328: 29–33
67 Pellock JM (1997) Overview of lamotrigine and the new antiepileptic drugs: The challenge. *J Child Neurol* 12 (suppl 1): S48–S52
68 Patry G, Lyagoubi S, Tassinari CA (1971) Subclinical electrical status epilepticus induced by sleep. An electroencephalographic study of six cases. *Arch Neurol* 24: 242–252
69 Deonna T, Peter CL, Ziegler A (1989) Adult follow-up of the acquired aphasia epilepsy syndrome in childhood: report of seven cases. *Neuropediatrics* 20: 132–138
70 Bednarek N, Motte J, Soufflet C, Plouin P, Dulac O (1998) Evidence of late-onset infantile spasms. *Epilepsia* 39 (1): 55–60
71 Dravet Ch, Bureau M, Roger J (1992) Benign myoclonic epilepsy in infants. In: Roger J, Bureau M, Dravet Ch, Dreifuss FE, Perret A, Wolf P (eds): *Epileptic syndromes in infancy, childhood and adolescence.* John Libbey, London-Paris, 67–74
72 Guerrini R, Dravet Ch, Gobbi G et al (1994) Idiopathic generalised epilepsy with myoclonus in infancy and childhood. In: A Malafosse, P Genton, E Hirsch et al (eds): *Idiopathic generalised epilepsy: clinical, experimental and genetics aspects.* John Libbey, London, 267–280
73 Dulac O, Plouin P, Shewmon A, Contributors to the Royaumont Workshop (1998) Myoclonus and epilepsy in childhood 1996 Royaumont meeting. *Epilepsy Research* 30: 91–106
74 Pons G, Dulac O, Ben-Ari Y (eds) (1994) *Les médicaments antiépileptiques chez l'enfant.* Springer-Verlag, Paris
75 Panayiotopoulos CP (1993) Interaction of lamotrigine with sodium valproate. *Lancet* 341: 445
76 Levy RH, Shen D (1995) Valproic acid: absortion, distribution, and excretion. In: Levy R, Mattson R, Meldrum BS (eds): *Antiepileptic drugs, Fourth edition.* Raven Press, New York, 605–619
77 Commission on Antiepileptic Drugs, International League Against Epilepsy. Guidelines for Therapeutic Monitoring on Antiepileptic Drugs. *Epilepsia* 34 (4): 585–587
78 Lundberg B, Nergardh A, Boreus L (1982) Plasma concentrations of valproate during maintenance therapy in epileptic children. *J Neurol* 228: 133–141
79 Wallace SJ (1998) Myoclonus and epilepsy in childhood: a review of treatment with valproate, ethosuximide, lamotrigine and zonisamide. *Epilepsy Res* 29: 147–154

Milestones in Drug Therapy
Valproate
ed. by W. Löscher

Treatment of epilepsies in adults

L. James Willmore

Department of Neurology Saint Louis University School of Medicine, 1402 S. Grand Blvd. Ste M226, St. Louis, Missouri 63104, USA

Introduction

Valproate is a broad-spectrum antiepileptic drug synthesized by Burton in 1882 [1]. Its effect on seizures was recognized by Meunier et al. in 1963 [2]. Clinical trials by Carraz et al. [3] followed the initial animal studies that showed efficacy against seizures caused by both pentylenetetrazol and maximal electroshock. Valproic acid was used to treat seizures in Europe almost 15 years before approval in the USA [4]. This drug was used initially as an adjunct, but was found to have value in treatment of primary generalized epilepsies [4–6], with efficacy when used as monotherapy reported as well [7].

Epilepsy in adults: Clinical patterns and classification

Epileptic seizures are classified based on clinical and electroencephalographic criteria that divide seizures into three major categories: partial, generalized, and unclassified (see Tab. 1) [8]. Partial seizures may be either simple or complex [9]. Simple partial seizures may arise from any neocortical region. The spectrum ranges from focal myoclonus of a limb or region of the hand or face to sensory experiences such as seeing colored spots or lines, as occurs with discharge in the primary visual cortex. A simple partial seizure may cause symptoms recalled by the patient; this often is referred to as an aura. If the patient remains fully conscious, the seizure is classified as simple partial. However, if the focal discharge involves brain regions subserving awareness or if the seizure spreads widely enough to cause the patient to lose conscious contact, the seizure is classified as complex partial.

Patients with complex partial seizures lose conscious contact regardless of the experience that may precede the seizure. Although the patient's level of consciousness is a key component of the definition, this information may be difficult to obtain. During altered conscious contact, patients are unable to respond to commands or to interact with their surroundings or recall events that occurred during the seizure. A report of behavior ob-

Table 1. International classification of epileptic seizures [59]

I. Partial (focal, local) seizures
   A. Simple partial
   B. Complex partial
      1. Impairment of consciousness at onset
      2. Simple partial at onset with subsequent impairment of consciousness
   C. Partial seizures becoming generalized tonic-clonic seizures (GTCS)
      1. Simple partial seizures with evolution to GTCS
      2. Complex partial seizures evolving to GTCS, as well as seizures with simple partial onset

II. Generalized seizures (nonconvulsive and convulsive)
   A. Absence
   B. Atypical absence
   C. Myoclonic
   D. Clonic
   E. Tonic
   F. Tonic-clonic
   G. Atonic

III. Unclassified epileptic seizures, including patterns of neonatal onset

served by a witness such as a family member is an important way to differentiate a simple from a complex seizure [10].

Simple and complex partial seizures may secondarily generalize and produce a tonic-clonic seizure or a convulsion. Spread from a focus of discharge within a local area to encompass the entire brain is common, but most partial seizures do not secondarily generalize. Because the focal manifestation of a secondarily generalized convulsion may not be identified by observers or recalled by the patient, it is best to assume that a newly diagnosed generalized tonic-clonic seizure originated from a focus until such a process can be excluded [11].

Generalized seizures cause a spectrum of behavior from the nonconvulsive pattern of simple absence through myoclonus to the fully developed generalized tonic-clonic seizure [8]. Absence seizures, part of an epilepsy syndrome beginning in childhood, are brief, usually lasting 10 s or less. The patient does not have an aura, and when the seizure ends, there are no lingering, or postictal effects. The patient has arrest of behavior and a blank stare. Occasionally, mild clonic manifestations such as subtle eyelid blinking or changes in postural tone occur. Patients with absence seizures have a specific electroencephalographic pattern of generalized spike-and-wave discharges at 3/s. This EEG abnormality is activated by forced overbreathing. These patients have normal intellectual and neurologic function. Atypical absence seizures differ from simple absence seizures in that onset occurs at an earlier age in childhood, the background EEG is abnormal, and the EEG discharges are slower than 3/s. These patients may also have atonic and myoclonic seizures and may be mentally retarded.

Motor movements of myoclonus are manifested as brief jerks or contractions of a specific muscle or group of muscles. Myoclonus associated with epilepsy commonly is symmetric. Myoclonus is a component of several epilepsy syndromes and occurs before or as a part of both absence and generalized tonic-clonic seizures. One important syndrome to identify is juvenile myoclonic epilepsy because of the specificity of treatment with valproate. Patients with this syndrome have a poor prognosis for abatement of treatment; they must be treated for life [12, 13]. Myoclonic seizures may be caused by disorders other than primary brain dysfunction; these include metabolic diseases or genetic brain disorders such as Lafora's disease.

Convulsions are the most common type of generalized seizures. These are characterized by loss of consciousness associated with apnea and violent contractions of the musculature of the trunk and extremities. Frequently, patients have mouth trauma and bladder incontinence. Salivation often increases, and both pulse rate and blood pressure rise during the seizure. Most generalized convulsions begin with a tonic phase, in which there is sustained contraction of all muscles with extended legs and either flexed or extended arms. This phase lasts for several seconds and is followed by a clonic phase, in which there are rhythmic contractions of the limbs that begin with high-frequency, low-amplitude movements and then gradually decrease in frequency for several seconds to a few minutes. Some patients may have only tonic seizures or only clonic seizures. A sequence of clonic, tonic, and then clonic movements may be observed in patients with primary generalized seizures. After the violent muscle contractions subside, the patient enters a postictal phase, in which breathing resumes and unresponsiveness is followed by gradual recovery of consciousness. The patient may remain confused for several minutes or longer and subsequently complain of muscle pain and headache. This sequence of behavior is stereotyped, occurring regardless of cause. If a structural cause cannot be found, the convulsion is an idiopathic or a primary generalized seizure. If the convulsion is preceded by a partial seizure, it is said to be a secondarily generalized seizure.

An epilepsy syndrome encompasses not only the behavior during a seizure but also the EEG changes, the patient's mental and motor development, and the family history (see Tab. 2). Defining a specific epilepsy syndrome often requires repeated assessment, evaluation of development, and review of responses to treatment. Syndromes are considered to be either benign or progressive. These terms are usually applied to ultimate outcome of intellectual function and survival. Although some syndromic seizures are benign in their impact upon intellectual function, lifelong treatment with antiepileptic drugs may be required.

Childhood absence epilepsy has a prevalence that is estimated to be 2 to 8%. The associated EEG abnormality of spike-and-wave discharges at 3/s occurs in the milieu of a normal background pattern [14]. Although remissions of 80% are reported, long-term follow-up shows that devel-

Table 2. International classification of epilepsies and epileptic seizures [8]

I. Localization-related epilepsies and epileptic syndromes (partial, focal local)
- A. Idiopathic with an age-related onset
  1. Benign childhood epilepsy with centrotemporal spikes
  2. Childhood epilepsy with occipital paroxysms
- B. Symptomatic

II. Generalized epilepsies and epileptic syndromes
- A. Idiopathic with an age-related onset
  1. Benign neonatal epilepsy
  2. Childhood absence epilepsy
  3. Juvenile myoclonic epilepsy
  4. Juvenile absence epilepsy with GTCS on awakening
- B. Secondary (symptomatic or idiopathic)
  1. Infantile spasms (West Syndrome)
  2. Lennox-Gastaut Syndrome
- C. Symptomatic
  1. Etiology nonspecific (early myoclonic encephalopathy)
  2. Specific syndromes that complicate diseases (Unverricht-Lundborg, Lafora body myoclonus)

opment of tonic-clonic seizures reduces the number of patients reaching medication-free remission. Only 30% of patients with absence have complete resolution without antiepileptic medication if they experience generalized tonic-clonic seizures. Also the outlook is not as optimistic for patients who are older than 15 years and at least 15 years after onset [12]. Remission is best with short duration of illness, which implies the need for early identification of patients and rapid institution of treatment. Factors associated with a good prognosis are normal intelligence quotient (IQ) and negative history of generalized tonic-clonic seizures. Unfortunately, tonic-clonic seizures complicate the course of absence epilepsy in about 50% of patients [13]. Risk factors for development of tonic-clonic seizures include later age at onset of absence, difficulty in controlling absence seizures with medication, and abnormal background activity on the EEG [14, 15].

Before the development of broad-spectrum drugs, some patients were given an additional drug to prevent tonic-clonic seizures. This dual-drug approach has been supplanted by the use of valproate, a broad-spectrum medication with combined efficacy against both absence and generalized tonic-clonic seizures. Discussion of treatment with parents and patients must include informing them about the possibility of development of tonic-clonic seizures [16].

Patients with juvenile myoclonic epilepsy have characteristic myoclonic jerks, generalized tonic-clonic seizures, a family history of seizures, and photosensitivity on the EEG. Precipitants of seizures include sleep deprivation, stress, and alcohol use. The myoclonic jerks are quite prominent in the morning and may involve the large muscles of the legs. The correct diagnosis is important because the seizures are usually well controlled with valproate. The prognosis for normal intellectual function is usually quite good; however, the outlook for drug-free remission is poor [12].

Drug selection in epilepsy treatment

Treatment should be directed at both controlling seizures and, if possible, correcting the underlying disease or disorder producing them. Patients who have recurrent seizures secondary to a treatable neurologic disease, such as brain tumor or intracranial infection, should be treated with AEDs and should be treated for the underlying problem. Patients with chronic recurrent seizures, regardless of etiology, should be treated with AEDs.

Treatment with AEDs should follow certain basic principles: Therapy should be started with a single suitable agent. Seizure control should be achieved, if possible, by increasing the dosage of the single initial agent rather than by adding a second one. If seizure control cannot be achieved with the first medication, a second, alternative agent should be considered. Two or more AEDs used in combination should be avoided whenever possible, but rational drug combinations may be useful when monotherapy fails. Changes in dosage should be guided by the patient's clinical response rather than by drug levels, with inadequate seizure control indicating the need for raising the dosage and with toxicity indicating the need to lower the dosage. Monitoring drug levels is usually not necessary for patients with good seizure control who are taking a well-tolerated medication. Doing so can be useful under some circumstances, however, such as to determine prescription compliance or unexplained changes in seizure control or drug toxicity.

The principal AEDs used to treat patients with epilepsy are carbamazepine, ethosuximide, gabapentin, lamotrigine, phenobarbital, phenytoin, primidone, and topiramate, tiagabine, and valproate (see Tab. 3). Some benzodiazepines, including clonazepam, diazepam, and lorazepam, are also used to treat seizures. With the exception of clonazepam, the benzodiazepines are used for short-term treatment of acute seizures or status epilepticus and are usually administered parenterally. Clonazepam can be used to treat epilepsy but is not recommended, because most patients develop tolerance to its antiepileptic effect. Felbamate is a newly available drug that should be reserved for use in a few patients with uncontrolled seizures; high incidence of serious adverse effects requires expert use best confined to experienced epilepsy centers [17]. Several additional drugs, including methsuximide, trimethadione, and acetazolamide, are used occasionally; none of these drugs will be considered in this subsection.

Most epileptologists agree that the drugs of choice for partial seizures are carbamazepine and phenytoin, which are highly effective and have acceptable adverse effects [18]. Valproate also has been demonstrated to be

Table 3. Principal drugs used to treat epilepsy

Drug	Uses	Comment
Carbamazepine	Potent efficacy for partial seizures and GTCS as well. A first choice drug for partial seizures.	Minimal cognitive or behavioral effects; induces metabolism making dosing a challenge, may worsen absence seizures.
Ethosuximide	Effective and well tolerated for absence epilepsy, a drug of first choice.	Specific for absence, will cause gastrointestinal side-effects. Not effective for convulsive seizures.
Felbamate	Broad spectrum drug, effective in Lennox-Gastaut.	GI symptoms, headache and insomnia prominent. Rare fatal aplastic anemia and hepatic injury limits use. Interacts vigorously with all other drugs.
Gabapentin	Demonstrated efficacy for partial seizures as add-on.	Renal excretion and lack of enzyme induction are useful characteristics.
Lamotrigine	Broad spectrum for partial and generalized seizures.	Skin reactions may be severe, requires slow titration.
Phenobarbital	Broad spectrum for partial and generalized seizures. Parenterial form available.	Causes sedation, adverse impact on behavior and cognition. Effective, but not a first choice drug.
Phenytoin	Quite effective for both partial and generalized seizures, a first choice drug for those types of seizures.	Enzymes for clearance saturated causing challenging kinetics; induces enzymes as well.
Primidone	Has effect for partial and GTCS seizures.	Side-effects limit use; behavioral changes a problem. Not a first choice drug.
Tiagabine	Effective as an add-on drug for partial seizures.	Brisk hepatic metabolism.
Topiramate	An add-on drug for partial seizures.	Potent drug with vigorous efficacy in clinical trials. Side-effects are specific, with word finding difficulty and weight loss particularly common.
Valproate	Broad spectrum with vigorous efficacy for generalized seizures and partial seizures as well.	First choice for idiopathic epilepsies. Teratogenicity is specific, rare hepatic dysfunction.
Vigabatrin	Partial seizures and GTCS including infantile spasms.	Psychiatric effects occur, some question of change in vision being reported.

effective in the treatment of partial seizures [19, 20]. Both phenobarbital and primidone are probably as effective as carbamazepine or phenytoin, but the two barbiturates are associated with a much higher incidence of adverse effects, particularly sedation and impaired cognition [18]. Newer drugs, gabapentin, lamotrigine, tiagabine and topiramate are effective against partial seizures, and at present, each is usually used in combination with one of the drugs mentioned above.

Generalized convulsive seizures may be controlled with carbamazepine, phenytoin, valproate, the barbiturates, gabapentin, or lamotrigine. Carbamazepine, phenytoin, and valproate seem to be approximately equally effective for generalized convulsive seizures. Carbamazepine, phenytoin, gabapentin, and lamotrigine are used for patients with secondarily generalized seizures because many of these patients often have partial seizures as well. On the other hand, valproate and lamotrigine are more often used for patients with primary generalized convulsions, because some of these patients may also have absence seizures, which may be controlled with these drugs. The barbiturates are not preferred in the treatment of generalized convulsive seizures, primarily because of their sedative effects. Generalized nonconvulsive seizures, particularly absence seizures, can be treated with either ethosuximide or valproate. For patients who have only absence seizures, ethosuximide is satisfactory. However, for patients who have absence seizures in conjunction with other types of seizures, such as generalized convulsions or myoclonic seizures, valproate is the drug of choice, lamotrigine may also be effective.

Pharmacokinetics of valproate in adults

Detailed review of pharmacokinetics will be found in the chapter by Shen. Briefly, absorption in adults showed valproic acid is available for oral administration as the acid, as a sodium salt, and complexed as divalproex sodium. All forms are highly bioavailable, but differ with time of onset of absorption and time to peak. Complete absorption of sodium valproate occurs within 60 min. Divalproex sodium, a formulation in common use in the US, begins absorption about 120 min after an oral dose. Peaks are reached for valproic acid and sodium valproate within 2 h; 3–4 h for divalproex sodium. The sprinkle formulation's peak is reached at 4 h after administration [21]. Intravenous valproic acid follows a distribution pattern similar to the oral preparation [22].

Binding and distribution studies in adults show the calculated volume of distribution of valproic acid is 0.1–0.4 L/kg. This drug distributes primarily within the extracellular space [23]. Valproic acid saturates protein binding at clinically relevant levels approaching 90% [23]. Because of binding saturation, free levels increase in a non-linear fashion with levels much above 100 µg/ml. Displacement is a problem in adults, especially the elderly, since other bound drugs, such as phenytoin, or even proprietary medications such as aspirin, will displace bound drug, causing increased tissue penetration and hepatic metabolism. Spinal fluid levels and brain tissues reflect the unbound circulating levels of drug.

Efficacy of valproate

Although initial studies and recommendations emphasized the impact of valproic acid on generalized seizure disorders, this drug was known to have a broad spectrum of effect that included seizures of focal onset. However, pivotal trials designed to gain FDA approval for use in the USA enrolled patients with absence seizures.

Initial studies in children, and in some adults, were designed to assess efficacy in patients with absence seizures. Numerous open and controlled trials demonstrated between 75% or greater reduction up to complete control in absence seizures in more than 50% of treated patients [4, 6]. Most studies found complete control could be achieved in 80 to 90% of patients with simple absence [24, 25]. Outcome of treatment if the patient had atypical absence or other combined seizures was less robust [26]. Comparison studies showed valproic acid was equally as effective as ethosuximide [27], but one advantage was overlapping efficacy in control of generalized tonic-clonic seizures. This effect was considered useful since patients with complex absence, or atypical absence, may develop generalized tonic-clonic seizures in 50% of cases.

With absence seizures control tended to correlate with control of EEG spike wave discharges [6]. Valproic acid was effective in reducing spike-wave discharges when given to patients with either atypical or typical absence seizures. Valproate not only reduced seizure frequency but caused a reduction in the total accumulated time a patient experienced spike-wave discharges [6].

In one pivotal trial conducted prior to drug approval in the USA, of 25 adults with absence seizures to whom valproage was given for 10 weeks, 19 had reduction in seizure frequency while 21 of these patients had lessening of the total time spent in spike-wave discharges, up to 75% in 11 patients [6]. Other studies corroborated these observations [28–30]. Telemetry EEG was used to provide objective data for both frequency and duration of seizures. These telemeter EEG studies showed valproate equal in efficacy to ethosuximide in controlling absence seizures [27]. This double-blind, crossover study treated 16 patients never exposed to AEDs and 29 patients with refractory seizures. Small series of patients showed complete control of simple absence seizures in more than 85% of patients [25, 31, 32].

Tonic-clonic seizures in adults seldom are a form of primary epilepsy, but occur following generalization from a focus [7]. Although generalized tonic-clonic seizures (GTCS) were assessed in adults, the precise classification was seldom clarified, since some reports of adult onset claimed to be enrolling patients with primary generalized seizures. Of patients evaluated in a broad range of studies, the 75–100% responder rate commonly exceeded 50% [4, 33]. In children, as many as 34% achieved complete control of GTCS. In adults, complete control with 2-year remission was achieved in 72% treated with valproate and 56% treated with phenytoin [7,

34]. Generally, approximately 70–85% of patients in a spectrum of age groups may become seizure free of tonic-clonic seizures with valproate treatment [25, 35].

Patients never treated before were enrolled in an open fashion to receive phenytoin or valproate. Two years of seizure remission were achieved in 65% of patients treated with valproate and in 56% of patients receiving phenytoin. One of the first studies of adults thought to have primary generalized epilepsy showed valproate caused complete seizure control in 73% [32].

Partial seizures are the most common form in the adult population [36]. Accumulated reports of open trials in patients with simple and complex partial seizures found 75–100% responder rates of 28%. Most of these studies were of add-on design. Valproate was compared to carbamazepine in a blinded, multicenter study of veterans with epilepsy [20]. This extensive assessment in 480 adults was conducted as a double-blind study in patients with complex partial seizures with or without secondarily generalized seizures [20]. Both drugs were equally efficacious in patients with generalized tonic-clonic seizures. At 12 months carbamazepine was favored in patients with complex partial seizures only, but that difference was no longer found following 24 months of treatment. Retention in the trial was no different for patients with secondarily generalized tonic-clonic treated with valproate or carbamazepine.

Broad efficacy of valproate to control complex partial seizures was suggested by data accumulated from numerous open studies. While simple partial seizures did not seem to respond to valproate [31]. 45% with CPS were seizure-free, albeit in combination with carbamazepine. While a small open study showed comparable efficacy between valproate and carbamazepine a larger series of 181 patients using untreated patients showed no difference in seizure control between carbamazepine, phenytoin and valproate [37]. Numerous other studies reported similar patterns of efficacy [7, 34, 38].

The only study meeting the usual criteria as evidence based medicine used a placebo control and valproate as a drug added to patients with intractable complex partial seizures receiving either carbamazepine or phenytoin in 137 patients in the intent-to-treat group [19]. Although median reduction in seizures was an outcome measure, data were expressed as responder rate of at least or more than 50% reduction in seizure frequency. Responder rate was 38% in patients treated with valproate, indicating efficacy in the treatment of complex partial seizures.

Some epilepsy syndromes that include myoclonus begin at the threshold of adulthood. One important syndrome beginning in young adulthood but then carried throughout life is juvenile myoclonic epilepsy (JME) [12]. Myoclonus in children responds to valproate, but similar efficacy has been observed in adults. Isolated myoclonus, or generalized seizures combined with myoclonus, respond to valproate [25, 31]. Valproic acid is specific therapy for juvenile myoclonic epilepsy [12]. Valproic acid will block postanoxic intention myoclonus [39]. For example, 70% of patients with myoclonic epilepsy of adolescence had complete control with valproate [31]. Most of those patients had abnormalities with photic stimulation.

Use of valproate in adults

Three forms for use in the USA are divalproex sodium, valproic acid, and valproic acid sodium for intravenous injection. Worldwide formulation used most commonly is sodium valproate. Depakote is an entantamer of sodium valproate and valproic acid.

Dose initiation of valproic acid in adults using sodium valproate or the divalproex sodium formulation is 15 mg/kg per day. Weekly dose escalation commonly is by 5 mg/kg per day. Ascension by 250 mg/day every 4–7 days to a target of 35 mg/kg/day is well tolerated. Monotherapy doses to achieve an adequate serum concentration range from 10 to 20 mg/kg per day [40, 41]. When valproate treatment was combined with other AEDs the doses needed to achieve an adequate level ranged from 30 to 60 mg/kg per day [24, 40]. Blood levels of valproate are not linear at higher doses. Saturation of protein binding sites with resultant increased free fraction that leads to brisk drug clearance accounts for the need for larger doses in patients receiving concomitant drug. Although plasma concentration and EEG alterations may not correlate, clinical efficacy is observed with levels above 50 µg/ml [6].

As with all drugs used to treat seizures, titration to efficacy balanced with side-effects should guide treatment. Blood levels of valproic acid are measured at the daily trough that occurs before the morning dose. Since divalproex sodium absorption is delayed from 1 to 2 h after ingestion, obtaining levels in that time window is appropriate. Random levels do little more than endorse compliance and may cause undue concern since peak blood levels are commonly well above the so-called upper end of the therapeutic range.

Common side effects

Dose-related side-effects may limit usefulness in adults. Although a change in formulation resulted in some abatement of these problems in adults, gastrointestinal effects commonly accompany initiation of valproic acid treatment; they include nausea, diarrhea, abdominal pain, and even vomiting [42]. Three dose-related effects require informing patients since they occur commonly. Tremor with sustension and at rest is dose-related [43]. Body weight gain is another common side-effect, with 20–54% of patients reporting this problem [44]. Patients report appetite stimulation. Weight change may require discontinuation of this drug.

Hair loss is common and transient in both children and adults. The transient alopecia that occurs in some patients taking valproate may be related to its tendency to chelate trace metals. Hair appears to be fragile; regrowth of the broken hair results in a curlier shaft [45]. Trace metal insufficiency can result in fragility of the hair shaft. Sequestration of selenium may play a role as well. Addition of a vitamin with zinc supplement usually results in an abatement of this hair loss.

Other more serious adverse effects and drug interactions are reviewed in the chapter by Schmidt. However, common adverse effects need to be kept in mind when informing patients about the use of valproate. Thrombocytopenia occurs in a pattern that appears to be dose related. Acute hemorrhage pancreatitis may develop in younger patients. Fatalities have been observed in children and adults during valproate treatment [46–48]. Early reports of changes in hepatic enzymes among treated adults [49] were soon followed by observation of fatal hepatoxicity [50]. Risk was found to be greatest in young patients receiving several medications. Valproate occasionally causes stupor or coma [51] and is known to affect mitochondrial function, causing elevations in serum levels of some branched chain fatty acids [52].

Laboratory monitoring does little to allow anticipation of serious adverse effects in either adults or children [53]. At initiation of treatment all patients must have assessment of complete blood cell count, including platelets, and measurement of enzymes derived from liver and of the products of liver metabolism. Hepatic studies should be repeated at 12 weeks after initiation of treatment, at the time when dose-related hepatic enzyme changes might be expected [49]. It is best to establish a clinical screening relationship with the patient in order to detect adverse events [53, 54].

Drug interactions are detailed by Prof. Schmidt in this volume. Although monotherapy use is a consistent theme for treatment, some patients require combination therapy leading to potential for drug interactions.

Phenobarbital is more commonly used in children, but finds occasional use in adults. Valproic acid given to patients treated with phenobarbital causes increases in the serum concentration of phenobarbital [55]. Phenobarbital dose must be reduced by about 40% at the inception of ascension dosing with valproic acid [56]. Valproate treatment causes phenytoin plasma levels to decline. This effect is caused by displacement of bound phenytoin by valproic acid with increased free phenytoin levels and subsequent enhanced clearance by hepatic metabolism [30]. Since CBZ is metabolized via an epoxide oxidase and valproate does inhibit that enzyme, toxicity with this drug combination may be related to accumulation of CBZ-10, 11-epoxide [57, 58]. Felbamate causes marked increase in blood levels of valproate.

Conclusions

Valproate is a broad spectrum anti-epilepsy drug with a unique molecular structure that has specific indications along with effects in many forms of epilepsy. While some side-effects in adults are specific and common, for the most part, this drug is both highly efficacious in forms of epilepsy affecting adults, and is well tolerated.

References

1 Burton BS (1882) On the propyl derivatives and decomposition products of ethylacetoacetate. *Am Chem J* 3: 385–395
2 Meunier H, Carraz G, Meunier V, Eymard M (1963) Proprietes pharmacodynamiques de l'acide n-propylacetique. *Therapie* 18: 435–438
3 Carraz G, Fau R, Chateau R, Bonnin J (1964) Essais cliniques sur l'activite anti-epileptique de l'acide n-dipropylacetique (sel de na). *Ann med Psychol* (Paris) 122: 577–584
4 Simon D, Penry JK (1975) Sodium di-n-propylacetate (DPA) in the treatment of epilepsy: a review. *Epilepsia* 22: 1701–1708
5 Richens A, Ahmad S (1975) Controlled trial of sodium valproate in severe epilepsy. *Brit Med J* 2: 255–256
6 Villarreal HJ, Wilder BJ, Willmore LJ, Bauman AW, Hammond EJ, Bruni J (1978) Effect of valproic acid on spike and wave discharges in patients with absence seizures. *Neurol* 28: 886–891
7 Turnbull DM, Rawlins MD, Weightman D, Chadwick DW (1982) A comparison of phenytoin and valproate in previously untreated adult epileptic patients. *J Neurol Neurosurg Psychiatry* 45: 55–59
8 Commission on classification and terminology of the International League Against Epilepsy (1989) Proposal for revised classification of epilepsies and epileptic syndromes. *Epilepsia* 30: 389–399
9 Theodore WH, Porter RJ, Penry JK (1983) Complex partial seizures: Clinical characteristics and differential diagnosis. *Neurol* 33: 1115–1121
10 Rocca WA, Sharbrough FW, Hauser WA, Annegers JF, Schoenberg BS (1987) Risk factors for complex partial seizures: a population-based case-control study. *Ann Neurol* 21: 22–31
11 Keranen T, Sillanpaa M, Riekkinen PR (1988) Distribution of seizure types in an epileptic population. *Epilepsia* 29: 1–7
12 Delgado-Escueta AV, Enrile-Bascal F (1984) Juvenile myoclonic epilepsy of Janz. *Neurol* 34: 285–294
13 Dreifuss F (1989) Pediatric epilepsy syndromes: an overview. *Cleve Clin J Med* 56: S166–S171
14 Penry JK, Porter RJ, Dreifuss FE (1975) Simultaneous recording of absence seizures with videotape and electroencephalography: A study of 374 seizures in 48 patients. *Brain* 98: 427–440
15 Sato S, Dreifuss FE, Penry JK, et al. (1983) Long-term follow-up of absence seizures. *Neurol* 33: 1590–1595
16 Wirrell EC, Camfield CS, Camfield PR, Gordon KE, Dooley JM (1996) Long-term prognosis of typical childhood absence epilepsy: remission or progression to juvenile myoclonic epilepsy. *Neurol* 47: 912–918
17 Pellock JM, Brodie MJ (1997) Felbamate: 1997 update. *Epilepsia* 38: 1261–1264
18 Mattson RH, Cramer JA, Collins JF, Smith DB, Delgado-Escueta AV, Browne TR (1985) Comparison of carbamazepine, phenobarbital, phenytoin, and primidone in partial and secondarily generalized tonic-clonic seizures. *N Engl J Med* 313: 145–151
19 Willmore LJ, Shu V, Wallin B, M88-194 Study Group (1996) Efficacy and safety of add-on divalproex sodium in the treatment of complex partial seizures. *Neurol* 46: 49–53
20 Mattson RH, Cramer JA, Collins JF, Department of Veterans Affairs Epilepsy Cooperative Study, No. 264 Group (1992) A comparison of valproate with carbamazepine for the treat-

ment of complex partial seizures and secondarily generalized tonic-clonic seizures in adults. *N Engl J Med* 327: 765–771

21. Carrigan PJ, Brinker DR, Cavanaugh JH, Lamm JE, Cloyd JC (1990) Absorption characteristics of a new valproate formulation: divalproex sodium-coated particles in capsules (Depakote Sprinkle). *J Clin Pharmacol* 30: 743–747
22 Devinsky O, Leppik I, Willmore LJ, Pellock JM, Dean C, Gates J, Ramsay RE, The Intravenous Valproate Study Team (1995) Safety of intravenous valproate. *Ann Neurol* 38: 670–674
23 Klotz U, Antonin KH (1977) Pharmacokinetics and bioavailability of sodium valproate. *Clin Pharmacol Ther* 21: 736–743
24 Henriksen O, Johannessen SL (1982) Clinical and pharmacokinetic observations on sodium valproate-a 5-year follow-up study in 100 children with epilepsy. *Acta Neurol Scand* 65: 504–523
25 Bourgeois B, Beaumanoir A, Blajev B, de la Cruz N, Despland PA, Egli M, Geudelin B, Kaspar U, Ketz E, Meyer C, Scollo-Larizzari G, Tosi C, Vassella F, Zagury S (1987) Monotherapy with valproate in primary generalized epilepsies. *Epilepsia* 28 (Suppl. 2): S8–S11
26 Erenberg G, Rothner AD, Henry CE, Cruse RP (1982) Valproic acid in the treatment of intractable absence seizures in children. A single-blind clinical and quantitative EEG study. *Am J Dis Child* 136: 526–529
27 Sato S, White BG, Penry JK, Dreifuss FE, Sackellares JC, Kupferberg HJ (1982) Valproic acid versus ethosuximide in the treatment of absence seizures. *Neurol* 32: 157–163
28 Bergamini L, Mutani R, Fulan PM (1975) The effect of sodium valproate (Epilim) on the EEG. *Electroencephalogr Clin Neurophysiol* 39: 429
29 Maheshwari MC, Jeavons PM (1975) The effect of sodium valproate (Epilim) on the EEG. *Electroencephalogr Clin Neurophysiol* 39: 429
30 Mattson RH, Cramer JA, Williamson PD, Novelly RA (1978) Valproic acid in epilepsy: clinical and pharmacological effects. *Ann Neurol* 3: 20–25
31 Covanis A, Gupta AK, Jeavons PM (1982) Sodium valproate: monotherapy and polytherapy. *Epilepsia* 23: 693–720
32 Feuerstein J (1983) A long term study of monotherapy with sodium valproate in primary generalised epilepsy. *Brit J Clin Pract Suppl.* 27: 17–25
33 Davis R, Peters DH, McTavish D (1994) Valproic acid: a reappraisal of its pharmacological properties and clinical efficacy in epilepsy. *Drugs* 47: 332–372
34 Turnbull DM, Howel D, Rawlins MD, Chadwick DW (1985) Which drug for the adult epileptic patient: phenytoin or valproate. *Br Med J* 290: 815–819
35 Wilder BJ, Ramsay RE, Murphy JV, Karas BJ, Marquardt K, Hammond EJ (1983) Comparison of valproic acid and phenytoin in newly-diagnosed tonic-clonic seizures. *Neurol* 33: 1474–1476
36 So EL (1995) Classifications and epidemiologic considerations of epileptic seizures and epilepsy. *Neuroimaging Clinics of North America* 5: 513–526
37 Callaghan N, Kenny RA, O'Neill B, Crowley M, Goggin T (1985) A prospective study between carbamazepine, phenytoin and sodium valproate as monotherapy in previously untreated and recently diagnosed patients with epilepsy. *J Neurol Neurosurg Psychiatry* 48: 639–644
38 Gupta AK, Jeavons PM (1985) Complex partial seizures: EEG foci and response to carbamazepine and sodium valproate. *J Neurol Neurosurg Psychiatry* 45: 131–138
39 Bruni J, Willmore LJ, Wilder BJ (1979) Treatment of post-anoxic intention myoclonus with valproic acid. *Can J Neurol Sci* 6: 39–42
40 Bruni J, Wilder BJ, Willmore LJ, Villareal HD, Thomas M, Crawford LEM (1978) Clinical efficacy of valproic acid in relation to plasma levels. *Can J Neurol Sci* 5: 385–387
41 Bruni J, Wilder BJ, Willmore LJ, Perchalski RJ, Villarreal HJ (1978) Steady-state kinetics of valproic acid in epileptic patients. *Clin Pharmacol Ther* 24: 324–332
42 Dreifuss FE, Langer DH (1988) Side effects of valproate. *Am J Med* (suppl 1A): 34–41
43 Hyuman NM, Dennis PD, Sinclar KG (1979) Tremor due to sodium valproate. *Neurol* 29: 1177–1180
44 Dinesen H, Gram L, Andersen T, Dam M (1984) Weight gain during treatment with valproate. *Acta Neurologica Scandinavica* 70: 65–69
45 Jeavons PM, Clark JE, Hirdme GA (1977) Valproate and curly hair. *Lancet* 1: 359

46 Camfield PR, Bagnell P, Camfield CS, Tibbles JAR (1979) Pancreatitis due to valproic acid. *Lancet* 1: 1198–1199
47 Rosenberg HK, Ortega W (1987) Hemorrhagic pancreatitis in a young child following valproic acid therapy. Clinical and ultrasonic assessment. *Clin Pediatr* 26: 98–101
48 Wyllie E, Wyllie R, Cruse RP, Erenberg G, Rothner AD (1984) Pancreatitis associated with valproic acid therapy. *Am J Dis Child* 138: 912–914
49 Willmore LJ, Wilder BJ, Bruni J, Villarreal HJ (1978) Effect of valproic acid on hepatic function. *Neurol* 28: 961–964
50 Suchy FJ, Balistreri WF, Buchino JJ, Sondheimer JM, Bates SR, Kearns GL, Stull JD, Bove KE (1979) Acute hepatic failure associated with the use of sodium valproate. *N Engl J Med* 300: 962–966
51 Sackellares JC, Lee SI, Dreifuss FE (1979) Stupor following administration of valproic acid to patients receiving other antiepileptic drugs. *Epilepsia* 20: 697–703
52 Coude FX, Grimer G, Pelet A, Benoit Y (1983) Action of the antiepileptic drug, valproic acid, on fatty acid oxidation in isolated rat hepatocytes. *Biochem Biophys Res Comm* 115: 730–736
53 Pellock JM, Willmore LJ (1991) A rational guide to routine blood monitoring in patients receiving antiepileptic drugs. *Neurol* 41: 961–964
54 Willmore LJ, Triggs WJ, Pellock JM (1991) Valproate toxicity: risk-screening strategies. *J Child Neurol* 6: 3–6
55 Schobben F, van der Kleijn D, Gabreels FJM (1975) Pharmacokinetics of di-N-propylacetate in epileptic patients. *Eur J Clin Pharmacol* 8: 97–105
56 Wilder BJ, Willmore LJ, Bruni J, Villarreal HJ (1978) Valproic acid: Interaction with other anticonvulsant drugs. *Neurol* 28: 892–896
57 Robbins DK, Wedlund PJ, Kuhn R, Baumann RJ, Levy RH, Chang S-L (1990) Inhibition of epoxide hydrolase by valproic acid in epileptic patients receiving carbamazepine. *Br J Clin Pharmacol* 29: 759–762
58 Tomson T, Bertilsson L (1984) Potent therapeutic effect of carbamazepine-10,11-epoxide in trigeminal neuralgia. *Arch Neurol* 41: 598–601
59 Commission classification and terminology of the International League Against Epilepsy. (1981) Proposal for revised clinical and electroencephalographic classification of epileptic seizures. *Epilepsia* 22: 498–501

Milestones in Drug Therapy
Valproate
ed. by W. Löscher

Valproate use in psychiatry: A focus on bipolar illness

Robert M. Post

Biological Psychiatry Branch, NIMH, NIH, Bldg. 10, Room 3N212, 10 Center Drive MSC 1272, Bethesda, MD 20892, USA

Introduction

Along with carbamazepine, valproate has emerged as one of the major mood stabilizer treatments for acute and prophylactic use in patients with bipolar affective illness. Early studies in France by Lambert and in Germany by Emrich presaged the widespread use of the drug in the 1990s [1, 2]. However, it was not until the publication of a series of double-blind controlled studies comparing valproate with lithium and placebo in the early 1990s that its use accelerated to the point at which it is now one of the leading medications used for the treatment of bipolar illness.

In addition, valproate has a variety of effects in other neuropsychiatric syndromes that make it particularly useful as they co-occur with bipolar illness. There is increasing recognition of the multiple comorbidities associated with bipolar illness including a very high incidence of comorbid alcohol and substance abuse as well as anxiety and a variety of personality disorders [3]. There is a moderate comorbidity with migraine and a variety of neurological symptoms including head trauma and pain. A modicum of data supports the efficacy of valproate in each of these syndromes and thus propels the wider use of valproate in bipolar illness when it occurs in conjunction with these entities.

Lithium, long considered the paradigmatic treatment of bipolar illness [7], is increasingly recognized to be an inadequate approach for many patients [8–15], with a low percentage of responsivity in specific subtypes of the illness. These include rapid cycling [16, 17], dysphoric mania, a variety of comorbidities including anxiety and personality disorder, and head trauma, as well as those with a pattern of depression followed by mania and then a well interval (D-M-I) as opposed to M-D-I, and those without a family history of bipolar illness in first-degree relatives [19–21]. Valproate and other mood stabilizing anticonvulsants appear to be useful in many of these subtypes that are inadequately responsive to lithium, even though valproate may not be preferentially responsive to these subtypes.

The primary focus of this chapter will be the use of valproate in bipolar illness, but I will also note reports of a variety of other psychiatric syndromes in which preliminary or substantial evidence exists for its efficacy. These include panic disorder, posttraumatic stress disorder (PTSD), migraine, and paroxysomal pain syndromes. The putatative mechanisms of action of valproate potentially relating to this range of psychotropic effects will be briefly outlined.

Valproate in acute mania

As illustrated in Table 1, Emrich et al. [2] reported that four of five patients showed a response to valproate in a double-blind B-A-B design. In a follow-up study they noted that several of these patients required lithium and valproate in combination to sustain their long-term responsiveness, however [22]. Similarly, Brennan et al. [23] reported acute antimanic response in six of eight patients, while Post et al. [24] reported seven of 13 patients for a 54% response rate.

Pope et al. [26] performed the first double-blind parallel group comparison with placebo, reporting nine of 17 (53%) patients responding to valproate, but only 11% to placebo.

In 1992, Freeman et al. [25] reported the first double-blind comparison of valproate to lithium, in which nine of 14 (64%) patients responded to valproate and 12 of 13 (92%) responded to lithium. This study provided evidence of valproate's effectiveness in patients with dysphoric mania. This study was followed by the largest double-blind, randomized study of valproate in the treatment of acute mania, in which patients were randomized to valproate, lithium, or placebo at a 2:1:2 ratio [27]. In this study, 33 of 69 (48%) patients showed a valproate response compared with 49% on lithium and only 25% on placebo.

Table 1. Controlled studies of valproate in acute mania

Investigators	Design	VPA responders	% VPA response	% Placebo response	% Lithium response
Emrich et al. 1980	DB, B-A-B	4/5	80%		
Brennan et al. 1984	DB, B-A-B	6/8	75%		
Post et al. 1989	DB, B-A-B	7/13	54%		
Freeman et al. 1992	DB vs Li	9/14	64%		92%
Pope et al. 1991	DB vs P	9/17	53%	11%	
Bowden et al. 1994	DB vs Li & P	33/69	48%	25%	49%
All studies		68/126	54%	22%	60%

VPA = valproate; Li = Lithium; P = Placebo; DB = Double-blind; B-A-B = Placebo-Active-Placebo.

These latter two studies [26, 27] formed the basis for the Food and Drug Administration's (FDA) approval of valproate for the treatment of mania in the United States. The time-course of response for the Bowden et al. [27] study is illustrated in Fig. 1, showing a relatively rapid onset of response, although not significantly faster than that of lithium. Patients with higher blood levels in the first 3 to 5 days of treatment showed better acute responsivity [27]. While valproate response was highly statistically significant, it is noteworthy that only about one-half of the patients showed responsivity at the criterion of 50% improvement by the end of 3 weeks. Moreover, when the criterion of response was reduced to 40% (i.e. moderate) and 30% (i.e. mild) improvement from baseline, a considerable number of patients still did not meet these lower thresholds (Tab. 2).

These double-blind monotherapy data thus converge with data from clinical settings, in which, under pressure to treat patients acutely for their

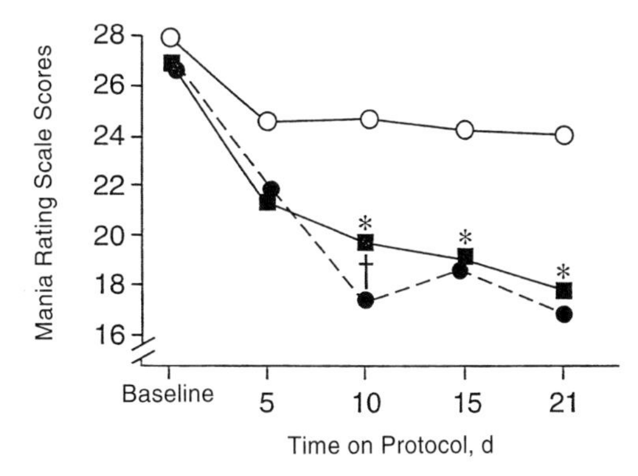

Figure 1. Changes from baseline to final evaluation in Mania Rating scale score, Schedule for Affective Disorders and Schizophrenia–Change version (SADS-C). Solid squares indicate divalproex; solid circles, lithium; and open circles, placebo. Numbers on the vertical axis indicate the sum of all items on this subscale of the SADS-C. Asterisks and dagger indicate time points at which a significant difference ($p < 0.05$) was observed between divalproex and lithium, respectively, and placebo treatment. Reprinted from [27], copyright 1994, American Medical Association.

Table 2. Degree of improvement at final evaluation

	% Patients			Fisher p Values	
% Improvement	Valproate	Lithium	Placebo	PLB vs. VPA	PLB vs. Li
30	66	51	35	$p < 0.001$	$p = 0.022$
40	54	49	28	$p < 0.001$	$p = 0.005$
50	48	46	25	$p = 0.004$	$p = 0.025$

PLB = placebo; VPA = valproate; Li = lithium.
Reprinted from [44], p. 135 by courtesy of Marcel Dekker, Inc.

manic syndrome, combination therapy is almost always used, as monotherapy is typically inadequate. Keck et al. [28] have indicated that loading doses of valproate can be achieved rapidly and safely in the treatment of acute mania. They used 20 mg/kg oral valproate on day-1 and saw a relatively rapid onset of clinical efficacy, although whether this rate is more rapid than that which occurs with nonloading doses was not systematically documented. Another group started with 30 mg/kg for 2 days and then continued with 20 mg/kg with fair tolerability [29].

Even with lithium and this loading dose strategy, valproate is often used in conjunction with high potency anticonvulsant benzodiazepines such as clonazepam or lorazepam for acute agitation and insomnia not initially responsive to valproate. The drug is also commonly used in conjunction with neuroleptics for psychotic mania. Formerly, such treatment options were limited to the typical neuroleptics, but more recently they have expanded to include those of the atypical variety, such as clozapine, risperidone, and olanzapine as well as quetiapine.

Valproate is widely used as the concomitant mood stabilizing anticonvulsant for patients treated with high dose clozapine when there is an increased risk of seizures. Given the substantial incidence of agranulocytosis with clozapine [30], most investigators are reluctant to combine it with carbamazepine, which has its own incidence of agranulocytosis and aplastic anemia [31]. Although the use of valproate is generally thought to be helpful for the prevention of clozapine-induced seizures and as a clinical adjunct for the acute manic syndrome, there is some suggestion from the data of Stevens, Denney, and colleagues [32, 33] that the efficacy of clozapine could be diminished with concomitant use of both drugs in acute psychosis and schizophrenia. These investigators speculate that the anticonvulsant effects of valproate may prevent some of the sub-epileptiform discharges induced by clozapine that could be important to its beneficial mechanisms of action.

Patients with dysphoric (or mixed) mania are known to be particularly unresponsive to lithium [8, 11, 34, 35]. Dysphoric mania includes the very large subgroup of approximately 40% of all manic patients who have substantial concomitant depressive or anxious symptomatology co-occurring with mania. Whereas the stereotype of mania is that of a euphoric, expansive mood progressing to grandiose or overtly delusional proportions, a substantial minority experience mania as uncomfortable, with an internal sense of pressure or being driven. This mania often includes large components of concomitant anxiety and even some depressive and hopeless elements to the point of suicidality. Moreover, full-blown panic attacks are not uncommon concomitants of the mixed manic syndrome. In contrast to lithium's poor responsivity to this syndrome, valproate has an approximately equal efficacy in those with euphoric and dysphoric presentations [25, 36–39].

Valproate in acute depression

There are consistent reports that valproate is less effective in the treatment of acute depression than acute mania [1, 36, 40–43]. There is some suggestion in the literature that valproate is more effective in patients with rapid cycling disorder (Tab. 3) [37, 38, 42, 44] or in patients with bipolar II disorder compared with those with bipolar I [43, 45]. Davis et al. [46] conducted an 8-week open trial of valproate in the treatment of acute non-bipolar depression. By week 4, 15 of the 28 completers (54%) demonstrated a response (defined by a 50% decrease in Hamilton rating scales for total score of nine or lower).

In 101 patients with rapid cycling bipolar disorder, Calabrese et al. [47] found a marked response (defined as the complete cessation of symptoms/cycling) in 12 of 58 acutely depressed patients (21%) and a moderate response (defined by any decrease in episode frequency, duration, or amplitude) in an additional 14 of the 58 (24%) for a total response rate of 45% (Tab. 3). This contrasts with a 72% moderate or marked response in the prophylaxis of depression, revealing a much better preventative than acute antidepressant response in this subgroup of patients.

Given the relative absence of controlled studies in the acute treatment of depression, there is much need for further work to define valproate's effi-

Table 3. Spectrum of outcome for rapid cycling bipolar patients

Mood and treatment	Total	Marked response		Moderate response		No response	
		N	%	N	%	N	%
Total cohort (N = 101)							
Mania							
Acute	58	37	64%	15	26%	6	10%
Prophylaxis	94	72	77%	16	18%	6	6%
Depression							
Acute	58	12	21%	14	24%	32	55%
Prophylaxis	94	36	38%	32	34%	26	28%
Mixed							
Acute	15	13	87%	0	0%	2	13%
Prophylaxis	18	16	89%	1	6%	1	6%
Monotherapy (N = 43)							
Mania							
Acute	19	14	74%	4	21%	1	5%
Prophylaxis	42	32	80%	8	20%	0	0%
Depression							
Acute	19	8	42%	6	32%	5	26%
Prophylaxis	42	19	45%	17	40%	6	14%
Mixed							
Acute	8	8	100%	0	0%	0	0%
Prophylaxis	9	8	89%	1	11%	0	0%

Reprinted from [47].

cacy in this area for both unipolar or bipolar depressed patients. However, it is perhaps noteworthy that despite multiple open and controlled clinical trials of lithium in the acute treatment of depression, its efficacy in this regard is still being debated. Thus, even with considerable further study, it may take time before an adequate definition of valproate's role in acute depression emerges.

Winsberg et al. [45] have observed a very high response rate (9 of 11, 82%) to valproate in patients with bipolar II depression who have not previously been exposed to other antidepressants. These data raise the possibility that the use of antidepressants in bipolar illness may make responsivity to valproate more difficult, as has previously been suggested for lithium carbonate [48]. However, it is common clinical practice to use antidepressants adjunctively with valproate or lithium, particularly in treating patients with breakthrough episodes of severe depression. For example, in the study of Calabrese et al. [44, 47], 43 of the 101 patients who stopped having hypomanic or manic episodes during treatment with valproate cycled into depression that required augmentation with an antidepressant. Of these 43, 60% experienced a marked antidepressant response and 40% did not respond. Only 25% exhibited drug-induced mania or hypomania and only 25% exhibited re-induction of rapid cycling from their "depressed state."

Valproate in the prophylaxis of manic and depressive episodes

While the currently published data on valproate in prophylaxis are almost entirely from open and uncontrolled studies, there is nonetheless considerable agreement in the literature about its clinical utility. This was apparent from the earliest studies of Lambert [1] reporting 13 of 141 (9%) marked and 45 of 141 (32%) partial response rates. Emrich et al. [22, 49] followed some of their acutely responsive patients over a considerable period of time, demonstrating responsivity to valproate or the combination of valproate and lithium in several instances, as illustrated in Fig. 2. In this patient, who was stabilized on combination therapy, discontinuation of valproate resulted in recurrence of depression, which then re-responded to combination therapy when valproate was resumed. Similarly, other patients relapsed when lithium was discontinued, indicating the need for both medications in some individuals.

Most of the initial series of studies were directed at patients inadequately responsive to lithium and carbamazepine, the other well-accepted or highly used mood stabilizing agents at the time (see review by McElroy et al. [50]; see also study by Hayes et al. [51]). The data of Schaff et al. [52] (see Tab. 4) are typical of many studies reporting a high prophylactic success rate in patients largely unresponsive to lithium and carbamazepine, including the rapid cycling patients of Calabrese et al. [47]. McElroy and

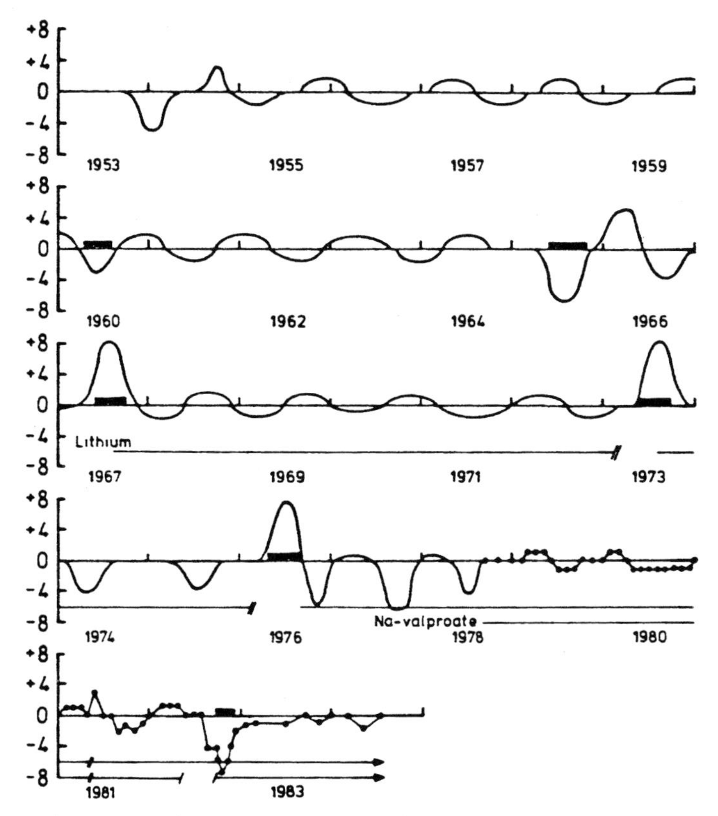

Figure 2. Time-course of the psychopathology of a patient with bipolar affective disorder, represented by use of the VBS (Verlaufs-Beurteilungs-Skala, i.e. course-assessment scale; cf. Emrich et al. 1977) under prophylactic long-term medication with valproate in combination with low doses of lithium. Symbols: [black bar], hospitalization; + 8, maximal mania; – 8, maximal depression. Reprinted from [22], p. 247, with permission from Elsevier Science.

Keck [40] reported a 65% response rate compared to Lambert et al. [1], who reported marked or moderate response in 58 of 141 bipolar patients (41%). In many of these sudies using global measures, results were more robust than those of Puzynski and Klosiewicz [43], who used careful lifecharting and reported some fragmentation of depressive episodes and actual exacerbation of cycling with the introduction of valproate prophylaxis.

The data of Lambert and Venaud [53] are open but randomized and thus provide an excellent comparative outcome with lithium. As illustrated in Table 5, there is a highly comparable degree of prophylactic efficacy of these two agents.

We are aware of two double-blind studies of valproate prophylaxis, one completed [54] and the other still in progress [55, 56]. The completed study [54] included patients who were well stabilized for a period of at least two consecutive biweekly appointments (1 month) on physician's choice medi-

Table 4. Response to divalproex in 63 patients with refractory affective disorders

Patient Subgroup	CGI Score[a]			Drug withdrawn due to side-Effects	N
	Good Response 1–2	Fair Response 3	Poor Response 4–7		
Total sample	47 (75%)	0	7	9	63
Patients who failed on carbamazepine	20 (69%)	0	7	2	29
Patients who failed on lithium	38 (84%)	0	0	7	45
Rapid-cycling patients	21 (81%)	0	2	3	26
Patients with psychotic episodes	10 (71%)	0	2	2	14
Patients diagnosed as schizoaffective	4 (80%)	0	0	1	5

[a] Clinical Global Improvement scores: 1–2 = very much or much improved = good response; 3 = minimally improved = fair response; 4–7 = no change to very much worse = poor response. Reprinted from [52].

Table 5. Comparative efficacy of valpromide and lithium in prevention of episodes[a]

	Valpromide (n = 78)		Lithium (n = 72)	
	Pre-Rx (2 years)	Post-Rx (1.5 years)	Pre-Rx (2 years)	Post-Rx (1.5 years)
Total No. episodes	321	39	282	42
Average No. episodes	4.12	0.51	3.92	0.61

[a] Adapted from [53].

cations and then randomized to lithium, valproate, or placebo for 1 year of prophylaxis. Many of the potential patients for this study did not enter it because of inability to meet stabilization criteria. Thus, a relatively "less ill" subgroup of subjects was recruited into the main phase of the study. This may account for some of the unconventional and unexpected findings that emerged.

For example, valproate, contrary to expectations, did not show significant prophylactic antimanic effects over that of placebo. However, valproate did show significant prophylaxis of depressive episodes compared with placebo, and, even more surprisingly, placebo was more effective than lithium in the prevention of depressive recurrences. While the reasons for this paradoxical lithium finding are not readily apparent, it is possible that the relatively high lithium levels intended and achieved contributed. It would be of particular interest to know the degree of lithium-induced suppression of the thyroid axis, since Frye and associates [57] reported that there was increased severity of depression in the year on lithium treatment as a function of the degree of lithium-induced suppression of free T_4. This relationship was evident from both increased Hamilton depression ratings as well as an increased frequency of cycling.

It should be noted that the data of Bowden et al. [54] are not convergent with the vast majority of controlled studies of lithium prophylaxis [21], which demonstrate that lithium is highly effective in both antimanic and antidepressant spheres. The data from this double-blind prophylactic trial are also at variance with the acute open studies that tend to demonstrate better antimanic than antidepressant efficacy. In the open studies of 101 rapid cycling patients (Tab. 3), Calabrese and associates [44, 47] found excellent antimanic and mixed-state prophylaxis. Obviously, further controlled trials will be required in order to delineate the consistency and relevance of these findings to the large majority of bipolar depressed patients.

Calabrese and associates [56, 55] are currently comparing the effects of lithium and valproate in a double-blind randomized trial of the long-term prophylaxis in bipolar patients initially stabilized on the combination of the two agents. It is noteworthy that in the intent-to-treat analysis, only a minority of the patients became eligible for randomization. A very substantial group of bipolar patients did not remain on their medication or in the study because of a high incidence of noncompliance in this difficult-to-treat population. A minority were adequately responsive to the combination of lithium and valproate.

In those who remained in the study, there was a good response rate to the combination of lithium and valproate. Although the prophylaxis phase is still blind, thus precluding analysis of efficacy of lithium *versus* valproate, Calabrese has reported that a number of the randomized patients have relapsed on monotherapy. These data, again, support those of Emrich et al. [49] and others indicating that for at least a subgroup of bipolar patients, combination therapy is required for long-term mood stability with lithium and valproate. These data mirror those of lithium and carbamazepine reported by Denicoff and associates [14] showing relatively low responses to either drug in a year of monotherapy (33.3% on lithium and 31.4% on carbamazepine), but then much better response to combination therapy (55.2%) in the third year of treatment. The efficacy of combination therapy was even clearer for the subgroup of rapid cycling patients in whom response to lithium and carbamazepine monotherapy was 28.0% and 19.0%, respectively, whereas response to the combination was 53.3%.

Those patients inadequately responsive to either of these three initial intended one-year phases of treatment (i.e. lithium, carbamazepine, or the combination) were offered the lithium-valproate combination in a fourth year with triple mood stabilizer therapy in the fifth year [58]. These data indicate that a subset of patients with confirmed inadequate response to lithium, carbamazepine, and the combination appear to respond to the addition of valproate to lithium (5 of 18, 28%). When carbamazepine was added to the lithium-valproate combination for seven patients, three responded to this treatment. Out of the original 42 evaluable patients, 26 (61.9%) responded to at least one of the five different intended 1 year prophylactic medication regimens.

These data are consistent with those of Keck et al. [59] and Ketter et al. [60], indicating that some patients respond to the carbamazepine-valproate combination (with or without adjunctive lithium) after failing to respond to monotherapy with either anticonvulsant. These data in bipolar illness thus mirror a substantial literature in the treatment-refractory epilepsies wherein patients inadequately responsive to either agent alone appear to be more highly responsive to the combination [61–65].

Since many actions of carbamazepine and valproate diverge as well as converge, as listed in Table 6 [66–87], these data bespeak the possible pharmacodynamic additive or synergistic effects of these two agents as opposed to pharmacokinetic effects. However, there are prominent pharmacokinetic effects as well, with valproate able to increase carbamazepine by both displacing it from its binding sites as well as increasing the 10,11-epoxide via inhibition of the epoxide hydroxylase enzyme in the conversion of the 10,11-epoxide to the inactive diol.

The ability of valproate to be effective in monotherapy in patients refractory to lithium and carbamazepine is illustrated in Fig. 3. This 32 year-old

Table 6. Comparative effects of two anticonvulsant mood stabilizers in bipolar illness*

	Carbamazepine	Valproate	Author
GABA turnover[a]	↓	↓	[67, 68]
$GABA_B$ receptors[a]	↑	↑	[69, 70]
GABA levels	(↑)	↑	[71]
GABA-mediated prepulse inhibition	0	↑	[72]
[^{3}H]-picrotoxinin binding	↓	↓	[73, 74]
Dopamine turnover[a]	↓	↓	[75, 76]
Na^+ influx	↓	↓	[77, 78]
K^+ efflux	↑	↑	[79–82]
NMDA-mediated currents	(↓)	↓	[72, 83, 84]
Release of aspartate	↓	↓	(personal communication), [72]
Somatostatin levels	↓	↓	[71, 85]
Substantia nigra lesions block CBZ's anticonvulsant effects on kindling	++	0	[86]
Decrease afterdischarge spread in:			
Hippocampus	–	↓	[87]
Cortex	↓	–	

* [66].
[a] Effects shared by lithium.

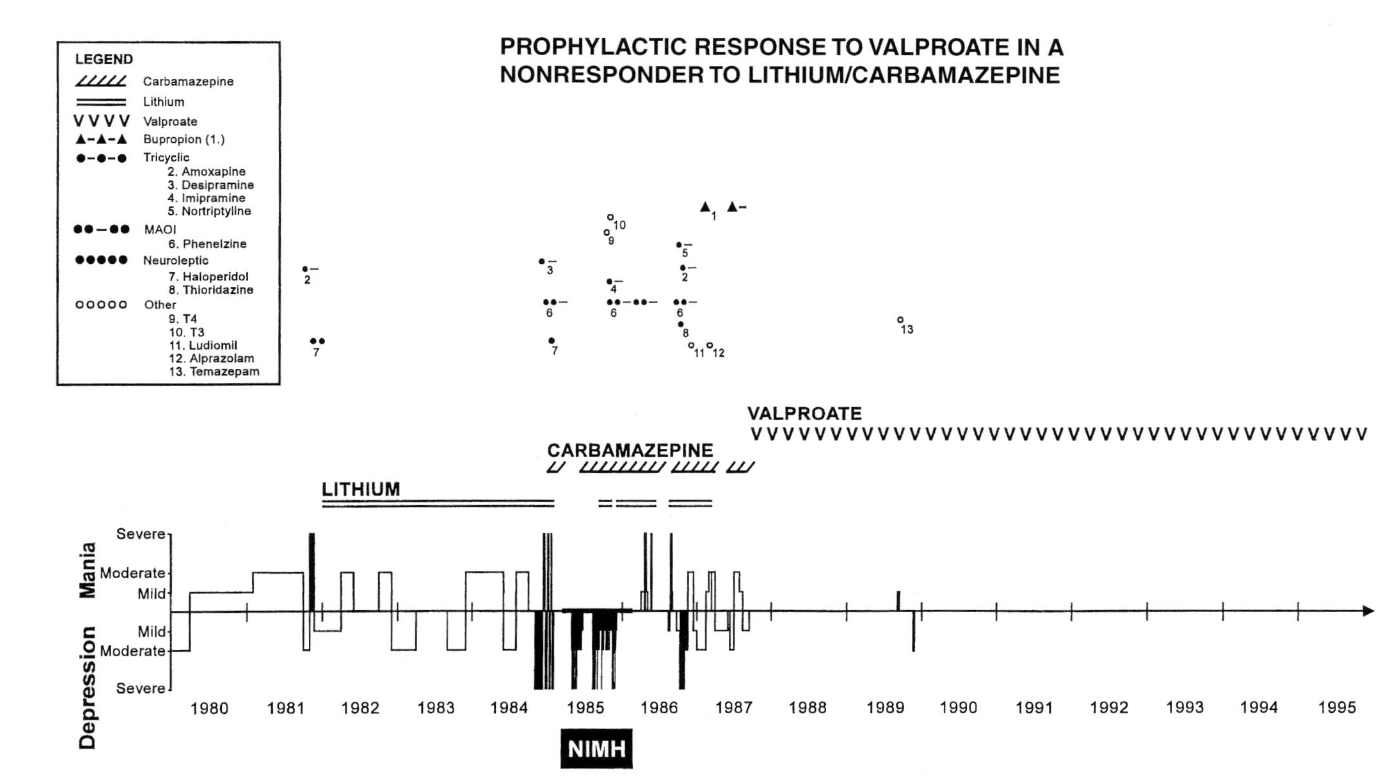

Figure 3. Prophylacatic response to valproate in a nonresponder to lithium/carbamazepine. Life chart illustrates a patients's response to valproate after nonresponse to lithium and carbamazepine.

business executive was fired from her job because of outrageous behavior during a manic episode. Although lithium moderated the severity of mania and carbamazepine was partially effective, the combination was inadequate to stabilize her mood and she felt mildly to moderately depressed. When tricyclic or monoamine oxidase antidepressants were utilized in an attempt to reach a euthymic mood, mania was exacerbated and cycling increased despite the presence of the two mood stabilizers. In contrast, affective episodes were minimal during valproate prophylaxis and this patient remained well on valproate monotherapy for more than a decade.

Tolerance to the Prophylactic Effects of Valproate

A subgroup of patients treated with any of the three major mood stabilizers, i.e. lithium [88, 89], carbamazepine [90, 91], or valproate [92] (McElroy, personal communication) appear to show lack of sustained efficacy during long-term prophylaxis. In lithium-refractory patients admitted to the clinical research unit of the NIMH, tolerance accounts for 34.9% of the lithium refractoriness, i.e. patients in this subgroup were initially responsive for a considerable number of years and then began to have breakthrough episodes [89]. We have observed similar degrees of tolerance prospectively with responders to carbamazepine with various adjunctive treatments. In these instances, carbamazepine combination therapy was associated with the loss of efficacy via the presumptive tolerance mechanism after an average of 2.8 years in approximately 45% of the patients followed for an average of 6.9 years [91]. There is some suggestion of a lesser incidence of loss of efficacy to valproate via such a tolerance mechanism, with 27% of our treatment-refractory patients showing loss of efficacy to valproate after an average of 3.6 years (Post et al. 1998, unpublished data). Because NIMH subjects are a highly selected subgroup of treatment-refractory patients, it is likely that the incidence of loss of efficacy via tolerance in the general population would be substantially less.

Loss of response to valproate via tolerance is exemplified in the life chart in Fig. 4a. After a period of approximately 4 years well on the lithium-valproate combination, this patient showed a gradual reemergence of manic and depressive episodes of increasing severity. Interestingly, upon discontinuation of valproate for a brief period of time and then its reinstitution, the patient appeared to show a re-responsivity to this drug regimen for a sustained period of time. This patient's experience is illustrative of a number of perspectives in the use of valproate in treatment-refractory bipolar patients in that he was not responsive to lithium monotherapy and had severe dysphoric, psychotic manias requiring seclusion and acute intermittent neuroleptic treatment. He also failed to respond to carbamazepine, but did respond to valproate in combination with lithium. However, despite the continued use of the combination of valproate and lithium, response was

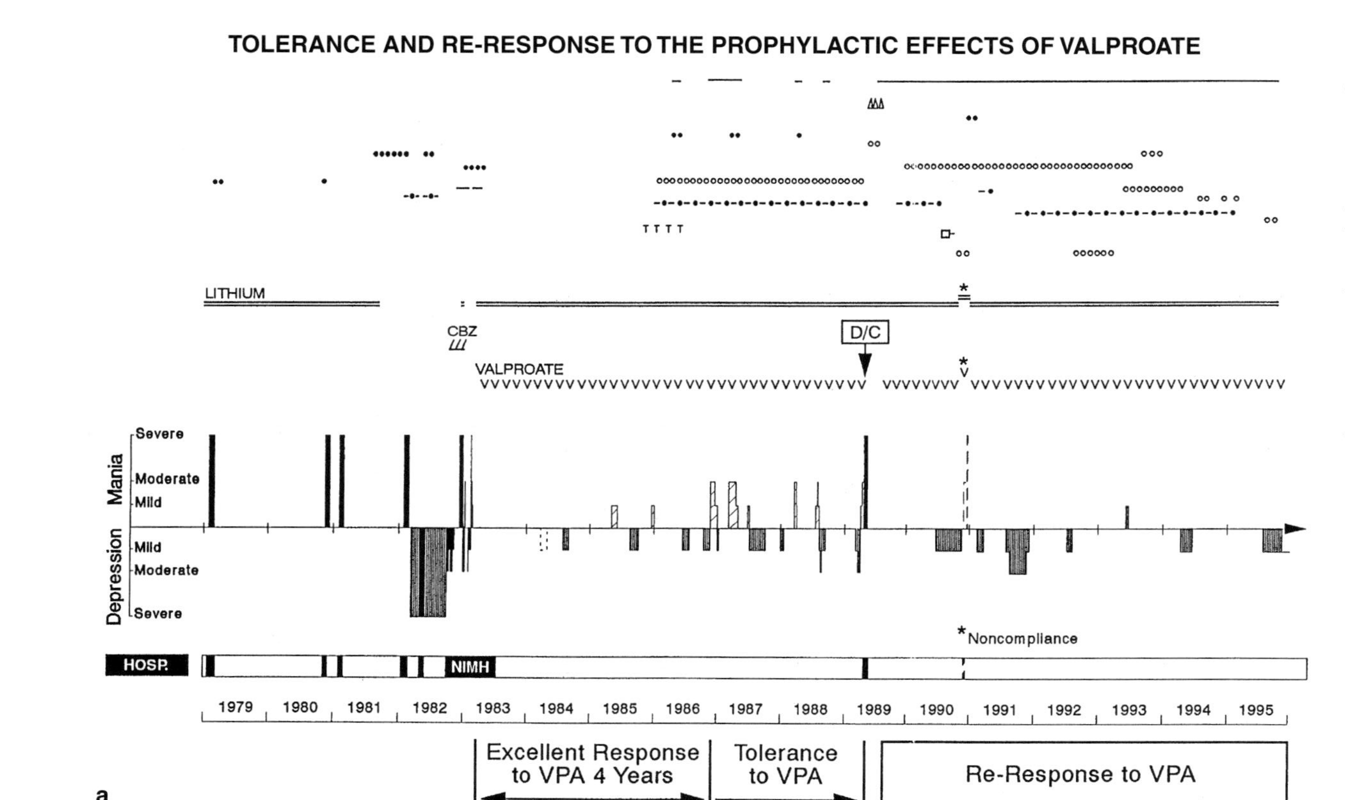

Figure 4 a and b. Loss of efficacy to valproate via tolerance: Comparative progressive evolution in syndrome breakthroughs. (a) Course of tolerance development in a patient. (b) Course of tolerance development in an amygdala-kindled rat. In both instances, altthough over vastly different time scales, there is a progression of episodic breakthroughs leading to complete tolerance. Interestingly, for both the patient and the kindled animal, drug efficacy was reinstated by a period of time off from the drug treatment (patient; right side ~ 1989, rat; right side ~ d 88–94).

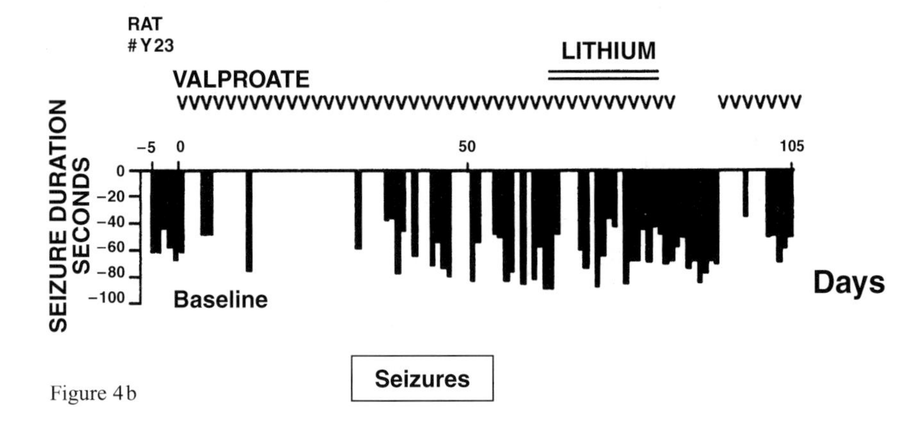

Figure 4b

not sustained, and manic and depressive episodes of generally increasing severity and/or duration began to breakthrough.

The potential for re-responsivity after a period of time off drug in the face of loss of efficacy via tolerance is illustrated in the right side of Fig. 4a. With the occurrence of a full-blown mania requiring hospitalization, valproate was initially discontinued in favor of treating with lithium alone, which had clearly been ineffective previously. We urged that valproate be restarted, and this was apparently sufficient to renew responsivity. The data from this naturalistic experiment are convergent with those of the preclinical model of amygdala-kindled seizures indicating that tolerance to anticonvulsants may be overcome and efficacy reinstated after a period of time off, when seizures occur in the absence of drug (Fig. 4b and 5) [93, 94].

Tolerance to the anticonvulsant effects of carbamazepine develops in a contingent fashion such that animals treated with the drug *prior* to amygdala-kindled stimulation become tolerant, but those that experience the kindled seizure and are treated with the same doses of drug but *after* the seizure has occurred do not become tolerant upon subsequent rechallenge [94, 95]. Unexpectedly, there was cross-contingent tolerance between carbamazepine and valproate despite presumptive different clinical spectrum of anticonvulsant effects and mechanisms of action between carbamazepine and those of valproate [66, 96].

A possible explanation for this cross-tolerance emerged in the finding that carbamazepine tolerance was associated with a failure of seizures to increase $GABA_A$ receptors and the mRNA for the alpha-4 subunit of the $GABA_A$ receptor, but not other subunits nor the benzodiazepine receptor itself [97]. This biochemical association was also demonstrated to be contingent, as it did not occur in animals who experienced equal amounts of carbamazepine and seizures but did not display tolerance since carbamazepine was administered *after* a seizure occurred. Subsequently, we have

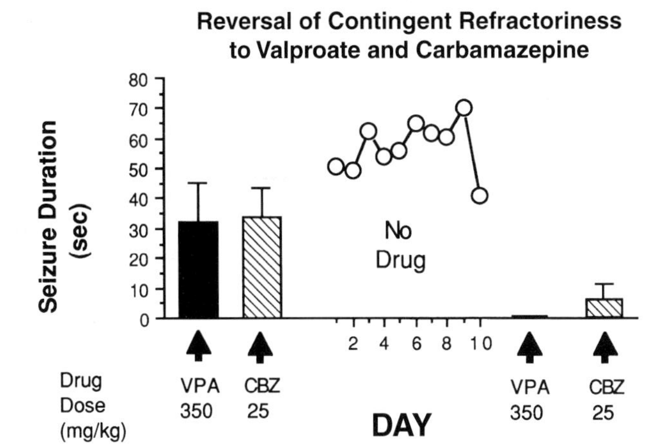

Figure 5 a and b. Reversal of contingent refractoriness to valproate and carbamazepine. With repeated daily drug treatments before (but not after) amygdala-kindled stimulation, contingent tolerance to the anticonvulsant effects developed. Contingent tolerance was reversed by inducing seizures without drug exposure (time-off drug), as measured by seizure duration (a); as well as afterdischarge duration (b).

observed that tolerance to diazepam is also associated with failure of upregulation of the alpha-4 subunit in a contingent fashion [98]. Since valproate is thought to exert part of its anticonvulsant effect through indirect actions of GABA enhancement at the $GABA_A$ receptor, the failure of seizure-induced upregulation at this site could be, in part, attributable to the cross-tolerance between carbamazepine and valproate [94, 97].

In the preclinical model we, at the same time, observed that it was relatively difficult to induce tolerance to valproate itself, and that with the administration of higher doses, such tolerance could typically be overcome [93, 94, 96]. In contrast, carbamazepine tolerance could not be readily prevented or overcome at higher doses that were not associated with behavioral toxicity and side-effects [94, 99]. Thus, it would appear from the preclinical model that valproate had unique properties compared with the relatively rapid induction of contingent tolerance to both carbamazepine [95] and diazepam [98, 100]; i.e. the demonstration of tolerance development to valproate was relatively difficult to achieve.

Nonetheless, when one used very low doses of valproate in order to propel such tolerance development and moderate doses of carbamazepine, one could demonstrate that the use of the combination of both drugs together resulted in a slower development of tolerance than either drug in monotherapy (Fig. 6) [93]. Thus, in addition to the additive effects of these two drugs on acute anticonvulsant efficacy and, presumptively, psychotropic efficacy [59, 60], these preclinical data would suggest that the two drugs in combination may be able to sustain a long-term efficacy better than either drug in monotherapy. The relevance of this to clinical seizure disorders and the psychotropic effects of these agents requires further direct and prospective observations in order to be confirmed.

Also of potential interest was the observation that amygdala-kindled seizing animals treated with valproate early in the course of their full-

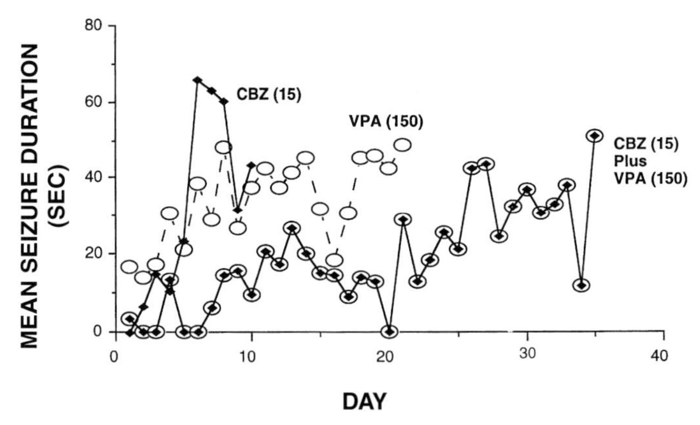

Figure 6. Enhanced anticonvulsant efficacy of carbamazepine plus valproate compared to monotherapy in amygdala-kindled rats. Group mean seizure duration is plotted over days in rats treated with carbamazepine (CBZ) alone (15 mg/kg) (filled symbols), valproate (VPA) alone (150 mg/kg) (open circles), or the combination (filled circles). Tolerance was slowest to develop and anticonvulsant efficacy was best and most prolonged in the group receiving the combined CBZ-VPA treatment.

blown seizure disorder were more responsive to the drug than those treated with similar doses of valproate late in the course of kindled seizures [101]. Whether valproate would be effective in the very latest stages of kindled seizure evolution, when it becomes spontaneous, also remains for systematic observation. In this regard it is noteworthy that diazepam, which is highly effective against the initial phases of kindling development and the full-blown completed kindled seizures, is not effective when these seizures become spontaneous [102]. Surprisingly, phenytoin, which is less effective in the initial phases of kindling in many instances, apparently is highly effective in the late spontaneous phases [102]. These data are examples of a now robust literature on different phases of kindled seizure evolution being differentially responsive to different agents. Whether valproate would remain effective in all stages of seizure evolution requires systematic experimental observation.

In this regard it is of interest that phenytoin is not effective in the prevention of post-traumatic epilepsy [103] and is also not effective in the early phases of the kindling model, as mentioned above. These data would suggest that valproate might be a potentially more effective agent in attempting to ward off the development of seizures in the post-traumatic situation, particularly if it proves that a kindling-like mechanism is involved in their late emergence. We raise these questions in the context of a chapter on the effects of valproate in psychiatric illness in relation to similar questions: would valproate treatment in the earliest phases of the illness also be effective in warding off the development of the full-blown syndrome in either the affective disorders or in anxiety and post-traumatic stress disorders? Similarly, one could ask a related question regarding early intervention in the prophylaxis of migraine as well: would the early institution of prophylaxis with a drug such as valproate help prevent the considerable tolerance development to many other acute therapies and also help ward off the late phase of its evolution and transformation into chronic daily headache?

Pharmacokinetics

The dose and blood level range for valproate in the treatment of affective and related psychiatric disorders is not well delineated. However, convention and unsystematic observations suggest that doses that achieve blood levels greater than 50 µg/ml are more effective for the most seriously ill patient compared with lower doses. The observations of Jacobsen [104] suggest that the milder forms of bipolar illness such as cyclothymia may be responsive to smaller doses of valproate, achieving blood levels in the range of 30–50 µg/ml, while bipolar II patients need higher levels, and the full-blown bipolar I syndrome may require the highest blood levels in the range 100 to 150 µg/ml [50].

As noted previously, Keck and associates [28] have used oral loading strategies of divalproex sodium (20 mg/kg on day one) with substantial success and a minimum of side effects. Since this dosing regimen tends to produce blood levels in the presumptive 50–100 μg/ml therapeutic range more rapidly than the usual or slow titration, it should theoretically be associated with more rapid onset of antimanic effects. Whether this is the case remains to be directly demonstrated. Some agents such as lithium or traditional antidepressants, even when rapidly loaded, do not produce a more immediate onset of therapeutic effects, and some delay (i.e. chronic administration of the drug) appears to be required before the appropriate biological pharmacodynamic effects are achieved that are sufficient for onset of antimanic and antidepressant efficacy [105, 106]. Whether this would also prove to be the case for loading doses of valproate remains to be clarified. Some of valproate's effects appear to require chronic administration in order to become evident, and it is possible that these may be pertinent to the antimanic and prophylactic efficacy of the drug.

A number of potential pharmacokinetic drug interactions of valproate are illustrated in Tab. 7 [107]. These include valproate increasing the blood levels of a number of anticonvulsant/psychotropic drugs, most notably carbamazepine and lamotrigine. Valproate's ability to increase lamotrigine

Table 7. Pharmacokinetic interactions of valproate (from [107])

Metabolic interactions

VPA → ↑ Drug	Drug → ↑ VPA	Drug → ↓ VPA
Benzodiazepines	ASA	CBZ
CBZ-E	Cimetidine	± Lamotrigine
Ethosuximide	Fluoxetine	Phenobarbital
Lamotrigine	Felbamate	Phenytoin
Phenobarbital	Erythromycin	Rifampin
Phenytoin	Phenothiazines	
TCAs		

Binding interactions

VPA → ↑ free drug	Drug → ↑ free VPA
CBZ	ASA
Diazepam	NSAIDs
Phenytoin	
Tiagabine	
Tolbutamide	
Warfarin	
Zidovudine	

VPA = valproate; CBZ-E = carbamazepine epoxide; TCAs = tricyclic antidepressants; ASA = aspirin; CBZ = carbamazepine; NSAIDs = nonsteroidal antiinflammatory drugs.

blood levels is of particular importance: the starting doses of lamotrigine and its rate of increase should be halved in the presence of valproate because of the risk of developing a severe rash, which appears to be partially dose-and rate of increase-related.

As listed in Tab. 7, a number of enzyme-inducing drugs, such as carbamazepine and phenobarbital, decrease the level of valproate. Potent enzyme-inhibiting drugs, such as fluoxetine, have been reported to increase levels of valproate in preliminary observations [108], and these remain to be more systematically documented. In contrast to the robust effects of carbamazepine on bupropion metabolism – decreasing bupropion levels and increasing the hydroxy bupropion active metabolite – valproate does not decrease bupropion levels substantially, although the metabolite is increased to some extent [109]. Similarly, in contrast to many other anticonvulsant agents, valproate does not interfere with the effects of birth control pills by inducing their metabolism [110, 111]. This is a particular liability with the mood stabilizing anticonvulsant carbamazepine [112].

Side-effects

Along with a similar spectrum of efficacy in bipolar affective illness, valproate also shares many of lithium's side effects (Tab. 8). For example, tremor, weight gain, and gastrointestinal upset are common to both agents, with some tendency for these side-effects to be additive when the two drugs are combined, as they often are. Whereas tremor appears to be dose-related, it is not as clear that this is the case for increased appetite and weight gain, which can be problematic [50, 113–116]. A potentially troublesome side-effect for a small percentage of people is the development of alopecia. Surprisingly, patients with straight hair may lose this, and when it grows back, be disconcerted by the fact that the regrowth is curly [117]. Clinical wisdom suggests the possibility that zinc and selenium vitamin supplements help prevent this side effect, although definitive studies have not been conducted [118].

Severe hematological side-effects are rare, with the most common being a reduction in platelet number or effectiveness in clotting; severe liver abnormalities have been observed, as well as pancreatitis [50, 114, 116, 119]. These occur in very low numbers, with the exception of patients under the age of 2 years on concomitant anticonvulsant medication wherein a number of lethalities have been recorded from valproate-related hepatitis [120, 121]. Transient fluctuations in liver enzymes are common with many of the mood stabilizing anticonvulsants, and conventional wisdom suggests that elevations in selective enzymes up to three times normal can be approached with caution without necessarily having to discontinue treatment. Clinical symptoms of hepatitis and an increasing progression of enzyme elevation above normal should trigger consideration of discontinuation, however.

Table 8. Comparative clinical and side-effects profiles of lithium, nimodipine, and the putative mood-stabilizing anticonvulsants (preliminary clinical impressions)

Clinical profile	Lithium (0.5–1.2 mEq/l)	Carbamazepine (4–12 µg/ml)	Valproate (50–120 µg/ml)	Nimodipine (12ng/ml)	Lamotrigine	Gabapentin	Topiramate
• *Acute episodes*							
Mania (M)	++	++	++	+	(+)	(++)	±
Dysphoric mania	+	(++)	++	+	?	(+)	±
Family history negative	±	++	±	?	?	?	?
Depression (D)	+	+	±	+	++	(+)	( )
• *Prophylaxis*							
Mania	++	++	++	+	(+)	(+)	(+)
Depression	++	++	+	+	(++)	(+)	(±)
Rapid cycling	+	++	+	++	++	(++)	+
Continuous cycling	++	++	(++)	(++)	?	?	+
• *Seizures*							
Generalized, Complex partial	0	++	++	±	++	++	++
Absence	0	– –	++	?	++		
• *Paroxysmal pain syndromes*	0	++	+	–	?	(+)	?
Migraine	±	±	++	+	?	(+)	?
Side effects profiles							
White blood count	↑↑*	↓↓	(↓)	–	–	–	–
Diabetes insipidus	↑↑*	↓	–	–	–	–	–
Thyroid hormones, T_3, T_4	↓↓	↓↓	↓	–	?	(↑)?	?
Thyroid stimulating hormone	↑↑*	–	?	–	–	–	?

Table 8 (continued)

Side effects profile	Lithium (0.5–1.2 mEq/l	Carbamazepine (4–12 μg/ml)	Valproate (50–120 μg/ml)	Nimodipine (12ng/ml)	Lamotrigine	Gabapentin	Topiramate
Serum calcium	↑	↓	?	?	?	?	–
Weight gain	↑↑	(↑)	↑↑[a]	–	–	↑	–
Tremor	↑↑	–	↑↑	–	↑	↑	–
Memory disturbances	(↑)	(↑)	(↑)	– –	(↑)	↑	↑
Diarrhea, Gastro-intestinal distress	↑↑	(↑)	↑	(↓)	–	(↑)	
Teratogenesis	(↑)	↑	↑	–	–	–	?
Psoriasis	↑	– –	–	–	–	–	–
Pruritic rash	– –	↑↑	(↑)	(↑)	↑↑	(↑)	–
Alopecia	(↑)	–	↑[a]	–	–	–	–
Agranulocytosis, aplastic anemia	– –	↑	–	–	–	–	–
Thrombocytopenia	–	(↑)	↑	–	–	–	–
Hepatitis	–	↑	↑	–	↑	–	(↑)?
Hyponatremia, water intoxication	–	↑	–	–	–	–	
Dizziness, ataxia, diplopia	–	↑	(↑)	–	↑	↑	↑
Hypercortisolemia, escape from examethasone suppression	–	↑↑	?	–	–	–	–

Clinical efficacy: 0 = none, ± = equivocal, + = effective, ++ = very effective, ( ) = very weak data, ? = unknown, – – = exacerbation.
Side-effects: ↑ = increase, ↓ = decrease, ( ) = inconsistent or rare, – = absent, – – = worse.
* Li = Effect of lithium predominates over that of carbamazepine when used in combination.
[a] = about 3 months after onset of VPA; prevent alopecia with Zinc and Selenium!?

Valproate has been associated with congenital malformations in a small percentage of patients with a particular risk for the induction of a spina bifida syndrome [122, 123]. Whether high doses of folate and minimal doses of valproate are able to substantially lessen this risk has not been proven but is thought likely [124, 125]. The clinical impact of valproate on cognitive functioning is typically minimal for most patients although occasional reports of word-finding difficulties are observed [116, 126, 127]. A confusional syndrome with a highly abnormal EEG has occasionally been reported in some epileptic patients [128]. Hyperammonemia is also a rare side-effect of valproate [129]. It may be more likely to occur when valproate and carbamazepine are used concomitantly and may also be associated with a marked asterixis. When hyperammonemia occurs with valproate alone, L-carnitine [130, 131] or citrulline [132] may be helpful.

Some of the relative risk and benefit ratios of choosing valproate among the various mood stabilizing and putative mood stabilizing options are listed in Table 9 (for rapid and ultra-rapid cyclers, see Fig. 7; see also Fig. 8 for mechanisms of action of these various mood stabilizers). In the absence

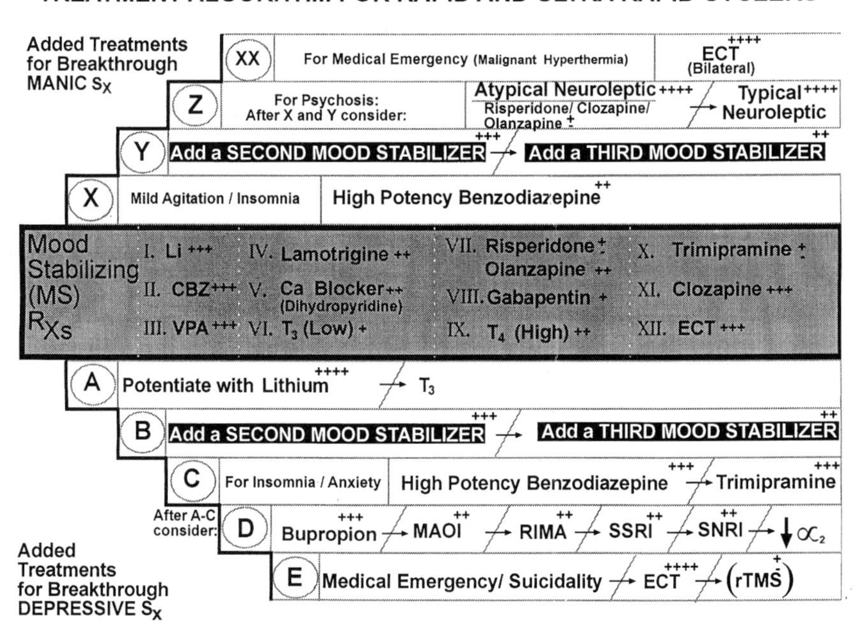

Figure 7. ECT indicates electroconvulsant therapy; Li, lithium; VPA, valproic acid; CBZ, carbamazepine; Ca, calcium; T_3, triiodothyronine; T_4, thyrocine; MAOI, monoamine oxidase inhibitor; RIMA, reversible inhibitor of monoamine oxidase; SSRI, selective serotonin reuptake inhibitor; SNRI, serotonin and noradrenaline reuptake inhibitor. Strength of evidence in the literature: + = weak, ++ = mild, +++ = moderate, ++++ = strong.

Table 9. Positive and negative selection factors for choice of mood stabilizer

Target symptoms and anxillary responsive syndromes	Lithium (Li)		Carbamazepine (CBZ)		Valproic acid (VPA)	
	Euphoric ++++ Family Hx positive +++ MDI pattern +++ Steroid-induced ++ Suicidal +++		Euphoric +++ Schizo-affective ++ Organic-affective +++ Aggressive +++ Dysphoric + Alcohol ++ Cocaine ± PTSD + Steroid-induced ± Pain syndromes ++++		Dysphoric +++ Rapid cycling +++ Organic-affective ++ Panic + Migraine ++++ Alcohol ± Cocaine ± PTSD + Pain syndromes ++	
	CHOOSE – Li	AVOID – Li	CHOOSE – CBZ	AVOID – CBZ	CHOOSE – VPA	AVOID – VPA
Side-effects profiles:	↑WBC (c)	Wt Gain (c)	Minimal Cognitive Δ's	Many drug interactions! (c)	Few drug interactions	Wt Gain (c)
Positive → CHOOSE Negative → AVOID	↑Ca^{++} (c)	Tremor (c) - DR	Little Wt Gain	↓ Potency of birth control pills (c)	Tolerated in O.D.	GI distress (c) Tremor (c) – DR
	Renal excretion	Subjective (c) – DR cog. slowing	Tolerated in O.D.		Minimal Cognitive Δ's	Alopecia (r) – I Pancreatitis (vr) – I
(c) = common (r) = rare (vr) = very rare	Non – sedating	↓ Thyroid (c) ↓ Renal – ↑ D.I. (c) – ↓ GFR (r) Toxic in O.D. – Cardiac (c) – Cerebellar (r)		Rash (10–15%) (c) Ataxia/Sedation (c) Hyponatremia (c) Agranulocytosis (r) – I Aplastic anemia (vr) – I Allergy – I		Polycystic ovaries (?) ↓ platelets (c) Liver failure – (Child < 2) (vr)
DR = Dose-related I = Idiosyncratic		Poor in M.S. & neurological illness				
MS = multiple sclerosis		Pregnancy? – Ebstein's Anomaly (vr)		Pregnancy? – Spina Bifida (1–3%) (r)		Pregnancy – Spina Bifida (2–6%) (r)

Table 9 (continued)

Target symptoms and anxillary responsive syndromes	Lamotrigine (LTG)		Gabapentin (GPN)		Topiramate (TOP)	
	Rapid cycling ++ R_x refractory ++ Pain syndromes ++		Parkinsonian symptoms ++ Rapid cycling ++ Insomnia ++ Anxiety +		Bulimia ++	
	CHOOSE – LTG	AVOID – LTG	CHOOSE – GPN	AVOID – GPN	CHOOSE – TOP	AVOID – TOP
Side-effects profiles	Non-sedating	Rash (c): 5–10%	Renal excretion	Inhibits own uptake –	Weight loss (c)	1% incidence renal
	Weight neutral to weight loss	Risk of severe rash: 1/500 (r)	Few interactions	requires T. I. D., Q. I. D. dosing		calculi Psychomotor slowing (c)
	Antidepressant; mood not set below baseline	Slow titration required	Helps essential tremor			Difficulty with word finding (c) Possible insomnia
Positive → CHOOSE Negative → AVOID		↑ levels (× 2) with VPA ↓ levels (× 2) with CBZ	Helps in pain syndromes Improves social phobia			
(c) = common (r) = rare (vr) = very rare						
DR = Dose-related I = Idiosyncratic MS = multiple sclerosis						

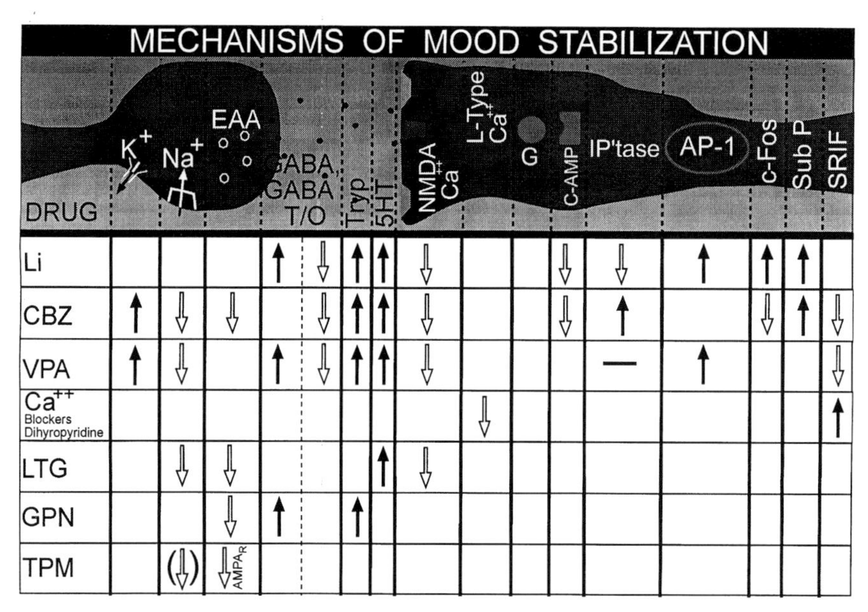

Figure 8. Depicted schematically at the top of the figure is a synapse with various types of channels, neurotransmitters, and proteins associated with the mechanisms of action of the mood stabilizers listed in the table below. Row headings: Li, indicates lithium; CBZ, carbamazepine; VPA, valproate; LTG, lamotrigine; GPN, gabapentin; TPM, topiromate. Column headings: K^+ efflux; Na^+ influx; EAA = excitatory amino acids; GABA; GABA turnover; tryptophan; 5HT = serotonin; NMDA; Ca^{++}; L-type Ca^{++}; G = G protein; C-AMP = cyclicAMP; IP'ase = Inositol phosphatase; AP-1 = activator protein 1; c-Fos; Sub P = substance P; SRIF = somatostatin. Arrows indicate increases or decreases in substance/activity.

of definitive differences in clinical efficacy profiles, choice of a given agent may also be dictated to a considerable degree by differential side-effects profiles, as listed in Table 8.

Possible clinical and biological correlates of clinical response to valproate sodium are outlined in Table 10 [22, 133, 134]. Several investigators have suggested that valproate is effective in many patients whose illness attributes and comorbidities make them prone to a low response rate to lithium, such as dysphoric mania. However, it is far from clear whether any of these is actually a positive clinical predictor of response and much work remains to identify robust and clinically useful predictors or markers of clinical responsivity. A number of concepts and principles for the use of valproate and related agents in bipolar illness are outlined in Table 11.

Table 10. Potential correlates of response to valproate*

Diagnosis-related factors:	
Diagnosis of bipolar disorder [vs. schizoaffective or schizophrenia disorder]	[22]
Dysphoric [or mixed mania]	[134]
Bipolar II disorder	[45]
Clinical course variables:	
Presence of rapid cycling	[47, 134]
Decreasing or stable episode frequency	[47]
Later age at onset of illness	[47]
Shorter duration of illness	[47]
Long and severe course of illness	[41]
Comorbid psychiatric diagnosis:	
Comorbid alcohol abuse	[37, 137]
Comorbid mental retardation	[138]
Comorbid panic attacks	[37]
Treated hypothyroidism	[137]
Marked sleep disturbance	[136]
Associated neurological factors:	
Presence of nonparoxysmal EEG abnormalities (not subsequently replicated)	[40]
Organic affective disorder	[139]
History of closed head trauma	[42, 140]
Family history:	
Positive family history of mood disorder	[135]
Treatment-related variables:	
Patients who are lithium naive	[45]
Higher serum valproate concentrations at days 3–6 of treatment	[26, 136, 141]
Higher serum valproate concentrations with more severe illness: cyclothymia < BPII < BPI	[104], 142]

* Modified from [133].
** Especially with addition of valproate to lithium (9/10).

Valproate in other psychiatric syndromes

Panic disorder

In an open study of valproate in panic disorder patients, Keck and associates [143] have observed that 10 of 12 patients (83%) who panicked on lactate while untreated failed to do so while they were actively being treated with valproate at an average dose of 20 mg/kg/day. In addition, 10 of the 14 patients completing the 28-day valproate trial experienced a 50% or greater reduction in the weekly frequency of panic attacks. Six (43%) had a complete remission of panic attacks, all by the end of the second week of treatment. In the 14 completers, there was a significant reduction in the mean frequency of full-blown panic attacks, but not those of the partial variety. Moreover, by the end of the 28-day trial, Hamilton Anxiety Scale

Table 11. Principles of treatment of bipolar illness with valproate

1. Dual treatment focus from the outset: acute episodes and prophylaxis
2. Mania: Treat first, chemistries later
3. Load valproate (VPA) and lithium (Li)
4. Potentiate VPA with Li and *vice versa*
5. Augment with other mood stabilizer instead of typical neuroleptics
6. Titrate using efficacy and side-effects rather than blood levels
7. Chart retrospective and prospective course of illness
8. Consider augmentation rather than substitution
9. Slow taper of Li, if at all
10. Educate and re-educate family about the illness and the need for long-term prophylaxis
11. Assess compliance and suicidality regularly (lithium is likely an antisuicide medicine)
12. Develop Early Warning System (EWS); identify the typical earliest symptoms that herald new episode
13. Develop specific contract regarding EWS, i.e. "augment with benzodiazepine if sleep by 1–2 h; call M.D. if sleep for 3 days"
14. Regular visits planned; monitor course and side effects closely; monitor blood levels and chemistry only as needed
15. Establish phone contact (p.r.n.) for emergencies
16. Develop "fire drill" for breakthrough mania/suicidality; when well, develop contract for hospitalization should it be needed in the future
17. Education about risks of comorbid alcohol and drug abuse; inquire about substance use; monitor if needed; treat aggressively
18. Use focused psychotherapy and medicalization illness; i.e. "you would not stop digitalis for congestive heart failure, why do it for manic depressive illness?"
19. Give statistics: 50% relapse in first 5 months off lithium; 80–90% within 1.5 years of lithium discontinuation
20. Recruit the patient as a co-principal investigator in the development of their optimal Rx (i.e. you, them, and meds together against the illness)
21. Be conservative and hold treatment constant in face of successful mood stabilization
22. Be radical and exploratory in the face of inadequate response

Scores dropped significantly from a baseline mean of 30.8 (±9.4) to 12.6 (±7). Other case [144, 145] and open studies [146, 147] as well as one small controlled study [148] have also suggested the potential efficacy of valproate in the treatment of panic disorder.

These preliminary observations are in need of further systematic study but are convergent with the view of valproate's efficacy in dysphoric mania [11, 39], which often has a substantial component of high anxiety levels, if not clear panic attacks. Moreover, a number of studies indicate the high comorbidity of a variety of anxiety disorders with bipolar illness [4, 5], making valproate a particularly attractive option because of its putative efficacy in this area. For example, in the Stanley Foundation's bipolar patient cohort, anxiety disorders were evident in 115 (44%) of the 261 patients [6].

Post-traumatic stress disorder (PTSD)

The efficacy of valproate has not been systematically evaluated in post-traumatic stress disorder, although some preliminary observations are promising. While carbamazepine appears particularly helpful in targeting some of the paroxysmal aspects of PTSD and its associated sleep disorder [149–151], preliminary assessment in an open study of Vietnam combat veterans with PTSD indicates that valproate may be helpful particularly in hyperarousal/hyperreactivity components of the disorder [152]. Several case studies have also reported beneficial results of valproate treatment in Vietnam veterans [153, 154] and a police officer [155] with PTSD, emphasizing the need for controlled studies.

Migraine

Although the effects of valproate on migraine have been dealt with in detail in another chapter in this volume, it is worth reemphasizing the high comorbidity of migraine with affective illness, as highlighted by Merikangas and colleagues [156]. As such, the use of valproate offers the possibility of a "two-for-one" benefit with a dual positive effect on both migraine prophylaxis and affective symptoms. Several controlled studies support the effects of valproate in migraine prophylaxis [157], although it is clear that use of the drug acutely for treatment of a given migraine episode is not effective. Average daily doses utilized in these migraine efficacy studies ranged from 500 mg/day to 2000 mg/day.

Paroxysmal pain syndromes

Several open observations [160–162] and one controlled study [163] suggest the potential efficacy of valproate for a variety of paroxysmal pain syndromes. While carbamazepine has been the treatment of choice for the prototypical lancinating pain syndromes, trigeminal [164] and glossopharyngeal [165] neuralgia, it is associated with a high degree of tolerance development [166]. Whether there is cross-tolerance to valproate for this targeted pain component, as there appears to be for anticonvulsant effects in the amygdala kindling preparation, remains to be shown. Pimozide appears to be effective in many patients who have lost responsivity to carbamazepine [168], and the potential efficacy of valproate in such patients should also be examined further.

Conclusion

Valproate has rapidly become one of the first-line treatments for bipolar illness, with proven efficacy in acute mania as well as in the prevention of manic and depressive recurrences, based on several open but extensive series. Valproate's broad spectrum of anticonvulsant effects is matched by a broad spectrum of psychotropic effects that make it particularly attractive in treating the many comorbidities of bipolar affective illness.

References

1 Lambert PA (1984) Acute and prophylactic therapies of patients with affective disorders using valpromide (dipropylacetamide). In: HM Emrich, T Okuma, AA Muller (eds): *Anticonvulsants in affective disorders*. Excerpta Medica, Amsterdam, 33–44
2 Emrich HM, Von Zerssen D, Kissling W, Moller H-J, Windorfer A (1980) Effect of sodium valproate in mania. The GABA hypothesis of affective disorders. *Arch Psychiatr Nervenkr* 229: 1–16
3 Regier DA, Farmer ME, Rae DS, Locke BZ, Keith SJ, Judd LL, Goodwin FK (1990) Comorbidity of mental disorders with alcohol and other drug abuse. Results from the Epidemiologic Catchment Area (ECA) Study. *JAMA* 264: 2511–2518
4 Kessler RC, Rubinow DR, Holmes C, Abelson JM, Zhao S (1997) The epidemiology of DSM-III-R bipolar I disorder in a general population survey. *Psychol Med* 27: 1079–1089
5 Weissman MM, Bruce ML, Leaf PJ, Florio LP, Holzer C (1991) Affective disorders. In: LN Robins, DA Regier (eds): *Psychiatric disorders in america. The epidemiologic catchment area study*. The Free Press, New York, 53–80
6 McElroy SL, Altshuler L, Suppes T, Keck PE, Frye MA, Denicoff KD, Nolen WA, Kupka R, Leverich GS, Rochussen J, et al. (1999) Axis I psychiatric comorbidity in a cohort of patients with bipolar disorder. *Am J Psychiatry*; submitted
7 Baastrup PC, Schou M (1967) Lithium as a prophylactic agent: its effect against recurrent depressions and manic-depressive psychosis. *Arch Gen Psychiatry* 16: 162–172
8 Prien RF, Himmelhoch JM, Kupfer DJ (1988) Treatment of mixed mania. *J Affective Disord* 15: 9–15
9 Harrow M, Goldberg JF, Grossman LS, Meltzer HY (1990) Outcome in manic disorders. A naturalistic follow-up study. *Arch Gen Psychiatry* 47: 665–671
10 Gitlin MJ, Swendsen J, Heller TL, Hammen C (1995) Relapse and impairment in bipolar disorder. *Am J Psychiatry* 152: 1635–1640
11 Swann AC, Bowden CL, Morris D, Calabrese JR, Petty F, Small J, Dilsaver SC, Davis JM (1997) Depression during mania. Treatment response to lithium or divalproex. *Arch Gen Psychiatry* 54: 37–42
12 Vestergaard P (1992) Treatment and prevention of mania: A Scandinavian perspective. *Neuropsychopharmacol* 7: 249–260
13 Goldberg JF, Harrow M, Sands JR (1996) Lithium and the longitudinal course of bipolar illness. *Psychiatric Ann* 26: 651–658
14 Denicoff KD, Smith-Jackson EE, Disney ER, Ali SO, Leverich GS, Post RM (1997) Comparative prophylactic efficacy of lithium, carbamazepine, and the combination in bipolar disorder. *J Clin Psychiatry* 58: 470–478
15 O'Connell RA, Mayo JA, Flatow L, Cuthbertson B, O'Brien BE (1991) Outcome of bipolar disorder on long-term treatment with lithium. *Br J Psychiatry* 159: 123–129
16 Calabrese JR, Delucchi GA (1989) Phenomenology of rapid cycling manic depression and its treatment with valproate. *J Clin Psychiatry* 50: 30–34
17 Dunner DL, Fieve RR (1974) Clinical factors in lithium carbonate prophylaxis failure. *Arch Gen Psychiatry* 30: 229–233
18 Dunner DL, Patrick V, Fieve RR (1977) Rapid cycling manic depressive patients. *Compr Psychiatry* 18: 561–566
19 Post RM (1990) Alternatives to lithium for bipolar affective illness. In: A Tasman, SM Goldfinger, CA Kaufmann (eds): *Review of Psychiatry*, Vol. 9, American Psychiatric Press, Inc., Washington, DC, 170–202
20 Post RM (1990) Prophylaxis of bipolar affective disorder. *Int Rev J Psychiatry* 2: 277–320
21 Goodwin FK, Jamison KR (1990) *Manic-depressive illness*. Oxford University Press, New York
22 Emrich HM, Dose M, Von Zerssen D (1985) The use of sodium valproate, carbamazepine and oxcarbazepine in patients with affective disorders. *J Affective Disorders* 8: 243–250
23 Brennan MJW, Sandyk R, Borseek D (1984) Use of sodium-valproate in the management of affective disorders: Basic and clinical aspects. In: HM Emrich, T Okuma, AA Muller (eds): *Anticonvulsants in Affective Disorders*, Excerpta Medica, Amsterdam, 56–65

24 Post RM, Trimble MR, Pippenger CE (eds) (1989) *Clinical use of anticonvulsants in psychiatric disorders*. Demos Publications, New York
25 Freeman TW, Clothier JL, Pazzaglia P, Lesem MD, Swann AC (1992) A double-blind comparison of valproate and lithium in the treatment of acute mania. *Am J Psychiatry* 149: 108–111
26 Pope HG, McElroy SL, Keck PE, Jr., Hudson JI (1991) Valproate in the treatment of acute mania. *Arch Gen Psychiatry* 48: 62–68
27 Bowden CL, Brugger AM, Swann AC, Calabrese JR, Janicak PG, Petty F, Dilsaver SC, Davis JM, Rush AJ, Small JG (1994) Efficacy of divalproex vs lithium and placebo in the treatment of mania. The Depakote Mania Study Group. *JAMA* 271: 918–924
28 Keck PE, Jr., McElroy SL, Tugrul KC, Bennett JA (1993) Valproate oral loading in the treatment of acute mania. *J Clin Psychiatry* 54: 305–308
29 Martinez JM, Russell JM, Hirschfeld RM (1998) Tolerability of oral loading of divalproex sodium in the treatment of acute mania. *Depress Anxiety* 7: 83–86
30 Alvir JM, Lieberman JA, Safferman AZ, Schwimmer JL, Schaaf JA (1993) Clozapine-induced agranulocytosis. Incidence and risk factors in the United States. *N Engl J Med* 329: 162–167
31 Sobotka JL, Alexander B, Cook BL (1990) A review of carbamazepine's hematologic reactions and monitoring recommendations. *DICP* 24: 1214–1219
32 Stevens JR, Denney D, Szot P (1997) Sensitization with clozapine: beyond the dopamine hypothesis. *Biol Psychiatry* 42: 771–780
33 Denney D, Stevens JR (1995) Clozapine and seizures. *Biol Psychiatry* 37: 427–433
34 Himmelhoch JM, Garfinkel ME (1986) Sources of lithium resistance in mixed mania. *Psychopharmacol Bull* 22: 613–620
35 Secunda SK, Swann A, Katz MM, Koslow SH, Croughan J, Chang S (1987) Diagnosis and treatment of mixed mania. *Am J Psychiatry* 144: 96–98
36 Hayes SG (1989) Long-term use of valproate in primary psychiatric disorders. *J Clin Psychiatry* 50, Supplement: 35–39
37 Calabrese JR, Delucchi GA (1990) Spectrum of efficacy of valproate in 55 patients with rapid-cycling bipolar disorder. *Am J Psychiatry* 147: 431–434
38 Calabrese JR, Markovitz PJ, Kimmel SE, Wagner SC (1992) Spectrum of efficacy of valproate in 78 rapid-cycling bipolar patients. *J Clin Psychopharmacol* 12: 53S–56S
39 Clothier J, Swann AC, Freeman T (1992) Dysphoric mania. *J Clin Psychopharmacol* 12: 13S–16S
40 McElroy SL, Pope HG, Jr., Keck PE, Jr., Hudson JI (1988b) Treatment of psychiatric disorders with valproate: A series of 73 cases. *Psychiatrie & Psychobiologie* 3: 81–85
41 McCoy L, Votolato NA, Schwarzkopf SB, Nasrallah HA (1993) Clinical correlates of valproate augmentation in refractory bipolar disorder. *Ann Clin Psychiatry* 5: 29–33
42 McElroy SL, Keck PE, Jr., Pope HG, Jr., Hudson JI (1988a) Valproate in the treatment of rapid-cycling bipolar disorder. *J Clin Psychopharmacol* 8: 275–279
43 Puzynski S, Klosiewicz L (1984) Valproic acid amide in the treatment of affective and schizoaffective disorders. *J Affective Disorders* 6: 115–121
44 Calabrese JR, Woyshville MJ, Rapport RT (1994) Clinical efficacy of valproate. In: RT Joffe, JR Calabrese (eds): *Anticonvulsants in mood disorders*. Marcel Dekker, Inc., New York, 131–146
45 Winsberg M, Degolia SG, Strong CM, Ketter TA (1997) Divalproex in medication-naive bipolar II depression. *APA New Research Program and Abstracts*, Abstract NR113: 97
46 Davis LL, Kabel D, Patel D, Choate AD, Foslien-Nash C, Gurguis GN, Kramer GL, Petty F (1996) Valproate as an antidepressant in major depressive disorder. *Psychopharmacol Bull* 32: 647–652
47 Calabrese JR, Woyshville MJ, Kimmel SE, Rapport DJ (1993b) Predictors of valproate response in bipolar rapid cycling. *J Clin Psychopharmacol* 13: 280–283
48 Kukopulos A, Reginaldi D, Laddomada P, Floris G, Serra G, Tondo L (1980) Course of the manic-depressive cycle and changes caused by treatment. *Pharmakopsychiatria Neuropsychopharmakol* 13: 156–167
49 Emrich HM (1990) Alternatives to lithium prophylaxis for affective and schizoaffective disorders. In: A Marneros, MT Tsuang (eds): *Affective and schizoaffective disorders*. Springer-Verlag, Berlin, 262–273

50 McElroy SL, Keck PE, Pope HG, Hudson JI (1992) Valproate in the treatment of bipolar disorder: Literature review and clinical guidelines. *J Clin Psychopharmacol* 12: 42S–52S
51 Hayes SG (1992) Long-term valproate prophylaxis in refractory affective disorders. *Ann Clin Psychiatry* 4: 55–63
52 Schaff MR, Fawcett J, Zajecka JM (1993) Divalproex sodium in the treatment of refractory affective disorders. *J Clin Psychiatry* 54: 380–384
53 Lambert PA, Venaud G (1995) Comparative study of valpromide vs lithium as prophylactic treatments in affective disorders. *Nervure J Psychiatrie* 1–9
54 Bowden CL, Calabrese JR, McElroy SL, Hirschfeld RM, Petty F, Gyulai L (1998) Maintenance treatment in bipolar disorder. *Syllabus and Proceedings Summary of the 151st Annual Meeting of the American Psychiatric Association*, Abstract No. 6C
55 Calabrese JR (1998a) Options for treatment-refractory rapid cycling. *Syllabus and Proceedings Summary of the 151st Annual Meeting of the American Psychiatric Association*, Abstract No. 12C: 287
56 Calabrese JR (1998b) Anticonvulsants in bipolar disorder. *Syllabus and Proceedings Summary of the 151st Annual Meeting of the American Psychiatric Association*, Abstract No. 40B: 324
57 Frye MA, Denicoff KD, Bryan A, Smith-Jackson E, Ali SO, Luckenbaugh D, Leverich GS, Post RM (1997) Longitudinal assessment of thyroid function and mood stability in manic depressive illness. *Proceedings of the Second International Conference on Bipolar Disorder*, Abstract 16
58 Denicoff KD, Smith-Jackson EE, Bryan AL, Ali SO, Post RM (1997) Valproate prophylaxis in a prospective clinical trial of refractory bipolar disorder. *Am J Psychiatry* 154: 1456–1458
59 Keck PE, Jr., McElroy SL, Vuckovic A, Friedman LM (1992) Combined valproate and carbamazepine treatment of bipolar disorder. *J Neuropsychiatry Clin Neurosci* 4: 319–322
60 Ketter TA, Pazzaglia PJ, Post RM (1992) Synergy of carbamazepine and valproic acid in affective illness: Case report and review of literature. *J Clin Psychopharmacol* 12: 276–281
61 Gupta AK, Jeavons PM (1985) Complex partial seizures: EEG foci and response to carbamazepine and sodium valproate. *J Neurol Neurosurg Psychiatry* 48: 1010–1014
62 Callaghan N, Goggin T (1988) Adjunctive therapy in resistant epilepsy. *Epilepsia* 29: S29-S35
63 Dean JC, Penry JK (1988) Carbamazepine/valproate therapy in 100 patients with partial seizures failing carbamazepine monotherapy: long term follow-up. *Epilepsia* 29: 687
64 Walker JE, Koon R (1988) Carbamazepine versus valproate vs combined therapy for refractory partial complex seizures with secondary generalization. *Epilepsia* 29: 693
65 Ashok PP, Maheshwari MC (1984) Role of combination of valproic acid with diphenylhydantoin and carbamazepine in the management of intrac Table seizures. *J Assoc Physicians India* 32: 565–567
66 Post RM, Weiss SRB, Chuang D-M (1992) Mechanisms of action of anticonvulsants in affective disorders: comparisons with lithium. *J Clin Psychopharmacol* 12: 23S–35S
67 Bernasconi R (1982) The GABA hypothesis of affective illness: influence of clinically effective antimanic drugs on GABA turnover. In: *Basic mechanisms in the action of lithium* – proceedings of a symposium held at Schloss Ringberg, Bavaria, October 4–6, 1981. Excerpta Medica, Amsterdam, 183–192
68 Bernasconi R, Hauser K, Martin P, Schmutz M (1984) Biochemical aspects of the mechanism of action of valproate. In: HM Emrich, T Okuma, AA Muller (eds): *Anticonvulsants in affective disorders*. Excerpta Medica, Amsterdam, 14–32
69 Motohashi N, Ikawa K, Kariya T (1989) $GABA_B$ receptors are up-regulated by chronic treatment with lithium or carbamazepine. GABA hypothesis of affective disorders. *Eur J Pharmacol* 166: 95–99
70 Motohashi N (1990) GABA receptor alterations after chronic lithium – in comparison with carbamazepine and sodium valproate. *Clin Neuropharmacol* 13: 207–208
71 Nagaki S, Kato N, Minatogawa Y, Higuchi T (1990) Effects of anticonvulsants and gamma-aminobutyric acid (GABA)-mimetic drugs on immunoreactive somatostatin and GABA contents in the rat brain. *Life Sci* 46: 1587–1595

72 Olpe HR (1991) Mechanism of action of antiepileptic drugs with special reference to carbamazepine. *Abstracts, World Congress of Biological Psychiatry*, 185
73 Leeb-Lundberg F, Snowman A, Olsen RW (1981) Some anticonvulsants interact with the GABA receptor-ionophore complex at barbiturate/picrotoxin receptor sites. *Fed Proc* 40: 309
74 Ticku MK, Davis WC (1981) Effect of valproic acid on [^{3}H]diazepam and [^{3}H]dihydropicrotoxinin binding sites at the benzodiazepine-GABA-receptor-ionophore complex. *Brain Research* 223: 218–222
75 Maitre L, Baltzer V, Mondadori C (1984) Psychopharmacological and behavioural effects of anti-epileptic drugs in animals. In: HM Emrich, T Okuma, AA Muller (eds): *Anticonvulsants in affective disorders*. Excerpta Medica, Amsterdam, 3–13
76 Waldmeier PC, Baumann PA, Fehr B (1984) Carbamazepine decreases catecholamine turnover in the rat brain. *J Pharmacol Exp Ther* 231: 166–172
77 McLean MH, Macdonald RL (1986) Sodium valproate, but not ethosuximide, produces use- and voltage-dependent limitation of high frequency repetitive firing of action potentials of mouse central neurons in cell culture. *J Pharmacol Exp Ther* 237: 1001–1010
78 Zona C, Avoli M (1990) Effects induced by the antiepileptic drug valproic acid upon the ionic currents recorded in rat neocortical neurons in cell culture. *Exp Brain Res* 81: 313–317
79 Zona C, Tancredi V, Palma E, Pirrone GC, Avoli M (1990) Potassium currents in rat cortical neurons in culture are enhanced by the antiepileptic drug carbamazepine. *Can J Physiol Pharmacol* 68: 545–547
80 Olpe H, Kolb CN, Hausdorf A, Haas HL (1991) 4-aminopyridine and barium chloride attenuate the anti-epileptic effect of carbamazepine in hippocampal slices. *Experientia* 47: 254–257
81 Slater GE, Johnston D (1978) Sodium valproate increases potassium conductance in aplysia neurones. *Epilepsia* 19: 379–384
82 Walden J, Altrup U, Reith H, Speckmann EJ (1991) Valproate induced increase of transmembraneous potassium currents. *Abstracts, World Congress of Biological Psychiatry*, 248
83 Lampe H, Bigalke H (1990) Carbamazepine blocks NMDA-activated currents in cultured spinal cord neurons. *NeuroReport* 1: 26–28
84 Zeise ML, Kasparow S, Zieglgansberger W (1991) Valproate suppresses N-methyl-D-aspartate-evoked, transient depolarizations in the rat neocortex in vitro. *Brain Research* 544: 345–348
85 Weiss SRB, Nguyen T, Rubinow DR, Helke CJ, Narang PK, Post RM, Jacobowitz DM (1987) Lack of effect of chronic carbamazepine on brain somatostatin in the rat. *J Neural Transm* 68: 325–333
86 Wahnschaffe U, Loscher W (1990) Effect of selective bilateral destruction of the substantia nigra on antiepileptic drug actions in kindled rats. *Eur J Pharmacol* 186: 157–167
87 Kubota T, Jibiki I, Hirose S, Yamaguchi N (1990) Comparative experimental study of antiepileptics-regional specificity of antiepileptic action of carbamazepine, phenobarbital and valproate sodium. *Brain Dev* 12: 503–508
88 Maj M, Pirozzi R, Kemali D (1989) Long-term outcome of lithium prophylaxis in patients initially classified as complete responders. *Psychopharmacology* 98: 535–538
89 Post RM, Leverich GS, Pazzaglia PJ, Mikalauskas K, Denicoff K (1993) Lithium tolerance and discontinuation as pathways to refractoriness. In: NJ Birch, C Padgham, MS Hughes (eds): *Lithium in Medicine and Biology*, 1 Ed., Marius Press, Lancashire, UK, 71–84
90 Post RM, Leverich GS, Rosoff AS, Altshuler LL (1990) Carbamazepine prophylaxis in refractory affective disorders: a focus on long-term follow-up. *J Clin Psychopharmacol* 10: 318–327
91 Leverich GS, Post RM, Frye M, Kimbrell T (1999) Long-term efficacy of carbamazepine in affective illness: The problem of tolerance. Unpublished manuscript
92 Post RM (1990) Prophylaxis of bipolar affective disorders. *Int'l Rev Psych* 2: 277–320
93 Post RM, Weiss SRB (1996) A speculative model of affective illness cyclicity based on patterns of drug tolerance observed in amygdala-kindled seizures. *Mol Neurobiology* 13: 33–60
94 Weiss SRB, Clark M, Rosen JB, Smith MA, Post RM (1995) Contingent tolerance to the anticonvulsant effects of carbamazepine: relationship to loss of endogenous adaptive mechanisms. *Brain Research Reviews* 20: 305–325

95 Weiss SRB, Post RM (1991) Development and reversal of contingent inefficacy and tolerance to the anticonvulsant effects of carbamazepine. *Epilepsia* 32: 140–145
96 Weiss SRB, Post RM, Sohn E, Berger A, Lewis R (1993) Cross tolerance between carbamazepine and valproate on amygdala-kindled seizures. *Epilepsy Res* 16: 37–44
97 Clark M, Massenburg GS, Weiss SRB, Post RM (1994) Analysis of the hippocampal GABA A receptor system in kindled rats by autoradiographic and *in situ* hybridization techniques: Contingent tolerance to carbamazepine. *Mol Brain Res* 26: 309–319
98 Post RM, Weiss SRB (1994) Kindling: Implications for the course and treatment of affective disorders. In: K Modigh, OH Robak, T Vestergaard (eds): *Anticonvulsants in Psychiatry*. Wrightson Biomedical Publishing Ltd., Stroud, England, 113–137
99 Weiss SRB, Post RM (1990) Contingent tolerance to the anticonvulsant effects of carbamazepine: implications for neurology and psychiatry. In: R Canger, E Sacchetti, GI Perini, MP Canevini (eds): *Carbamazepine: A bridge between epilepsy and Psychiatric disorders*. Ciba-Geigy Edizioni, Origgio, 7–32
100 Mana MJ, Kim CK, Pinel JP, Jones CH (1992) Contingent tolerance to the anticonvulsant effects of carbamazepine, diazepam, and sodium valproate in kindled rats. *Pharmacol Biochem Behav* 41: 121–126
101 Weiss SRB, Post RM (1995) Caveats in the use of the kindling model of affective disorders. *J Toxicol Industr Health* 10: 421–447
102 Pinel JPJ (1983) Effects of diazepam and diphenylhydantoin on elicited and spontaneous seizures in kindled rats: a double dissociation. *Pharmacol Biochem Behav* 18: 61–63
103 Young B, Rapp RP, Norton JA, Haack D, Walsh JW (1983) Failure of prophylactically administered phenytoin to prevent post-traumatic seizures in children. *Childs Brain* 10: 185–192
104 Jacobsen FM (1993) Low-dose valproate: a new treatment for cyclothymia, mild rapid cycling disorders, and premenstrual syndrome. *J Clin Psychiatry* 54: 229–234
105 Rey AC, Jimerson DC, Post RM (1979) Lithium and electrolytes in cerebrospinal fluid of affectively ill patients during acute and chronic lithium treatment. *Commun Psychopharmacol* 3: 267–278
106 Post RM, Uhde TW, Rubinow DR, Huggins T (1987) Differential time course of antidepressant effects following sleep deprivation, ECT, and carbamazepine: clinical and theoretical implications. *Psychiatry Res* 22: 11–19
107 Ketter TA, Frye MA, Cora-Locatelli G, Kimbrell TA, Post RM (1999) Metabolism and excretion of mood stabilizers and new anticonvulsants. *Cell Mol Neurobiol*; submitted
108 Sovner R, Davis JM (1991) A potential drug interaction between fluoxetine and valproic acid [letter]. *J Clin Psychopharmacol* 11: 389
109 Ketter TA, Jenkins JB, Schroeder DH, Pazzaglia PJ, Marangell LB, George MS, Callahan AM, Hinton ML, Chao J, Post RM (1995) Carbamazepine but not valproate induces bupropion metabolism. *J Clin Psychopharmacol* 15: 327–333
110 Crawford P, Chadwick D, Cleland P, Tjia J, Cowie A, Back DJ, Orme ML (1986) The lack of effect of sodium valproate on the pharmacokinetics of oral contraceptive steroids. *Contraception* 33: 23–29
111 Perucca E, Grimaldi R, Gatti G, Pirracchio S, Crema F, Frigo GM (1984) Pharmacokinetics of valproic acid in the elderly. *Br J Clin Pharmacol* 17: 665–669
112 Rapport DJ, Calabrese JR (1989) Interactions between carbamazepine and birth control pills [letter]. *Psychosomatics* 30: 462–464
113 McElroy SL, Keck PE (1994) Experience with valproate in affective disorders. In: K Modigh, OH Robak, P Vestergaard (eds): *Anticonvulsants in psychiatry*. Wrightson Biomedical Publishing Ltd., Great Britain, 57–72
114 Rimmer EM, Richens A (1985) An update on sodium valproate. *Pharmacotherapy* 5: 171–184
115 Corman CL, Leung NM, Guberman AH (1997) Weight gain in epileptic patients during treatment with valproic acid: a retrospective study. *Can J Neurol Sci* 24: 240–244
116 Smith MC, Bleck TP (1991) Convulsive disorders: toxicity of anticonvulsants. *Clin Neuropharmacol* 14: 97–115
117 Jeavons PM, Clark JE, Harding GF (1977) Valproate and curly hair [letter]. *Lancet* 1: 359
118 Hurd RW, Van Rinsvelt HA, Wilder BJ, Karas B, Maenhaut W, De Reu L (1984) Selenium, zinc, and copper changes with valproic acid: possible relation to drug side effects. *Neurology* 34: 1393–1395

119 Asconape J, Penry JK, Dreifuss F, Riela A, Mirza W (1993) Valproate-associated pancreatitis. *Epilepsia* 34: 177–183
120 Dreifuss FE, Santilli N, Langer DH, Sweeney KP, Moline KA, Menander KB (1987) Valproic acid hepatic fatalities: a retrospective review. *Neurology* 37: 379–385
121 Dreifuss FE, Langer DH, Moline KA, Maxwell JE (1989) Valproic acid hepatic fatalities. II. US experience since 1984. *Neurol* 39: 201–207
122 Omtzigt JGC, Los FJ, Grobbee DE, Pijpers L, Jahoda MGJ, Brandenburg H, Stewart PA, Gaillard HLJ, Sachs ES, Wladimiroff JW, et al. (1992) The risk of spina bifida aperta after first-trimester exposure to valproate in a prenatal cohort. *Neurology* 42: 119–125
123 Jeavons PM (1984) Non-dose-related side effects of valproate. *Epilepsia* 25 Suppl 1: S50-S55
124 Anonymous (1991) From the Centers for Disease Control. Use of folic acid for prevention of spina bifida and other neural tube defects 1983–1991. *JAMA* 266: 1190–1191
125 Delgado-Escueta AV, Janz D (1992) Consensus guidelines: preconception counseling, management, and care of the pregnant woman with epilepsy. *Neurology* 42: 149–160
126 Beghi E (1986) Adverse reactions to antiepileptic drugs: a multicenter survey of clinical practice. *Epilepsia* 27: 323–330
127 Vining EP (1987) Cognitive dysfunction associated with antiepileptic drug therapy. *Epilepsia* 28 Suppl 2: S18–S22
128 Pedersen B, Juul-Jensen P (1984) Electroencephalographic alterations during intoxication with sodium valproate: a case report. *Epilepsia* 25: 121–124
129 Hulsman J (1989) Hyperammonemia and use of antiepileptic drugs including valproate. *Epilepsia* 30: 647–648
130 Bohles H, Sewell AC, Wenzel D (1996) The effect of carnitine supplementation in valproate-induced hyperammonaemia. *Acta Paediatr* 85: 446–449
131 Gidal BE, Inglese CM, Meyer JF, Pitterle ME, Antonopolous J, Rust RS (1997) Diet- and valproate-induced transient hyperammonemia: effect of L-carnitine. *Pediatr Neurol* 16: 301–305
132 Stephens JR, Levy RH (1994) Effects of valproate and citrulline on ammonium-induced encephalopathy. *Epilepsia* 35: 164–171
133 West SA, McElroy SL, Keck PE (1996) Valproate. In: PJ Goodnick (ed): *Predictors of treatment response in mood disorders*. American Psychiatric Press, Inc., Washington, DC, 133–146
134 Bowden CL (1995) Predictors of response to divalproex and lithium. *J Clin Psychiatry* 56 Suppl 3: 25–30
135 Calabrese JR, Rapport DJ, Kimmel SE, Reece B, Woyshville MJ (1993 a) Rapid cycling bipolar disorder and its treatment with valproate. *Can J Psychiatry* 38: S57–S61
136 McElroy SL, Keck PEJ, Pope HGJ, Hudson JI, Morris D (1991b) Correlates of antimanic response to valproate. *Psychopharmacol Bull* 27: 127–133
137 Fogelson DL, Jacobson S, Sternbach H (1991) A retrospective study of valproate in private psychiatric practice. *Ann Clin Psychiatry* 3: 315–320
138 Sovner R (1989) The use of valproate in the treatment of mentally retarded persons with typical and atypical bipolar disorders. *J Clin Psychiatry* 50: 3 (Suppl): 40–43
139 Kahn D, Stevenson E, Douglas CJ (1988) Effect of sodium valproate in three patients with organic brain syndromes. *Am J Psychiatry* 145: 1010–1011
140 Pope HGJ, McElroy SL, Satlin A, Hudson JI, Keck PEJ, Kalish R (1988) Head injury, bipolar disorder, and response to valproate. *Compr Psychiatry* 29: 34–38
141 Bowden CL, Janicak PG, Orsulak P, Swann AC, Davis JM, Calabrese JR, Goodnick P, Small JG, Rush AJ, Kimmel SE, et al. (1996) Relation of serum valproate concentration to response in mania. *Am J Psychiatry* 153: 765–770
142 McElroy SL, Keck PEJ (1993) Treatment guidelines for valproate in bipolar and schizoaffective disorders. *Can J Psychiatry* 38: S62–S66
143 Keck PEJ, Taylor VE, Tugrul KC, McElroy SL, Bennett JA (1993 b) Valproate treatment of panic disorder and lactate-induced panic attacks. *Biol Psychiatry* 33: 542–546
144 Roy-Byrne PP (1989) Anticonvulsants in anxiety and withdrawal syndromes: hypotheses for future research. In: HG Pope, S McElroy (eds): *Use of anticonvulsants in psychiatry: recent advances*, Raven Press, New York, 155–168
145 McElroy SL, Keck PEJ, Lawrence JM (1991a) Treatment of panic disorder and benzodiazepine withdrawal with valproate [letter]. *J Neuropsychiatry.Clin Neurosci* 3: 232–233

146 Primeau F, Fontaine R, Beauclair L (1990) Valproic acid and panic disorder. *Can J Psychiatry* 35: 248–250
147 Woodman CL, Noyes RJ (1994) Panic disorder: treatment with valproate. *J Clin Psychiatry* 55: 134–136
148 Lum M, Fontaine R, Elie R, Ontiveros A (1991) Probable interaction of sodium divalproex with benzodiazepines. *Prog Neuropsychopharmacol Biol Psychiatry* 15: 269–273
149 Lipper S, Davidson JRT, Grady TA, Edinger JD, Hammett EB, Mahorney SL, Cavenar JO, Jr. (1986) Preliminary study of carbamazepine in post-traumatic stress disorder. *Psychosomatics* 27: 849–854
150 Wolf ME, Alavi A, Mosnaim AD (1988) Posttraumatic stress disorder in Vietnam veterans clinical and EEG findings; possible therapeutic effects of carbamazepine. *Biol Psychiatry* 23: 642–644
151 Stewart JT, Bartucci RJ (1986) Posttraumatic stress disorder and partial complex seizures [letter]. *Am J Psychiatry* 143: 113–114
152 Fesler FA (1991) Valproate in combat-related posttraumatic stress disorder. *J Clin Psychiatry* 52: 361–364
153 Berigan TR, Holzgang A (1995) Valproate as an alternative in post-traumatic stress disorder: a case report. *Mil Med* 160: 318
154 Szymanski HV, Olympia J (1991) Divalproex in posttraumatic stress disorder [letter]. *Am J Psychiatry* 148: 1086–1087
155 Ford N (1996) The use of anticonvulsants in posttraumatic stress disorder: case study and overview. *J Trauma Stress* 9: 857–863
156 Merikangas KR, Merikangas JR, Angst J (1993) Headache syndromes and psychiatric disorders: association and familial transmission. *J Psychiatr Res* 27: 197–210
157 Jensen R, Brinck T, Olesen J (1994) Sodium valproate has a prophylactic effect in migraine without aura. *Neurol* 44: 647
158 Klapper J (1997) Divalproex sodium in migraine prophylaxis: a dose-controlled study. *Cephalalgia* 17: 103–108
159 Kaniecki RG (1997) A comparison of divalproex with propranolol and placebo for the prophylaxis of migraine without aura. *Arch Neurol* 54: 1141–1145
160 Peiris JB, Perera GL, Devendra SV, Lionel ND (1980) Sodium valproate in trigeminal neuralgia. *Med J Aust* 2: 278
161 Raftery H (1979) The management of post herpetic pain using sodium valproate and amitriptyline. *Ir Med J* 72: 399–401
162 Karlov VA, Savitskaia ON (1980) Comparative effectiveness of antiepileptic preparations in the treatment of patients with trigeminal neuralgia [in Russian]. *Zh Nevropatol Psikhiatr In S S Korsakova* 80: 530–535
163 Desai N, Shah K, Gandhi I (1991) Baclofen sodium valproate combination in carbamazepine resistant trigeminal neuralgia: a double blind clinical trial. *Cephalalgia* June: 321–322
164 Green MW, Selman JE (1991) Review article: the medical management of trigeminal neuralgia. *Headache* 31: 588–592
165 Dalessio DJ (1991) Diagnosis and treatment of cranial neuralgias. *Med Clin North Am* 75: 605–615
166 Pagni CA (1993) The origin of tic douloureux: a unified view. *J Neurosurg Sci* 37: 185–194
167 Taylor JC, Brauer S, Espir ML (1981) Long-term treatment of trigeminal neuralgia with carbamazepine. *Postgrad Med J* 57: 16–18
168 Lechin F, Van der Dijs B, Lechin ME, Amat J, Lechin AE, Cabrera A, Gomez F, Acosta E, Arocha L, Villa S (1989) Pimozide therapy for trigeminal neuralgia. *Arch Neurol* 46: 960–963

Milestones in Drug Therapy
Valproate
ed. by W. Löscher

Valproate in the treatment of headache

Stephen D. Silberstein

Thomas Jefferson University School of Medicine, Jefferson Headache Center, 111 South Eleventh Street, Gibbon Building, Suite #8130, Philadelphia, PA 19107, USA

Introduction

Valproic acid is commercially available in the United States in three preparations: valproic acid, sodium valproate (Depakene syrup), and divalproex sodium (Depakote), an enteric-coated, stable coordination compound containing equal proportions of valproic acid and sodium valproate [1]. In this chapter the term valproate will be used for these formulations [2]. The amide of valproic acid, valpromide (Depamide), is available in Europe. Additionally, an extended-release form (Depakote-SR) has just become available; this should allow once-daily dosing and decrease peak related side-effects such as tremor and dyspepsia.

Valproate is an anticonvulsant agent approved for use either alone or in combination with other antiepileptic drugs for simple and complex absence seizures and as adjunctive therapy in patients with multiple seizure types, including absence seizures [3]. It is useful and well-tolerated in the manic phase of bipolar and schizoaffective disorder, even in patients who have been unable to tolerate or failed to respond to lithium [4]. Divalproex is approved for mania and migraine in the United States by the Food and Drug Administration. Valproate use in epilepsy and mania are discussed in other chapters.

Four double-blind placebo-controlled studies [5–8] have confirmed that divalproex sodium/valproate is an effective migraine treatment. Some investigators have suggested that divalproex sodium is especially useful for treating migraineurs who also have mania, seizures, aggressive behavior, or an anxiety disorder [4, 9]. In addition, studies have shown that it is effective for chronic daily headaches [10] and cluster headaches [11]. The most common, early, self-limited adverse events (AEs) were nausea and vomiting. Tremor, weight gain, and hair loss may occur after longer use. The most serious, although rare, idiosyncratic reaction is hepatotoxicity. Valproate is also teratogenic. This paper will review the pharmacology of valproate, as well as the clinical studies that have been performed, and the associated AEs. It will suggest monitoring procedures and present an approach to treating migraine with valproate.

Migraine treatment

The pharmacologic treatment of migraine may be acute (abortive, symptomatic) or preventive (prophylactic) [12], and patients who are experiencing frequent severe headaches often require both approaches. Symptomatic treatment attempts to abort (reverse or stop the progression of) a headache once it has started. Preventive therapy is given, even in the absence of a headache, to reduce the frequency and severity of anticipated attacks. Symptomatic treatment is appropriate for most acute attacks, even in patients who are on preventive medication, and should be used 2 to 3 days a week at most. Preventive treatment is used more selectively. For example, treatment strategies should emphasize decreasing the frequency of frequent attacks. Preventive treatments include a broad range of medications, most notably beta-blockers, calcium channel blockers, antidepressants, serotonin antagonists, anticonvulsants, and nonsteroidal antiinflammatory drugs (NSAIDs).

Table 1. Choices of preventive treatment in migraine: Influence of comorbid conditions

Drug	Efficacy*	Side-effects*	Comorbid condition	
			Relative contraindication	Relative indication
Beta-blockers	4+	2+	Asthma, depression, CHF, Raynaud's disease, diabetes	HTN, angina
Antiserotonin				
Methysergide	4+	4+	Angina, PVD	Orthostatic hypotension
Calcium channel blockers				
Verapamil	2+	1+	Hypotension, Constipation	HTN, angina, asthma Migraine with aura
Antidepressants				
TCAs	4+	2+	Mania, urinary retention, heart block	Other pain disorders, depession, anxiety disorders, insomnia
SSRIs	2+	1+	Mania	Depression, OCD
MAOIs	3+	4+	Unreliable patients	Refractory depression
Anticonvulsants				
Divalproex	4+	2+	Liver disease, bleeding disorders	Mania, epilepsy, anxiety disorders
NSAIDs				
Naproxen	2+	2+	Ulcer disease, gastritis	Arthritis, other pain disorders

* Ratings are on a scale from 1+ (lowest) to 4+ (highest).
CHF, Congestive heart failure; HTN, Hypertension; MAOIs, Monoamine oxidase inhibitors; NSAIDs, Nonsteroidal anti-inflammatory drugs; OCD, Obsessive compulsive disorder; PVD, Peripheral vascular disease; SSRI, Serotonin specific reuptake inhibitor; TCA, Tricyclic antidepressants.

Preventive medication is usually given daily for months or years; however, treatment can be episodic, subacute, or chronic. Episodic treatment is used when there is a known headache trigger, such as exercise or sexual activity, and these patients can be instructed to pretreat prior to the exposure or activity. Patients who are undergoing a time-limited exposure to a trigger, such as ascent to a high altitude, or a reduced migraine threshold such as menstruation, can be treated subacutely by having them medicate before and during the exposure. Preventive medication can also be taken on a regular basis (chronic treatment) to decrease the frequency of migraine attacks [13].

Circumstances that warrant chronic treatment include: (1) two or more attacks a month that produce disability that lasts 3 or more days; (2) contraindication to, or ineffectiveness of, symptomatic medications; (3) the use of abortive medication more than twice a week; or (4) special circumstances, such as hemiplegic migraine or rare headache attacks producing profound disruption or risk of permanent neurologic injury. These rules are more strict during pregnancy, during which time severe disabling attacks accompanied by nausea, vomiting, and possibly dehydration are required for chronic treatment to be prescribed [14].

The major medication groups for preventive migraine treatment include β-adrenergic blockers, antidepressants, calcium channel antagonists, serotonin (5-HT) antagonists, anticonvulsants, and NSAIDs. If preventive medication is indicated, the agent should be chosen from one of the major categories, based on side-effect profiles and coexistent comorbid conditions (Tab. 1) [15].

Migraine mechanisms

There is no certainty about how migraine medications work. While there are no true animal models for migraine, a number of innovative models are available for the development of drugs for acute migraine treatment. These models were developed based on either the presumed pathophysiology of the migraine attack or the presumed mechanism of action of an existing migraine drug. Most migraine preventive medications were designed to treat other disorders. In contrast, methysergide was developed as a migraine preventive, based on the concept that migraine is a serotonin excess disorder and methysergide is a serotonin antagonist. Some have suggested that down-regulating the 5-HT_2 receptor or modulating the discharge of serotonergic neurons may be involved in migraine prevention [16, 17].

Antidromic stimulation of the trigeminal nerve releases substance P, calcitonin gene related peptide, and neurokinin A from the sensory C-fibers and results in neurogenic inflammation (NI). The released neuropeptides interact with the blood vessel wall, producing dilatation, plasma extravasation, and sterile inflammation. Electron micrographs of the interior of these blood vessels show platelet activation [18].

The development of NI results in the breakdown of the blood-brain barrier in the dura mater. Sumatriptan and dihydroergotamine, drugs that are agonists at the presynaptic inhibitory 5-HT_{1D} heteroreceptor, prevent the leakage. Neither drug blocks the production of inflammation induced by direct application of neuropeptides to the dural vessels. Methysergide, following long-term but not acute administration, worked in this model consistent with its clinical usefulness as a migraine preventive.

NI can be prevented by blocking neuronal transmission. Acute specific headache medications may work by blocking nerve fiber transmission in the trigeminal system rather than through vascular constriction. Elevated calcitonin gene related peptide levels have been found in the jugular blood during a human migraine attack. 5-HT_{1D}-receptor agonists abort the headache and reduce calcitonin gene related peptide to control levels [19]. Other drugs that block NI include divalproex, neurosteroids, NSAIDs, and neuropeptide Y.

Diener's group [20] has recently reported brainstem activation in spontaneous migraine attacks. Using positron emission tomography to measure regional cerebral blood flow, nine patients with migraine without aura with right-sided headaches were studied within hours of migraine onset. High regional cerebral blood flow values compared with the headache-free interval were found in the cerebrum bilaterally in the cingulate cortex, the auditory association cortex, the visual association cortex, and the inferior anterocaudal cingulate cortex (on the left side only). There was increased regional cerebral blood flow lateralized to the left in the brainstem anterior to the aqueduct and posterior to the corticospinal tract. Sumatriptan relieved the headache and associated symptoms and reversed the cerebral, but not the brainstem, increase in regional cerebral blood flow.

This is the first report of brainstem activation during a spontaneous attack of migraine without aura, most likely a result of activation of brainstem gray matter including the dorsal raphe nucleus (a serotonergic nucleus) and the locus ceruleus (a noradrenergic nucleus).

Migraine may be due to the abnormal activation of a network that has normal physiological functions. The strong familial association of migraine and the association of some varieties of migraine with chromosome 19 strongly suggest an underlying genetic basis whose biological basis is still unknown. It is still uncertain whether migraine with aura and migraine without aura are distinct entities.

A preventive migraine drug could raise the threshold to activation of the migraine process either centrally or peripherally. Drugs could conceivably decrease activation of the migraine generator, enhance central antinociception, raise the threshold for spreading depression, or stabilize the more sensitive migrainous nervous system by changing sympathetic or serotonergic tone. Preventive drugs most likely work by more than one mechanism. The drugs could, in part, have a peripheral mechanism of action, similar to specific acute medications. Since the prolonged use of most acute medications can cause daily headache and can block the effect of preventive drugs, a theoretical problem is created. A clue to this problem may be found in the exceptions, long-acting NSAIDs and dihydroergotamine, which are both acute and preventive medications. Saxena [21] has postulated a primary peripheral mechanism for methysergide, which may work by closing cerebral AVAs, and Moskowitz [18] has shown that both methysergide and valproate block the development of NI in his model.

The mechanism of action of valproate in migraine prophylaxis may be related to facilitation of GABAergic neurotransmission [24–26; see the chapter by Löscher in this volume]. Valproate also attenuates plasma extravasation in the Moskowitz model of NI [22] by interacting with the $GABA_A$ receptor. The relevant receptor may be on the parasympathetic nerve fibers projecting from the sphenopalatine ganglia; in so doing, it attenuates nociceptive neurotransmission [23]. In addition, valproate-induced increased central enhancement of $GABA_A$ activity may enhance central antinociception [18]. Valproate also interacts with the central 5-HT system and it reduces the firing rate of midbrain serotonergic neurons [18].

Valproate pharmacokinetics

With the exception of divalproex sodium, an enteric coated formulation whose absorption may be delayed 2 to 4 h, valproate preparations are rapidly absorbed [24]. Their bioavailability is nearly complete [24]. Peak serum concentrations occur in 1 to 4 h, although absorption may be delayed if the drug is taken with food [24]. Valproate is rapidly distributed, reaching the central nervous system within minutes of administration, and is highly bound (90%) to plasma proteins, mainly albumin when serum levels are less than 100 μg/ml. The extended release formulation exhibits a peak serum concentration in approximately 15 h in nonhepatically induced patients. Cerebrospinal fluid concentrations are in equilibration with free drug in the plasma [24]. (See Shen, this volume.)

The plasma elimination half-life of valproate ranges from 8 to 17 h [24]. Concurrent use of hepatic microsomal enzyme-inducing drugs may shorten it. Less than 3% of the drug is excreted unchanged in the urine or feces [24], most is metabolized in the liver, primarily by conjugation with glucuronide, as well as by beta, omega, and omega-1 oxidation. This results in the production of many metabolites, some of which are active. (See Abbott and Anari, in this volume.)

The suggested valproate plasma concentration is 50 to 100 μg/ml for epilepsy and up to 125 μg/ml for mania, although the correlation between concentration and response is poor [24, 25]. There may be a response threshold (especially for epilepsy) at 50 μg/ml, the approximate concentration at which plasma albumin sites begin to saturate [26]. Serum concentrations higher than 50 μg/ml may be more effective than lower concentrations. Valproate inhibits hepatic metabolism rather than inducing its own metabolism or the metabolism of other hepatically cleared agents (e.g. oral contraceptives) [25].

Clinical Trials (Tab. 2)

In 1988, in a prospective open trial of valproate, Sorenson studied 22 patients with severe migraine resistant to previous prophylactic treatment [27]. Seventeen patients had common migraine and five had classic migraine. The attack frequency ranged from four to 16 a month in 21 patients (one had daily headache). The dose of valproate was 600 mg twice a day adjusted upward to a serum level of about 700 µmol/l. Follow-up in 3 to 12 months revealed that 11 patients were migraine-free, six had had a significant reduction in frequency, one had had no change, and four had dropped out. Sorenson's clinical observations of the benefit of valproate prompted him to do this prospective open trial.

In 1991, Viswanathan et al. [28] conducted an open study on 16 patients who had resistant migraine and paroxysmal EEG changes. The patients were treated with sodium valproate, 200 mg three times a day as an add-on medication. After 2 weeks, 12 patients were headache-free and the rest had 50% relief. The drug was continued for 3 months in the totally headache-free patients, while the other four patients increased their dose to 800 to 1000 mg a day. Two of these dropped out, and the other two had complete relief.

Moore, in 1992, reported a retrospective analysis of 207 patients with refractory headache who were treated with valproate 750 to 2000 mg a day [29]. Patients with either migraine (n = 125) or chronic daily headache (n = 82) were treated with the drug. Sixty-five percent of patients with migraine without aura and 50% of patients with migraine with aura had a good or excellent response (criteria not defined). Patients with mixed headache had a 52% response, while those with chronic tension-type headache had a 73% response. The average duration of treatment was 246 days.

In 1992, Hering and Kuritzky evaluated sodium valproate's efficacy in migraine treatment, in a double-blind randomized crossover study [7]. Thirty-two patients were divided into two groups and given either 400 mg of sodium valproate twice a day or placebo for 8 weeks. The patients were then crossed over to the opposite treatment for an additional 8 weeks. Three patients dropped out. Sodium valproate was effective in preventing migraine or reducing the frequency, severity, and duration of attacks in 86.2% of the remaining 29 patients, whose attacks were reduced from 15.6 to 8.8 a month. The drug was a well-tolerated, effective migraine treatment.

In 1993, Sianard-Gainko et al. [30] assessed the prophylactic effects of sodium valproate over 6 months in 56 patients with severe primary headaches. Thirty-two patients had migraine without aura, 3 had migraine with aura, 14 had both frequent migraine attacks and tension-type headache

Table 2. Clinical trials

Patient population (diagnostic criteria)	No.	Design	Dosage (mg/d)/ other medication	Plasma levels	Duration	Results	Study
Resistant common or classic migraineurs	22	Open; prospective	600 mg bid adjusted to serum level	700 µmol/l	3 to 12 Months	1 Headache free; 6 significant improvement; 4 dropped out; 1 no effect	[27]
Migraine with EEG changes	16	Open	600 mg/Valproate	??	14 Days	12 Headache-free	[28]
Migraine; CDH	20, 7	Retrospective	750 to 2000 review	??	Average 246 days	Migraine with aura 50%; migraine 65%; CDH, mixed 52%; CTTH 73%	[29]
Migraine	29	Double-blind/	800 mg (400 mg bid) placebo-controlled crossover	31.1 to 91.9 µg/ml	8 Weeks each; total of 16 weeks	86.2% of patients responded better to valproate	[48]
Migraine with or without aura	35	Open	928.5 mg/Valproate mean dosage	70 to 90 µg/ml	6 Months	60%: 75–100% reduction; 20%: 50–75% reduction; improvement correlated to plasma level	[30]
Migraine with or without aura	62	Open	400 mg/Valproate	Mean 30 µg/ml	12 Weeks	69.8% substantial benefit; no correlation to plasma level	[31]
Migraine without aura	43	Double-blind/	1000 to 15000 mg/ placebo-controlled crossover	Mean 73.4 µg/ml Sodium valproate	32 Weeks	50% valproate 18% placebo	[32]
Intractable headache; migraine; TM; TTH	75	Prospective; open	500 mg bid/Dival-proate titrate	< 150 µg/ml	10 Weeks	TTH 27%; TM 61%; Frequent migraine 85%	[35]
Migraine with or without aura	107	Double-blind/ placebo controlled	500 to 1500 mg/ Divalproex	70 to 120 µg/ml	16 Weeks	48% divalproex 14% placebo	[6]

Table 2 (continued)

Patient population (diagnostic criteria)	No.	Design	Dosage (mg/d)/ other medication	Plasma levels	Duration	Results	Study
Migraine without aura	32	Open; 2 groups	600 mg,* titrate to 1422 mg**/Valproate	* < 50 µg/ml **70 to 100 µg/ml	26 Weeks	Better results with higher dose	[34]
Migraine with or without aura	17 6	Double-blind/ placebo-controlled	500 to 1600 or 1500 mg/Divalproex	??	10 Weeks	43% divalproex 21% placebo	[8]
Migraine without aura	37	Single blind cross-over comparison	Mean dose divalproex 1414 mg; propranolol to propranolol	Mean 68.5 µg/ml 168 mg	28 Weeks	Placebo 19%; propanolol 66%; divalproex 66%	[35]
Migraine with or without aura	12	Open label crossover	Highest dose tolerated	??	4.5 Months	92% had fewer headache on divalproex compared to propanolol	[36]
Cluster, chronic episodic	15	Open	600 to 2000 mg/ Valproate	31 to 94 µg/ml	Until end of cluster	73.3% responded; 9/15 complete disappearance; 2/15 marked improvement	[10]
Chronic daily headache	30	Open	1000 to 2000 mg Divalproex	75 to 100 µg/ml	3 Months	2/3 Significant improvement	[9]

(TTH), and 7 had only chronic tension-type headache. Twenty-nine percent of the patients overused analgesics and/or ergotamine. The mean daily dose of sodium valproate (given twice a day) was 928.5 mg. Efficacy was assessed by comparing the number of headache days before treatment and during the sixth month of therapy. In the migraine-only group (n=35), 60% had a 75 to 100% and 20% a 50 to 75% reduction in headache days. In the mixed or TTH group, only 33% had a greater-than-50% reduction in headache days. Clinical improvement significantly correlated with sodium valproate blood levels. These results confirmed that sodium valproate is an effective prophylactic treatment for severe migraine, but is of limited value for TTH at least when compounded by overuse of symptomatic medication. This study suggested aiming for a sodium valproate plasma level between 70 and 90 µg/ml.

Coria, in 1994, conducted an open label trial of sodium valproate in which they assessed migraine prophylaxis in 62 consecutive patients with severe migraine with and without aura [31]. The patients were given 200 mg of sodium valproate twice a day for 3 months and then asked to withdraw the drug for 3 months. The therapeutic response was measured with a scale of 15 items, which included the frequency and severity of migraine attacks. Substantial benefit was obtained by 69.8% of patients; it lasted 3 months after drug withdrawal in 67.6% of cases, perhaps due to a positive carry-over effect. No significant correlation was found between sodium valproate levels and the therapeutic response as measured by the migraine assessment scale.

Jensen et al., in 1994, studied 43 patients with migraine without aura in a triple-blind, placebo- and dose-controlled, crossover study of slow-release sodium valproate [32]. After a 4-week medication-free run-in period, the patients were randomized to sodium valproate (n=22) or placebo (n=21). Thirty-four patients completed the trial. Patients randomized to valproate received 1000 mg a day for the first week. Patients with serum levels below 50 µg/ml were blindly adjusted to 1500 mg of sodium valproate a day and those with serum levels above 50 µg/ml were continued on 1000 mg a day. The number of migraine days was 3.5 per 4 weeks during valproate treatment and 6.1 during placebo ($p=0.002$) treatment. The severity and duration of the migraine attacks that did occur were not affected by sodium valproate when compared with placebo. Fifty percent of the patients had a reduction in migraine frequency to 50% or less for the valproate group compared to 18% for placebo. During the last 4 weeks of valproate treatment 65% responded. The mean serum sodium valproate concentration was 73.4 µg/ml after 8 days and 64.2 µg/ml after 12 weeks of treatment. The most common side-effects (33% valproate, 16% placebo) were intensified nausea and dyspepsia, tiredness, increased appetite, and weight gain and were usually mild or moderate. Fifty-eight percent of the patients had no side-effects. It was concluded that sodium valproate was an effective and well-tolerated prophylactic medication for migraine without aura.

Rothrock et al., in 1994, consecutively recruited 75 patients with intractable headache syndromes [33]. They divided these patients into three groups (frequent migraine [FM] (n = 18), transformed migraine (n = 43), and TTH [n = 14]) and treated all 75 with divalproex sodium. Thirty-six patients (48%) reported a ≥ 50% reduction in headache frequency. Significantly different treatment response rates were found in the three groups: FM patients improved the most (61%); transformed migraine patients less (51%); and TTH patients the least (21%). They concluded that prophylactic treatment with divalproex sodium may be effective in selected patients with intractable headache syndromes and that identification of clinically distinct headache subtypes may assist in predicting response to treatment.

In 1995, in a multicenter, double-blind, randomized, placebo-controlled investigation, Mathew et al. [6] compared the effectiveness and safety of divalproex sodium and placebo in migraine prophylaxis. A 4-week, single-blind, placebo-baseline phase was followed by a 12-week treatment phase (4-week dose adjustment, 8-week maintenance). One-hundred seven patients were randomized to divalproex sodium or placebo (2 : 1 ratio), with 70 receiving divalproex sodium and 37 receiving placebo. Divalproex sodium and placebo dosages were titrated in blinded fashion during the dose adjustment period to achieve actual/sham trough valproate sodium concentrations of approximately 70 to 120 µg/ml. During the treatment phase, the mean migraine headache frequency per 4 weeks was 3.5 in the divalproex sodium group and 5.7 in the placebo group ($p \leq 0.001$), compared with 6.0 and 6.4 respectively during the baseline phase. Forty-eight percent of the divalproex sodium-treated patients and 14% of the placebo-treated patients showed a 50% or greater reduction in migraine headache frequency from the baseline phase ($p \leq 0.001$). Among those with migraine headaches, the divalproex sodium-treated patients reported significantly less functional restriction than the placebo-treated patients and used significantly less symptomatic medication per episode. No significant treatment-group differences were observed in average peak severity or duration of individual migraine headaches. (A nonsignificant decrease in duration was noted.) Treatment was stopped in 13% of the divalproex sodium-treated patients and 5% of the placebo-treated patients because of intolerance (p, not significant). The authors concluded that divalproex sodium is an effective prophylactic drug for patients with migraine headaches and is generally well tolerated.

Czapinski, in 1995, assessed the effect of valproic acid in a 6-month open study of 32 patients (24 women and 8 men) with migraine without aura [34]. For the first month valproate was slowly increased to a total divided dose of 600 mg daily. Blood serum valproate levels were obtained, and the patients were divided into two groups: A) those with valproate levels below 50 µg/ml (n = 14); and B) those with valproate levels above 50 µg/ml (n = 18). In Group B, the valproate dose was increased until the concentration reached 70 to 100 µg/ml (mean dose 1422 mg), while Group

A received the original dose. The outcome measure was the number of days with migraine before and after 2, 4, and 6 months of treatment. Valproate was found to be effective, with its beneficial effect increasing during the course of therapy. Efficacy correlated to serum valproate concentrations. AEs occurred in 10 patients (gastrointestinal complaints, loss of hair, weight gain), but did not necessitate the discontinuation of therapy. (No data were provided in the abstract.)

Klapper et al. evaluated the efficacy and safety of divalproex sodium as prophylactic monotherapy in a multicenter, double-blind, randomized, placebo-controlled study [8]. During the 4-week, single-blind, baseline-placebo phase, patients completed a headache diary. Patients with two or more migraine attacks during the baseline phase were randomized to a daily divalproex sodium dose of 500 mg, 1000 mg, or 1500 mg, or placebo. The experimental phase lasted 12 weeks, the first 4 weeks for dose escalation and the remaining 8 weeks for dose maintenance. The primary efficacy variable was 4-week headache frequency during the experimental phase. During the experimental phase, the mean reduction in the combined daily divalproex sodium groups was 1.8 migraines per 4 weeks compared to a mean reduction of 0.5 attacks per 4 weeks in the placebo group. Overall, 43% of divalproex sodium-treated patients achieved $\geq 50\%$ reduction in their migraine attack rates, compared to 21% of placebo-treated patients. A statistically significant ($p \geq 0.05$) dose-response effect across the dose range placebo, 500 mg, 1000 mg, 1500 mg, was observed for both overall reduction in attack frequency and a $\geq 50\%$ reduction in attack frequency. A nonsignificant decrease in headache duration was noted. With the exception of nausea, AEs were similar in all groups (divalproex sodium 24%, placebo 7%, $p = 0.015$) and most AEs were mild or moderate in severity. This study showed divalproex sodium to be an effective migraine prophylactic agent.

In 1995, Kaniecki enrolled 37 patients with International Headache Society-diagnosed migraine without aura in a randomized, single-blind, placebo-controlled, crossover study comparing divalproex sodium and propranolol prophylaxis [35]. After a 4-week placebo phase, patients were randomized to either divalproex sodium or propranolol for 12 weeks. After a 4-week washout phase they were crossed over to the opposite treatment. Thirty-two patients completed the study. Responders were defined as those patients achieving a greater than 50% reduction in either mean migraine frequency (events per month) or mean number of days with migraine (days per month). Assessment of migraine frequency revealed a significant response to divalproex sodium in 66% of patients, to propranolol in 63% of patients, and to placebo in 19% of patients. Similar results were seen in assessment of migraine days per month.

Klapper [36] compared divalproex sodium to propranolol in an open label crossover study in which migraineurs randomly received either divalproex sodium or propranolol. The dose was titrated to the highest level tolerated, and the patients recorded their headaches for 2 months. They were then withdrawn from their first medication over a period of 2 weeks and the process was repeated for the second drug. Twelve patients completed both arms of the study. The number of headaches in the divalproex sodium arm was 10.9 per 2 months compared to 20.4 per 2 months in the propranolol arm, with 92% having fewer headaches when taking divalproex sodium. Nine of the patients who took divalproex sodium (38%) did not complete the study due to side-effects, compared with 3 patients (13%) that took propranolol. In this study, migraine patients had significantly fewer headaches when taking divalproex sodium than when taking propranolol. However, the dropout rate was higher in the divalproex sodium group.

Silberstein et al. reported on 163 patients [37] who, following completion of either of the two placebo-controlled divalproex sodium trials [6, 8], were enrolled in an open-label extension study. This represented 198 patient-years of divalproex exposure, with an average dose of 974 mg/day. Forty-nine percent of patients experienced a ≥ 50% reduction in migraine headache rates from days 1 to 90. This increased to 70% from days 901 to 1080, in part due to selective dropout.

A subset of patients who were treated for at least 12 months was analyzed. Patients who discontinued divalproex after 1 year failed to show this improvement up to days 361–500 in contrast to patients who remained in the study. Overall, 67% of the patients discontinued divalproex. Reasons included administrative problems (31%), drug intolerance (21%), and treatment ineffectiveness (15%). The most common AEs were nausea (42%), infection (39%), alopecia (31%), tremor (28%), asthenia (125%), dyspepsia (25%), and somnolence (25%). No unexpected AEs or safety concerns were found in the prophylactic treatment of migraine with divalproex.

Studies of other headache types

Hering and Kuritzky, in 1989, treated 15 cluster patients (2 chronic and 13 episodic) with sodium valproate in an open pilot study [11]. The dose ranged between 600 and 2000 mg a day. Based on clinical response 11 patients (73.3%) improved: 9 reported complete cessation of attacks and 2 reported marked improvement. There was no correlation between efficacy and valproate plasma levels.

Mathew and Ali, in 1991, reported on 32 chronic daily headache patients who were unresponsive to prior treatment [10]. After a baseline observation period of 1 month, the patients were given divalproex sodium 1000 to 2000 mg a day for 3 months. Blood valproic acid levels were between 75 and 100 µg/ml. Two-thirds of the patients showed a significant improvement in headache index and headache-free days. The most common side-effects were weight gain, tremor, hair loss, and nausea. No liver function abnormalities were noted.

In all of the clinical studies, whether open, retrospective or placebo-controlled and double-blind, valproate was an effective preventive treatment for migraine. There was a reduction in the number of migraine attacks, and migraine duration and intensity were also reduced in some instances. There was some evidence that symptomatic medication use could be decreased. Some studies have suggested that clinical efficacy is correlated to serum concentration [11, 30]. Preliminary reports [35, 38] suggest that divalproex sodium is as effective as the beta-blocker propranolol. The data from two multicenter, double-blind, placebo-controlled studies were combined and demonstrated that divalproex sodium was equally as effective in migraine with aura as in migraine without aura [38]. It also appears to be effective in cluster headache and chronic daily headache but does not seem to be as effective for pure TTH. It is equally as effective in patients with severe frequent migraines as in those with less severe migraines. In clinical trials, the most frequent AEs reported by patients treated with divalproex sodium were nausea, asthenia, dyspepsia, dizziness, somnolence, and diarrhea, with most AEs being mild to moderate in severity.

Thomas [39] looked at the long-term safety of divalproex sodium in the prophylactic treatment of migraine. A total of 248 patients were treated with divalproex sodium for up to 35 months. The average daily dose was 750–1000 mg per day. Nausea was the most frequently reported AE (37% of patients); it was transient and declined in frequency to 3 to 6% after 6 months. In contrast, tremor and weight gain generally increased in frequency over time (maximum 34% and 24% per 6 months respectively). Alopecia declined after month 12 (maximum 12% per six months). Discontinuations due to AEs were caused by alopecia (6%), nausea (4%), weight gain (2%), and tremor (2%). The long-term safety profile of divalproex sodium in migraine is similar to that in other indications and to the cumulative postmarketing experience.

Adverse effects

There are two types of adverse drug reactions: drug-induced disease and dose-related toxicity. The latter is more common and is best identified by measuring the plasma drug concentration [40]. Routine blood level monitoring in a stable patient is performed to excess, however, and blood levels should be obtained only to answer specific questions about compliance, toxicity, and level-to-dose relationships or when dosage alteration is necessary [41].

Valproate has a favorable side-effect profile when compared with other anticonvulsants [42] and is often better tolerated than either lithium [43], neuroleptics, or TCAs. The most common side-effects include gastrointestinal complaints (anorexia, nausea, vomiting, dyspepsia, diarrhea), asymptomatic serum hepatic transaminase elevations, neurologic symptoms (tremor, sedation, and, less frequently, ataxia), increased appetite and weight gain, alopecia, and, less commonly, rashes and hematologic dysfunction [24, 25]. Valproate is teratogenic and can produce neural tube defects if taken during the first trimester of pregnancy [44, 45]. These will be discussed in detail in the chapter by Schmidt in this volume.

Migraine Treatment

The goals of treatment are to relieve or prevent the pain and associated symptoms of migraine and to optimize the patient's ability to function normally. The medications used to treat migraine can be divided into two major categories: 1) First-line choices of high efficacy, which include β-blockers, tricyclic antidepressants (TCAs), and valproate, and of lower efficacy, which include selective serotonin reuptake inhibitors (SSRIs), calcium channel antagonists, and NSAIDs, and, 2) second-line choices of possible high efficacy, which are difficult to use or have significant AEs, which include methysergide and monoamine oxidase inhibitors, and second-line choices of unproven or low efficacy, which include cyproheptadine, lithium, and phenytoin. Choose a first-line drug based on the patient's profile and the presence or absence of coexisting or comorbid disease (Tabs. 1 and 3). Use the drug with the best risk-to-benefit ratio for the individual patient and take advantage of the side-effect profile of the drug. An underweight patient would be a candidate for one of the medications that commonly produce weight gain, such as a TCA; in contrast, one would avoid these drugs in the overweight patient. Sedating tertiary TCAs would be useful at bedtime for patients with insomnia. The older patient with cardiac disease may not be able to use TCAs or calcium channel or β-blockers but could easily use divalproex. In the athletic patient β-blockers should be avoided. Medication that can impair cognitive functioning should be avoided in patients who are dependent on their wits.

Comorbid and coexistent diseases have important implications for treatment. The presence of a second illness provides therapeutic opportunities

Table 3. Migraine preventive drugs

First-line	High efficacy	β-blockers, TCAs, Valproate
	Low efficacy	Calcium-channel blockers, NSAIDs, SSRIs
Second-line	High efficacy	Methysergide* MAOIs*
	Low efficacy	Cyproheptadine Phenytoin Lithium*

* Significant adverse effects.

Table 4. Drug Combinations for Migraine Prevention

Suggested	Antidepressants	– β-blocker – calcium channel blocker – valproate – methysergide
	Methysergide	– calcium channel blocker
	SSRI	– TCA's
Caution	β-blocker	– calcium channel blocker – methysergide
	MAOIs	– amitriptyline or nortriptyline
Contraindications	MAOIs	– SSRIs – most TCAs (except amitriptyline or nortriptyline) – carbamazepine
	NSAIDs	– lithium

but also imposes certain therapeutic limitations. In some instances, two or more conditions may be treated with a single drug. When migraine and hypertension and/or angina occur together, β-blockers or calcium channel blockers may be effective for all conditions [46]. For the patient with migraine and depression, TCAs or SSRIs may be especially useful [47]. For the patient with migraine and epilepsy [6, 48], or migraine and manic depressive illness [49, 50], divalproex sodium is the drug of choice. The pregnant migraineur who has a comorbid condition that needs treatment should be given a medication that is effective for both conditions and has the lowest potential for adverse effects on the fetus. In individuals with more than one disease, certain categories of treatment may be relatively contraindicated. For example, β-blockers should be used with caution in the depressed migraineur, while TCAs, neuroleptics, or sumatriptan may lower the seizure threshold and should be used with caution in the epileptic migraineur.

Drug combinations are commonly used for patients with refractory headache disorders. Some combinations, such as antidepressants and β-blockers, are suggested; others, such as β-blockers and calcium channel blockers, should be used with caution; and some, such as monoamine oxidase inhibitors and SSRIs, are contraindicated because of potentially lethal interactions (Tab. 4). Many clinicians use the combination of an antidepressant (such as a TCA or SSRI) and a β-blocker, and find that they act synergistically [14, 51]. Lance has advocated combining methysergide with a vasodilator such as a calcium channel blocker to decrease side effects. Divalproex, used in combination with antidepressants, is a logical choice to treat refractory migraine that is complicated by depression or bipolar disease. Some clinicians cautiously use the combination of phenelzine and amitriptyline in refractory headache patients.

Recommendations

1. Before initiating valproate, perform a physical examination and take a thorough medical history, with special attention to hepatic, hematologic, and bleeding abnormalities. Inform the patient about possible hair loss, weight gain, and teratogenic effects and the signs and symptoms of hepatic and hematologic dysfunction.

Obtain screening baseline laboratory studies to help identify risk factors that could influence drug selection. Suggested studies include: complete blood count, differential, and platelets; prothrombin time and partial thromboplastin time; serum chemistry, including glucose, BUN, electrolytes, calcium, potassium, magnesium, creatinine, urate, cholesterol, bilirubin, alkaline phosphatase, AST, ALT, total protein, and albumin (SMA profile).

2. To minimize gastrointestinal side-effects, use enteric-coated formulations of valproate. Begin with a dose of 250 mg at bedtime. If nausea still occurs use the sprinkle formulation (125 mg) and very slowly increase the dose. Slowly increase the dose to 500 to 750 mg a day (in 2 to 3 divided doses) to limit gastrointestinal side-effects. Higher doses are needed at times.

3. Obtain follow-up divalproex levels to test for compliance, toxicity, and drug reactions as needed. Rigid adherence to the therapeutic range for epilepsy (50 to 100 μg/ml) is not likely to benefit the patient; we often push to trough levels of 125 μg/ml.

4. See the patient on a regular basis (every 1 to 2 months) during the first 6 to 9 months of therapy.

5. It is not necessary to monitor blood and urine in otherwise healthy and asymptomatic patients on monotherapy, despite the manufacturer's recommendation that liver function tests be performed at frequent intervals, especially during the first 6 months. Identify patients who belong to one of the high-risk groups at the inception of treatment. Obtain follow-up chemistry, if needed, at most 2 and 6 months following the onset of treatment, particularly in patients on polypharmacy.

6. If mild hepatic transaminase elevation occurs, continue divalproex sodium at the same dose or a lower dose until the enzymes normalize. If the hepatic transaminase elevations are much higher (e.g. two to three times the upper limit of normal), discontinue valproate and restart it a lower dose once the abnormalities have resolved. If severe abnormal pain suggestive of pancreatitis occurs, check serum amylase and lipase.

7. Avoid valproate for headache in children under the age of 10 years unless routine treatments have failed, in which case use it as monotherapy. Avoid valproate in patients with preexistent liver disease. Avoid valproate for headache in women who are pregnant or attempting to become pregnant.

8. Tremor may occur in 10% of treated patients. If this is bothersome, decrease the dose of divalproex sodium or use propranolol, a β-blocker that is also an effective migraine preventive [52].

9. Excessive weight gain can occur with an incidence of up to 44%. Advise the patient to exercise regularly, obtain a dietary consultation, and avoid using other medications, such as TCAs, which can produce weight gain. Hair loss occurs with an incidence of 26 to 12% [25]. Anecdotally, multivitamins and zinc supplements has been reported to control hair loss. We routinely have our patients take 220 mg a day of zinc (mega zinc) and a multivitamin.

References

1 Physicians' desk reference, 49th ed (1998) New Jersey: Medical Economics Company
2 McElroy SL, Keck PE, Pope HG, Hudson JI (1989) Valproate in psychiatric disorders: literature review and clinical guidelines. *J Clin Psychiatry* 50(3): 23–29
3 Bourgeois FD (1995) Valproic acid. In: RH Levy, RH Mattson, BS Meldrum (eds): *Antiepileptic drugs*, 4th ed. Raven Press, New York, 633–639
4 Balfour JA, Bryson HM (1994) Valproic acid. A review of its pharmacology and therapeutic potential in indications other than epilepsy. *CNS Drugs 2*(2): 144–173
5 Jensen R, Brinck T, Olesen J (1994) Sodium valproate has a prophylactic effect in migraine without aura. *Neurology* 44: 647–651
6 Mathew NT, Saper JR, Silberstein SD, Tolander LR, Markley H, Solomon S, Rapoport A, Turkewitz LJ, Silber CJ, Deaton R (1995) Prophylaxis of migraine headaches with divalproex sodium. *Arch Neurol* 52: 281–286
7 Hering R, Kuritzky A (1992) Sodium valproate in the prophylactic treatment of migraine: a double-blind study versus placebo. *Cephalalgia* 12: 81–84
8 Klapper J, on behalf of the Divalproex Sodium in Migraine Prophylaxis Study Group (1997) Divalproex sodium in migraine prophylaxis: a dose-controlled study. *Cephalalgia* 17: 103–108
9 Roy-Byrne PP, Ward NG, Donnelly PJ (1989) Valproate in anxiety and withdrawal syndromes. *J Clin Psychiatry* 50(3): 44–48
10 Mathew NT, Ali S (1991) Valproate in the treatment of persistent chronic daily headache. An open label study. *Headache* 31: 71–74
11 Hering R, Kuritzky A (1989) Sodium valproate in the treatment of cluster headache: an open clinical trial. *Cephalalgia* 9: 195–198
12 Silberstein SD, Lipton RB (1994) Overview of diagnosis and treatment of migraine. *Neurology* 44(7): 6–16
13 Silberstein SD, Saper J (1993) Migraine: diagnosis and treatment. In: D Dalessio, SD Silberstein (eds): *Wolffis headache and other head pain* (6th ed). Oxford University Press, New York, 96–170
14 Silberstein SD (1997) Migraine and pregnancy. *Neurol Clin 15*: 209–231
15 Tfelt-Hansen P, Welch KMA (1993) Prioritizing prophylactic treatment. In: J Olesen, P Tfelt-Hansen, KMA Welch (eds): *The headaches*. Raven Press, New York, 403–404
16 Silberstein SD (1994) Review: serotonin (5-HT) and migraine. *Headache* 34: 408–417
17 Peroutka SJ (1990) Developments in 5-hydroxytryptamine receptor pharmacology in migraine. *Neurol Clin* 8: 829–838
18 Moskowitz MA (1992a) Neurogenic vs vascular mechanisms of sumatriptan and ergot alkaloids in migraine. *Trends Pharmacol Sci* 13: 307–311
19 Goadsby PJ, Edvinsson L (1993) Sumatriptan reverses the changes in calcitonin gene-related peptide seen in the headache phase of migraine. *Ann Neurol* 33: 48–56
20 Weiller C, May A, Limroth V, Faiss JH, Timmann D, Mueller SP, Diener HC (1995) Brainstem activation in spontaneous human migraine attacks. *Nature Medicine* 1(7): 658–660
21 Saxena PR (1995) Cranial arteriovenous shunting: an *in vivo* animal model for migraine. In: J Olesen, MA Moskowitz (eds): *Experimental headache models*. Lippincott-Raven, Philadelphia, 189
22 Limmroth V, Lee WS, Cutrer FM, Waeber C, Moskowitz MA (1995) Meningeal $GABA_A$ receptors located outside the blood brain barrier mediate sodium valproate blockade of

neurogenic and substance P-induced inflammation: possible mechanism in migraine. Proceedings of the 7th International Headache Congress, Canada, September 16–20, 1995. *Cephalalgia* 15(14): 102
23 Cutrer FM, Limmroth V, Moskowitz MA (1997) Possible mechanisms of valproate in migraine prophylaxis. *Cephalalgia* 17: 93–100
24 Rall TW, Schleifer LS (1990) Drugs effective in the therapy of epilepsies (Chapter 19). In: AG Giman, TW Rall, AS Nies, P Taylor (eds): *Goodman and Gilmansí the pharmacologic basis of therapeutics* (8th ed). Pergamon Press, New York, 436–462
25 Rimmer EM, Richens A (1985) An update on sodium valproate. *Pharmacotherapy* 5: 171–184
26 Chapman A, Keane PE, Meldrum BS, et al. (1982) Mechanism of anticonvulsant action of valproate. *Prog Neurobiol* 19: 315–359
27 Sorensen KV (1988) Valproate: a new drug in migraine prophylaxis. *Acta Neurol Scand* 78: 346–348
28 Viswanathan KN, Sundraram N, Rajendiran C, Manohar DS, Balaraman VT (1995) Sodium valproate in therapy of intractable headaches with EEG changes. *Cephalalgia* 11(11): 282–283
29 Moore KL (1992) Valproate in the treatment of refractory recurrent headaches: a retrospective analysis of 207 patients. *Headache* 3(3): 323–325
30 Sianard-Gainko J, Lenaerts M, Bastings E, Schoenen J (1993) Sodium valproate in severe migraine and tension-type headache: clinical efficacy and correlations with blood levels. *Cephalalgia* 13(13): 252
31 Coria F, Sempere AP, Duarte J, Claveria LE, Cabezas C, Bayon C (1994) Low-dose sodium valproate in the prophylaxis of migraine. Clin Neuropharmacol 17: 569–573
32 Jensen R, Brinck T, Olesen J (1994) Sodium valproate has a prophylactic effect in migraine without aura: a triple-blind, placebo-controlled crossover study. *Neurology* 44: 647–651
33 Rothrock JF, Kelly NM, Brody ML, Golbeck A (1994) A differential response to treatment with divalproex sodium in patients with intractable headache. *Cephalalgia* 14(3): 241–244
34 Czapinski P (1995) Valproic acid in preventive treatment of migraine. *Cephalalgia* 15(14): 283
35 Kaniecki RG (1995) A comparison of divalproex sodium to propranolol hydrochloride and placebo in the prophylaxis of migraine without aura. *Investigator* 1(2):
36 Klapper JA (1995) An Open label crossover comparison of divalproex sodium to propranolol HCI in the prevention of migraine headaches. *Investigator* 1(2):
37 Silberstein SD, Deaton R, Collins SD (1998) The safety of divalproex sodium (Depakote) in headache prophylaxis. An open-label, long-term study. *Headache* 38(5): 405 (abs)
38 Deaton RL, Thomas JR (1995) The efficacy of divalproex sodium prophylactic treatment in patients experiencing migraine with or without aura. Proceedings of the 7th International Headache Congress. Toronto, Canada, September 16–20, 1995. *Cephalalgia* 15(14): 268
39 Thomas JR (1995) The long-term safety of divalproex sodium prophylactic treatment of patients with migraine. Proceedings of the 7th International Headache congress. Toronto, Canada, September 16–20, 1995. *Cephalalgia* 15(14): 266
40 Pellock JM, Pippenger CE (1993) Adverse effects of antiepileptic drugs. In: EW Dodson, JM Pellock (eds): *Pediatric epilepsy diagnosis and therapy*. Demos Publications, New York,
41 Silberstein SD, Willmore LJ (1996) Divalproex sodium: migraine treatment and monitoring. *Headache* 36: 239–242
42 Beghi E, DiMascio R, Tognoni G (1986) Adverse effects of anticonvulsant drugs: a critical review. *Adverse Drug Reactions & Acute Poisoning Rev* 5(2): 63-86
43 Vencovsky E, Soucek K, Zatecka I (1983) Comparison of side effects of lithium and dipropylacetamide (Depamide). *Cesk Psychiatr* 4: 223–227
44 Jeavons PM (1982) Sodium valproate and neural tube defects. *Lancet 2*: 1282–1283
45 Centers for Disease Control (1983) Valproate: a new cause of birth defects – Report from Italy and followup from France. *MMWR* 32: 438–439
46 Solomon GD (1989) Management of the headache patient with medical illness. *Clin J Pain* 5: 95-99
47 Silberstein SD, Lipton RB, Breslau N (1995) Migraine: association with personality characteristics and psychopathology. *Cephalalgia* 15: 337–369

48 Hering R, Kuritzky A (1992) Sodium valproate has a prophylactic effect in migraine: a double-blind study vs placebo. *Cephalalgia* 12: 81–84
49 Bowden CL, Brugger AM, Swann AC, Calabrese JR, Janicak PG, Petty F, Dilsaver SC, Davis JM, Rush AJ, Small JG et al (1994) Efficacy of divalproex vs lithium and placebo in the treatment of mania. *JAMA* 271: 918–924
50 Curran DA, Hinterberger H, Lance JW (1967) Methysergide. Res Clin Stud *Headache* 1: 74–22
51 Silberstein SD, Lipton RB, Goadsby PJ (eds): *Headache in clinical practice.* Isis Medical Media Ltd, Oxford, 1998
52 Karas BJ, Wilder BJ, Hammond EJ, Bauman AM (1983) Treatment of valproate tremors. *Neurology* 33: 1380–1382

Milestones in Drug Therapy
Valproate
ed. by W. Löscher

Adverse effects and interactions with other drugs

Dieter Schmidt

Epilepsy Research Group, Goethestrasse 5, D-14163 Berlin, Germany

Introduction

The ascertainment of valproate-induced adverse effects is fraught with a number of specific problems. Valproate was introduced in Europe in the early 1960s when few clinical trials were required, and those performed have simply used incidence reporting following passive inquiry. The prevalence and incidence of adverse effects has not been reported at different points during these early trials and, apart from withdrawal from the trial, no quantification of the severity and the impact of adverse effects was possible. Furthermore, the pharmaceutical formulation of valproate was changed several times over the years. More recently, enteric coated and slow release preparations were introduced which incidentally reduced the number of adverse reactions [1]. In addition, the indications which were initially limited to idiopathic generalized seizures could be widened to include not only virtually all types of epileptic seizures but also into other therapeutic areas such as migraine and acute mania. Despite the wide and well accepted use of valproate, it took decades to discover altogether rare but serious adverse effects. These include acute liver failure [2] and an increased rate of congenital malformations, especially neural tube defects in the offspring of women who had taken valproate during pregnancy [3]. Even after being on the market for so many years, new adverse effects continue to be discovered. A recent example is the contribution of valproate to polycystic ovaries syndrome in women [4, 4a]. Apart from the problems of ascertainment, the classification and nomenclature of adverse effects is difficult and often tenuous [5, 6]. For this review, adverse effects are broadly classified in those occurring early, i.e. during the first weeks of treatment and in those usually seen months and often years thereafter. Drug interactions involving valproate may cause adverse effects and therefore are discussed here. The adverse effects of valproate have been reviewed earlier [6–9] and this update mainly deals with the more recent literature.

An overview of common and less common adverse effects of valproate

A wide range of adverse events has been reported with valproate from controlled trials, spontaneous reports and other sources (Tab. 1). In most patients the adverse effects of valproate are mild to moderate in intensity [7]. Sedation, ataxia, acute impairment of cognitive function and hypersensitivity are usually no or minimal problems in the daily use of the compound.

Characteristic and mostly rare adverse effects of valproate that are not shared by antiepileptic drugs with a similar spectrum of efficacy such as lamotrigine and topiramate include subacute encephalopathy with or without hyperammonemia, polycystic ovaries, hyperinsulinism and hyperandrogenism, reversible Parkinsonism with or without insidious cognitive impairment and reversible pseudo brain atrophy, acute liver failure due to metabolic idiosyncrasy, teratogenicity with predominantly neural tube effects, and platelet dysfunction with or mostly without coagulation problems (Tab. 1). Common adverse effects of valproate are transient nausea, indigestion and vomiting, which incidentally have become less frequent with the introduction of enteric coated preparations of valproate and are reported in no more than 3–6% of patients [7]. Tremor, weight gain and hair loss remain the most common dose related adverse effects, while acute liver failure and the teratogenicity of valproate, although altogether uncommon, have proved the most devastating adverse effects of valproate (Tab. 2).

Although we are currently far from being able to explain how valproate induces common adverse effects such as weight gain or tremor, we are starting to elucidate the putative mechanism of some of the enigmatic adverse effects of valproate. The reasons for the variability of the intensity and the incidence of these adverse effects need to be addressed in future studies. Despite these shortcomings, putative mechanisms have been identified including the inhibition of enzymes and the induction of mitochondrial dysfunction through valproate (Tab. 3).

Although these mechanisms do not explain all characteristic adverse reactions to valproate, they provide a very modest measure of understanding for the mechanism of a number of adverse reactions of valproate. From a clinical perspective, adverse effects are an important criterion for choosing a particular drug for the individual patient. It may therefore be of interest to briefly assess the comparative tolerability and safety of valproate in relation to other antiepileptic drugs.

Comparative tolerability and safety of valproate

When valproate is given with antiepileptic drugs, it is often not possible to determine whether the adverse events are those of valproate or the co-medication. However in recent years several controlled trials have compared the

Table 1. Adverse effects of valproate compared with those reported with the use of other antiepileptic drugs

	VPA	CBZ	PHT	PB	PRM	CLB	ESM	FBM	GBP	LTG	OXC	TGB	TPM	VGB	ZON
Early Adverse Effects															
Drowsiness				+++	+++	++			+	+		+	+++	+	++
Dizziness		++	++		+++	+			+	+	++		++	+	++
Stupor	+														
Induction of seizures		+	+			+			+			+		++	
Gastro-intestinal disorders	+	+			++		++	+	+		+				+
Acute liver failure	+							+							
Hypersensitivity reactions		+	+	+	+			+		++	+		+		
Late Adverse Effects															
Sedation				++	+++	++	+								+
Subacute encephalopathy	+		+												
Visual disturbances		+	+						+?			+?		+	
Movement disorders	+++	+	++	+	+				+	+					
Behavioral changes	+		+	+++	+++								++	++	+
Depression			+	+			+					+		+	
Psychotic reactions	+	+	+	+	+		++	+	+	+		+	+	++	
Cerebellar deficits			+												
Peripheral neuropathy		+?	+	+?	+?										
Leukopenia	+	++	+	+	+		+	+			+				
Aplastic anemia		+	+	+	+		+	+++							
Thrombocytopenia	++							+							
Macrocytic anemia		+?	+	+	+										
Pancreatitis	+								+						
Kidney													+		+

CBZ = Carbamazepine; CLB = Clobazam or Clonazepam; ESM = Ethosuximide; FBM = Felbamate; GBP = Gabapentin; LTG = Lamotrigine; OXC = Oxcarbazepine; PB = Phenobarbital; PHT = Phenytoin; PRM = Primidone; TGB = Tiagabine; TPM = Topiramate; VPA = Valproate; VGB = Vigabatrin; ZON = Zonisamide; AE = Adverse effect and possible causal relationship has been reported. + = small risk; ++ = intermediate risk; +++ = high risk; i.e. 30% or more at high doses [modified from 7, 10].

Table 1 (continued)

	VPA	CBZ	PHT	PB	PRM	CLB	ESM	FBM	GBP	LTG	OXC	TGB	TPM	VGB	ZON
Heart		+	+												
Connective tissue			++	++	++										
Skin	+		+												
Osteomalacia			+	+	+										
Hyponatremia		+									++				
Weight problems	+++								+					+++	
Cognitive deficits		+	+	+++	+++	++						+	+++		+
Teratogenicity	++	++	+	+	+		+								
Reproductive disorders (PCOS)	+														
Immunological disorders	+	+	++		+		++								
Interactions of valproate on other drugs	++	+++	+++	+++	+++			+++			++				+
Interactions of other drugs on valproate		+++	+++	++	+++	++	+	++		+++					+

Table 2. Adverse effects of valproate

Clinical features	Percent
Single-drug therapy	
Increased appetite	1.4
Drowsiness	1.4
Hair loss	1.3
Tremor	1.0
Paresthesia	1.0
Weight gain	0.7
Adjunctive therapy	
Drowsiness	14.4
Anorexia	4.8
Hair loss	4.0
Nausea	3.2
Weight gain	3.0
Gastrointestinal symptoms	2.9
Lymphopenia	2.3
Vomiting	2.0
Tremor	1.5
Hypersalivation	0.7
Ataxia	0.6
Increased appetite	0.6
Thrombocytopenia	0.4
Loss of weight	0.3
Coma	0.3
Headache	0.3
Behavioral disturbances	0.3
Hyperactivity	0.2
Vertigo	0.2
Diarrhea	0.2
Leukopenia	0.1
Bleeding	0.1
Increased spasticity	0.1
Tinnitus	0.1
Peripheral edema	0.1
Macular rash	0.1

Results of 16 trials published until 1982 in a total of 1140 patients (modified from [7]). Adverse effects were noted in 300 patients (26%). Treatment was discontinued in 2% of the 300 patients.

adverse effects of valproate to those of other anticonvulsants, mostly carbamazepine. Mattson et al. [16] noted that weight gain, hair loss and tremor were more often seen on valproate while rash was more common on carbamazepine. It is of special interest that 12 months after starting treatment weight gain was the only persistent adverse effect, all other side-effects disappeared in most patients during ongoing treatment (Tab. 4).

In a multicenter comparative trial of sodium valproate and carbamazepine in pediatric epilepsy, appetite increase was more common in patients receiving valproate, while somnolence, and dizziness were more often seen

Table 3. Putative mechanisms of adverse effects of valproate

Characteristic adverse effects of valproate	Putative mechanism(s)
Hyperammonemia with or without subacute encephalopathy	Variable inhibition of the urea cycle produces hyperammonemia, plus a variable CNS effect [11].
Polycystic ovaries or hyperandrogenism, obesity of androgen type, hyperinsulinemia, low HDL-cholesterol/total cholesterol ratio, elevated serum triglycerides	The mechanism of valproate-induced insulin-resistance and polycystic ovaries is presently unknowm but may involve valproate-induced decrease of insulin-like growth factor 1 [4] .
Reversible Parkinsonism with or without insidious cognitive impairment	Production of mitochondrial respiratory chain dysfunction by valproate [12].
Acute liver failure unrelated to hypersensitivity	Interaction of valproate with an unrecognized idiosyncrasy in fatty acid metabolism (see Nau, this volume).
Teratogenicity with predominantly neural tube effects	Effect of valproate on an unrecognized idiosyncrasy in folate metabolism (see Nau, this volume).
Valproate-induced increase of the serum concentration of other drugs e.g. phenytoin, phenobarbital, or warfarin, and to a lesser degree that of carbamazepine	Mainly by inhibition of the P-450 isoenzyme CYP2C9 [13].
Platelet dysfunction with or without clotting problems	Inhibition of the second phase of platelet aggregation (REF).
Weight gain	Valproate-induced impairment in fatty acid beta oxidation may be of possible relevance [14, 15]

Table 4. Incidence of systemic adverse effects during treatment with valproate as compared to carbamazepine (modified from [16])

Effect	% of pts.		p Value	% at 12 month follow-up		p Value
	Valproate	Carbamazepine		Valproate	Carbamazepine	
Gastrointestinal symptoms	33	29	0.36	2	6	0.11
Rash	1	11	< 0.001	0	1	0.31
Hepatic toxicity	3	4	0.56	0	0	1.00
Weight gain	43	32	0.02	20	9	0.01
Weight gain (> 5.5 kg)	20	8	< 0.001	13	3	0.01
Hair change or loss	12	6	0.02	4	1	0.06
Impotence	10	7	0.30	1	2	0.29
No of patients	240	231		136	130	

Percentage of patients in whom each type of adverse effects occurred at any time during the trial.

Table 5. Adverse events reported by four or more patients during treatment with sodium valproate and carbamazepine in pediatric epilepsy [17]

Adverse event	N patients reporting event	
	Valproate	Carbamazepine
Somnolence	11	25*
Fatigue	6	13
Weight increase	12	5
Headache	6	9
Rash	4	8
Nausea/vomiting/dyspepsia	6	5
Abdominal pain	6	5
Appetite increase	11*	0
Dizziness	1	8*
Concentration impaired	6	3
Aggressive reaction	4	4
Anorexia	5	3
Alopecia	5	2
Diplopia	0	5
Gait abnormality / ataxia	0	5
Insomnia	0	5
Other events	29	58
Total	112	133

* p = 0.05 between treatments (Fisher exact test; two-tailed).
The analysis of adverse events was based on reports from 118 patients treated with sodium valproate (110 for more than 3 months) and 126 patients treated with carbamazepine (117 for more than 3 months).

in those on carbamazepine (Tab. 5). In another comparative trial in 300 patients, the overall incidence of adverse effects was similar for valproate (46%) and carbamazepine (50%). Most adverse effects were seen within the first 12 weeks of treatment [18]. More patients withdrew from therapy with carbamazepine (23 of 152) than with valproate (14 of 149), predominantly due to rashes (14 vs 1). Neurological adverse effects (ataxia, fatigue, headache, somnolence, dizziness, tremor) were more frequent in patients receiving carbamazepine than on valproate, while weight gain was the most frequent side effect of valproate. Gastrointestinal disturbances (nausea, vomiting, dyspepsia) had a similar frequency in both groups.

When slow release formulations of valproate and carbamazepine were compared in an open randomized trial, both drugs had similar and satisfactory tolerability and safety (Tab. 6). In a trial comparing valproate with oxcarbazepine, adverse effects leading to discontinuation of valproate included hair loss, amnesia, aggressiveness, agitation, dizziness and nausea (Fig. 1).

When divalproex, a 1:1 molar compound of sodium valproate and valproic acid, was added to the medication of patients with refractory partial seizures, the adverse effects using the COSTART nomenclature [21] could be compared to those during the double-blind addition of placebo. Gastrointestinal adverse events occurring significantly more often in the dival-

Table 6. Adverse effects of slow release preparations of valproate and carbamazepine requiring discontinuation [19]

Adverse effects	Valproate CR	Carbamazepine CR
Tremor	1	1
Tremor plus weight gain	2	0
Tremor, abdominal pain, drowsiness	1	0
Drowsiness	0	1
Rash, angioedema	1	0
Rash	0	5
Rash, lower platelets	1	0
Hair loss, irritability	1	0
Pruritus	0	1
Leucopenia	0	1
Unknown	1	0
Total of patients	8/78 (10%)	9/75 (12%)

Valproate and carbamazepine appeared to have satisfactory safety and tolerability in this open randomized trial in partial epilepsy. CR = controlled release formulations.

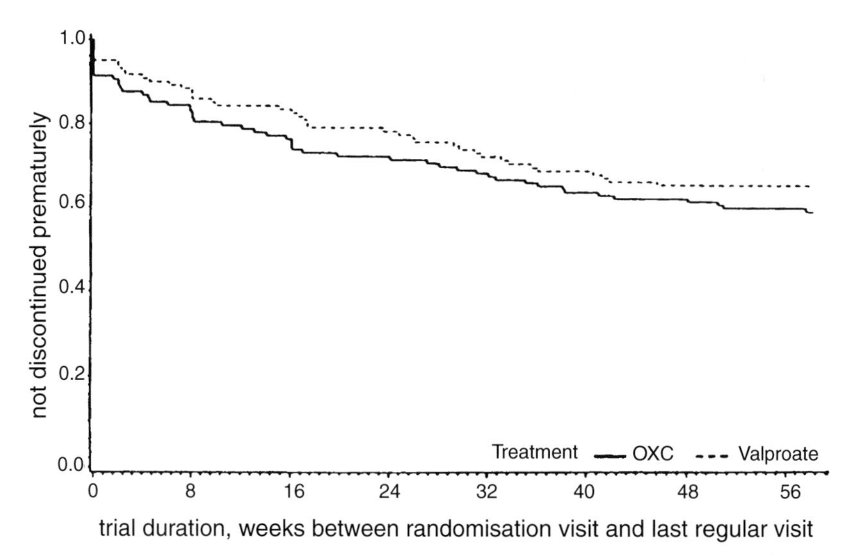

Figure 1. Comparative tolerability of valproate and the new antiepileptic drug oxcarbazepine. Shown are the number and percentage of patients with adverse reactions thought to be related to the treatment in this double-blind trial comparing valproate with oxcarbazepine in the treatment of previously untreated patients with epilepsy. Patients may have had more than one adverse reaction, and only adverse reactions noted in at least 5% in a treatment group are listed (Taken from [2a] with permission).

proex group versus the placebo group were nausea (48 vs 14%), vomiting (27 vs 75), abdominal pain (23 vs 6%), and anorexia (12 vs 0%). In addition, asthenia (27 vs 7%), somnolence (27 vs 115), and tremor (25% vs 6%) were more common on divalproex comedication than during placebo [22]. Except for tremor, the incidence of the adverse events decreased over the 4 months of treatment.The side-effects and adverse events of valproate were modest. In recent years valproate has also been used in an intravenous formulation [23]. The most frequent adverse effects of intravenous valproate were headache, reaction at the injection site, nausea (2.2% each), somnolence (1.9%), vomiting (1.6%), and dizziness and taste perversion (1.3% each). No persistent hematological or serum chemistry abnormalities were found. Vital signs were not significantly affected by the i.v. infusion of valproate [24, 25].

Finally, it may be worthwhile to review the adverse effects when valproate is employed for the treatment of disorders other than epilepsy including migraine and acute mania. Based on two placebo-controlled clinical trials and their long-term extension [26], divalproex was generally well tolerated with most adverse events rated as mild to moderate in severity. Of the 202 patients, 17% discontinued for intolerance.This compares to a rate of 5% for the patients on placebo. Reasons for discontinuation were alopecia (6%), nausea/vomiting (5%), weight gain (2%), tremor (2%), somnolence (1%), elevated SGOT and /or SGPT (1%), and depression (1%). In two placebo controlled trials of divalproex for the treatment of acute mania, nausea (22 vs 15%), somnolence (19 vs 12%), dizziness (12 vs 4%), vomiting (12 vs 3%), accidental injury (11 vs 5%) and asthenia (10 vs 7%) were reported by more than 10% of those on divalproex compared to placebo. Vomiting was the only event that was reported by significantly more patients receiving divalproex compared to placebo [26]. A total of 8% discontinued for intolerance to divalproex compared to 4% on placebo and 11% on lithium.

Although a direct comparison of the adverse effects of valproate with those of other anticonvulsants is difficult, the survey of individual adverse effects of commonly used antiepileptic drugs in Table 1 appears to suggest that valproate is causing fewer neurological adverse effects and fewer skin rashes than phenytoin, phenobarbital and primidone, and its tolerability and safety appears to be similar to that of carbamazepine. A comparison with newer antiepileptic drugs is even more difficult since the new drugs have considerably less exposure than valproate. Nevertheless, some of the newer drugs which are similarly effective such as lamotrigine may become a challenge for valproate. Lamotrigine may turn out to be less teratogenic, and has been shown to partly reverse some of the reproductive problems associated with valproate (see below). Although only effective for treatment of partial seizures, gabapentin has probably the best safety profile of all new drugs. Acute liver failure, reproductive dysfunction, teratogenic effects have not been described, and sedation or hypersensitivity reactions are not a problem. It should be noted however that valproate currently is and will remain for the foreseeable one of the best tolerated and, on a world wide basis, probably the most commonly used antiepileptic drugs with a broad spectrum of efficacy. The mostly mild to moderate individual side-effects will be discussed in more detail in the following section starting with early adverse effects seen in the first weeks of treatment.

Early adverse effects

Early adverse effects to the administration of valproate include mostly transient and modest gastro-intestinal side-effects at the initiation of therapy and mostly asymptomatic laboratory changes, while acute liver failure, pancreatitis, hypersensitivity reactions and acute encephalopathy, although altogether rare, are serious adverse reactions of valproate and require immediate medical attention (Tab. 1).

Gastro-intestinal side-effects

The most commonly reported adverse effects to valproate have been transient anorexia, indigestion, abdominal discomfort, nausea and vomiting occurring in up to 42%, the higher figures are seen in children [10]. In most cases they will disappear however within a few days after starting valproate. As discussed above, they are reported less often when enteric coated formulations of valproate are used [1]. In addition, anorexia and weight loss, diarrhea, abdominal cramps, and constipation have been reported in individual patients. Avoiding acidic drinks, e.g. orange or apple juice and coffee, attenuates indigestion and nausea in some patients. Although gastrointestinal side-effects are usually transient and mild to modest in intensity and rarely require discontinuation of therapy, it should be noted that in rare cases nausea and vomiting may herald the onset of valproate-associated acute liver failure. Severe abdominal pain may be a warning symptom of valproate-induced pancreatitis as will be discussed below. Much more common than these rare but serious gastro-intestinal complications are asymptomatic elevations of liver enzymes.

Elevated liver enzymes

Most traditional anticonvulsants including valproate induce minor elevations of the concentration in plasma of enzymes that are used as liver function tests. The elevations are usually dose-related. Up to fourfold increases of gamma-glutamyl-transferase (GGT) are commonly seen while elevated levels of SGOT and SGPT are seen in no more than 25% and 5%, respectively. In a controlled study, 22% of patients with normal values before receiving valproate developed an elevated serum aspartate aminotransferase during the first year on valproate [16]. In that respect it is important to note that, as a rule, bilirubin is not increased unless there is significant hepatic disease. How valproate induces elevation of liver function tests is not well known. Since these changes do not indicate impairment of hepatic function and may result in part from induced synthesis of the enzymes or "spilling" out of normal hepatocytes during physiological functioning [27] they do not necessitate withdrawal of the anticonvulsant. If the patient is asymptomatic without any evidence for aggravated liver dysfunction, moderate elevations do not require any modification of the therapeutic regimen. Unwarranted abrupt discontinuation of anticonvulsants out of ill-advised fear may precipitate a flurry of seizures, and infrequently, status epilepticus [28]. In addition, drug compliance may suffer and considerable anxiety may arise if physicians offer ill-founded conclusions that the medication may have caused liver problems [28].

Furthermore, changes in hepatic enzymes such as GGT, LDH, SGOT, SGPT and muscle serum creatine kinase may be elevated in the days following a tonic-clonic seizure; these changes may erroneously be attributed to incipient hepatic disease [5]. Finally, abnormal liver function tests may of course indicate hepatic disorders unrelated to drug treatment. Alkaline phosphatase may signal the development of osteomalacia [5]. The potential for erroneous interpretation is a further concern about the clinical utility of indiscriminate monitoring of liver function tests. If, on the other hand, the patient is showing symptoms and signs of hepatic disease, and liver function test rise beyond a moderate increase, e.g. three times the normal, continue to rise, or include abnormal bilirubin, aggravated liver dysfunction and potentially serious anticonvulsant-induced hepatotoxicity is more likely. Still, critics maintain that careful clinical monitoring may be just as effective or even more useful than indiscriminate laboratory controls in detecting early stages of valproate-induced acute hepatic failure [29].

Acute liver failure

Valproate, among other traditional anticonvulsants, has been implicated to contribute or to cause hepatic disease including acute hepatic failure. Among over one million patients receiving new prescriptions for valproate during the years 1987 to 1993, 29 patients developed fatal hepatotoxicity [2]. It is now well recognized that the early stages of valproate-associated serious or fatal hepatic failure may be heralded by unspecific presenting symptoms and signs such as decreased alertness, jaundice, vomiting, hemorrhage, increased seizures, anorexia, and edema [2, 30, 31]. Patients should be closely monitored for these presenting signs. It is these early warning symptoms and signs that allow the recognition of incipient hepatic failure in many, if not most patients. Early investigations suggested periodic routine monitoring of liver function in patients receiving anticonvulsants [32]. These tests were perceived by their advocates to protect patients from anticonvulsant-induced hepatic injury [33, 34]. Liver function tests were soon routinely recommended for monitoring the use of most anticonvulsants [26]. The inconvenience and the cost involved in repeated sampling and subsequent laboratory analysis, and a growing concern that frequent routine liver function testing may not reliably identify serious hepatotoxicity [35–39], prompted a reevaluation of the research evidence on the role of liver function tests in monitoring anticonvulsant use. What then, is the contribution of abnormal liver function tests for early recognition of valproate-associated hepatic failure ?

According to a careful recent review by Koenig et al. [29], the clinical symptomatology and liver function tests were described in sufficient detail for metaanalysis in 101 of the total of 177 published cases with fatal valproate-associated hepatic failure [29]. The following clinical symptoms and signs were leading to the diagnosis in 80 of 101 patients (79%) [29].These were: apathy, somnolence, nausea, vomiting, and coma. Jaundice and status epilepticus were the leading signs in only 3 patients each, respectively [29]. Hepatic failure was curiously preceded by infection and fever in more than 50% [29]. In 14 further patients a concomitant increase of transaminases or coagulation deficits was noted [29]. In two patients valproate was discontinued before liver function tests were performed, one patient was in status epilepticus [30], and one had a subarachnoidal hemorrhage [40]. As a consequence, in only 5 of the 101 patients available for analysis, abnormal liver function tests were preceding clinical symptoms and signs of hepatic failure [29]. These five patients are discussed in some detail. First, it should be noted that pre-existing abnormal liver function tests were documented in 3 of the 5 patients [41, 42] before valproate was started. One patient with Friedreich's ataxia in addition developed abnormal transaminases on valproate without any clinical symptomatology. Valproate was discontinued in this patient and phenytoin was given instead. Four days after changing to phenytoin, the 19-year-old patient developed somnolence and jaundice. Phenytoin was discontinued, yet the patient died the next day [43]. Finally, it is worthwhile to note that 3 of the 5 patients suffered from neurodegenerative disorders or had a familial history of unexplained fatal hepatic failure with so called Reye-like symptomatology [33, 41]. It is highly debatable if these 5 out of 101 patients do not justify routine laboratory monitoring during treatment with valproate. Furthermore, transaminases may not be elevated in severe cases of valproate toxicity, such as those with Reye's syndrome [44]. A major deficiency of these tests is that they do not assess how well the liver is performing at precisely the time they are measured; instead they are essentially indicators of liver dysfunction over the previous days and weeks [45]. This may explain in part why these tests are not invariably abnormal in patients with acute hepatic failure at the very recent onset. Given the absence of any prospective or randomized evaluation of screening procedures and the inherent flaws of retrospective analysis, it is of course difficult to be certain. Furthermore, such a clinical trial would have to include a very large sample size given the rare occurrence of acute hepatic failure. What can be said, however, is that current evidence, summarized above, does not support the notion that monitoring by liver function tests is of proven value for early recognition of valproate-associated hepatic failure. Instead, it is becoming quite clear that careful clinical monitoring appears to be much more effective than measuring liver function tests in detecting early cases of hepatic failure associated with the use of valproate. Based on the available evidence, it may be appropriate to re-consider recommendations for liver function tests in patients on anticonvulsants including valproate.

Periodic liver function tests are routinely recommended during treatment with valproate, e.g. "liver function tests should be performed prior to therapy and at frequent intervals thereafter, especially during the first 6 months" [26]. The two main goals of monitoring liver function tests are screening of patients for pre-existing hepatic disease prior to starting valproate, and early detection of serious hepatotoxicity. The premise to screen patients for hepatic disease prior to exposing them to valproate is two-fold. First, if preexisting disease is found, potentially hepatotoxic anticonvulsants such as valproate can be avoided, making room for newer antiepileptic drugs such as clobazam, vigabatrin, gabapentin or tiagabine which have not been implicated to cause or to contribute to hepatic injury [26, 28, 46, 47]. Second, in patients with normal liver function tests, a reference is established for later comparison, should it become necessary. It is for these reasons that routine liver function tests prior to starting valproate therapy are generally seen as clinically useful. Once the decision has been made to assess liver function for screening purposes, the test parameters need to be determined. Most commonly, GGT, SGOT, SGPT, and bilirubin are chosen. In developmentally retarded or handicapped children, additional screening parameters may be appropriate, such as blood glucose, lactate, ammonia, amino acids, and organic acids [29].

Although the relative contribution of liver function tests is not always easy to disentangle from the value of careful history taking and clinical examination, there is a consensus that patients with a history or a familial history of hepatic disease or non-endocrine pancreatic disease should not receive valproate. The same is true for children with a host of metabolic disorders, especially mitochondrial, peroxisomal and urea cycle disorders [29, 48]. Children under the age of 2 years who receive valproate as polytherapy were at the greatest risk (1:600) of developing fatal hepatotoxicity [2, 30, 31]. Above the age of 2 years experience shows that the incidence of fatal hepatotoxicity decreases considerably progressively in older patients [26]. Additional risk factors for valproic acid hepatic fatalities include polytherapy, developmental delay, and coincidental metabolic abnormalities [2, 30, 31]. Similarly, those with a family history of hepatic failure or unexplained death should avoid valproate [29]. As a consequence, valproate should only be given after careful consideration of all relevant aspects in children below the age of 2 years receiving other enzyme-inducing anticonvulsants, and especially in those children with severe mental retardation, neuro-metabolic or hereditary metabolic disorders with involvement of the liver [2, 30, 31]. Metabolic disorders such as urea cycle disorders, familial spongiotic glio-neuronal dystrophy, progressive familial cerebellar degeneration, progressive myoclonus epilepsies, and subacute leukodystrophy [29] were reported in patients with valproic acid hepatic fatalities. In fact, the development of valproate-associated hepatic failure may be seen as a catastrophic interaction of valproate with underlying preexisting disease such as progressive neuronal degeneration of childhood [41]. In a typical clinical course of progressive neuronal degeneration of childhood (Alpers' disease) with liver disease early developmental delay is followed by intractable epilepsy leading rapidly to death, in some cases in liver failure [42]. In these children, valproate is thought to aggravate an already existing liver disorder [42]. In several young children with liver failure ascribed to valproate, the association is thought to be coincidental and secondary to previously undiagnosed Alpers' disease [49]. Dodson [50] recently noted that several children who underwent liver transplantation for treatment of valproate-associated hepatic failure died subsequently of progressive brain disease. He speculated that valproate may have accelerated the fulminant hepatic disorder, but was not its primary cause [50]. Valproate-induced liver failure was recently described in one of two siblings with Alpers' disease [51]. The authors suggest that a diagnosis of Alpers disease should be considered in children with unexplained early developmental delay, cerebellar signs, or partial seizures, including epilepsia partialis continua. When Alpers disease is strongly suspected, the use of valproate should be avoided. Children with Reye syndrome or those with hepatic porphyrias may suffer exacerbations when exposed to valproate [2,30,31]. Valproate-induced acute liver failure was reported in a boy with medium chain acyl-CoA dehydrogenase deficiency 4 months after starting treatment with valproate [52]. The authors suggest that valproate should not be given in the treatment of seizures in patients with possible medium chain acyl-CoA dehydrogenase deficiency. As a matter of further precaution co-administration of salicylates should be avoided in children below 3 years since they employ similar metabolic pathways. Once the mechanism of valproate-associated hepatic failure is better understood, screening for patients for high risk will be more effective. Nevertheless, it is quite clear that liver function tests are valuable for screening of pre-existing acute symptomatic liver disease before exposing patients to potentially hepatotoxic anticonvulsants such as valproate, especially now, when better tolerable alternatives such as gabapentin are available.

One of the major challenges when monitoring the hepatotoxic effects of valproate is to distinguish between common, asymptomatic elevations in liver enzymes and rare but deleterious acute valproate-induced hepatic injury.

Practical guidelines for monitoring liver function tests

Given the lack of randomized prospective investigations on the clinical utility of liver function test to detect anticonvulsant hepatotoxicity, as discussed above, it is difficult to give general recommendations (Tab. 7).

Nevertheless, the available experience indicates that valproate hepatotoxicity may, at least in part, be prevented by excluding exposure to patients with preexisting hepatic disease and, in the case of valproate, those with a host of rare diseases, outlined above, where valproate may precipitate fatal hepatic failure. Monotherapy is recommended in children under the age of 3 years when prescribing valproate and the potential benefit of valproate therapy should be weighed against the risk of liver damage in these patients prior to starting the treatment. Liver function tests, i.e. transaminases, bilirubin and alkaline phosphatases should therefore be performed on every patient before any anticonvulsant is started. Knowing liver function test results has immediate therapeutic consequences because better tolerated anticonvulsants such as clobazam, or some of the newer agents such as gabapentin, vigabatrin, tiagabine may then be used. Among the newer drugs, felbamate [53] and progabide [54] should be avoided because they have been implicated to be hepatotoxic. As discussed above, routine liver function testing in patients has not been shown to be able to detect incipient hepatic disease that has not already been suspected based on careful history and clinical examination. No prospective studies have been undertaken of the predictive value of liver function tests for clinical hepatotoxicity. Based on the available evidence which is incomplete and retrospective in nature,

Table 7. Recommendations for monitoring liver function tests in patients receiving valproate

- Screening for preexisting overt liver disease with clinical examination and monitoring transaminases, alkaline phosphatase and total bilirubin before starting valproate. Patients with active liver disease, a history of significant hepatic dysfunction or a family history of liver disease should not receive anticonvulsants associated with acute liver failure or severe hepatotoxicity such as valproate. Patients with increased risk for acute liver failure should receive the drug only after careful consideration of all circumstances involved.
- Careful and repeated clinical monitoring for presenting symptoms of acute hepatic failure, especially during the first 6 months of treatment. Presenting symptoms may be unspecific, and include malaise, lethargy, weakness, somnolence, facial edema, vomiting, abdominal discomfort, aversion to food, and worsening of seizure control. Whenever hepatic disease is suspected, a complete examination including appropriate liver function tests is of course required immediately. In case of doubt, valproate should be discontinued at once. Indiscriminate repeated liver function tests are not recommended because patients with acute liver failure may not invariably have abnormal values at the onset.

it would be difficult to rationally justify any repeat laboratory testing. Nevertheless, given the fact that most cases of acute anticonvulsant-induced hepatic failure occur within the first 6 months of exposure, testing of SGOT at one of the clinical visits early in the course of treatment (weeks after starting treatment) may be considered. Other laboratory tests, such as blood ammonia, are much more complicated; protein ingestion will increase the concentration of ammonia, there is much more intraindividual variation, and the tests are not practical to use. Most patients will however be asymptomatic and have moderate and isolated changes in transaminases and normal bilirubin that bear no relationship to incipient hepatic disease, and consequently have no therapeutic consequences. In otherwise healthy patients with epilepsy or migraine [39], clinicians do not need to be overly preoccupied with monitoring liver function tests [39]. The current data strongly suggest that careful clinical supervision within the first 6 months and alerting patients, relatives and caregivers to the warning symptoms and signs of incipient hepatic failure is more effective. In patients suspected of developing hepatotoxicity, the incriminated drug should be immediately discontinued, and, if necessary, newer drugs such as vigabatrin, tiagabine, topiramate, gabapentin or benzodiazepines at reduced doses tailored to the degree of impaired hepatic biotransformation can be given to maintain seizure control [26]. Carnitin supplementation has recently been claimed to contribute to successful recovery after immediate withdrawal of valproate [29]. If standard conservative care fails, liver transplantation is an option.

Pancreatitis

The use of valproate has been associated with pancreatitis [26, 55]. Fatal outcome has been reported. Pancreatitis may occur as early as week 6 [56]. However individual cases have been reported as late as 12–18 months of treatment [16]. Serum concentrations were at the therapeutic range in two adolescents developing pancreatitis [56]. Asymptomatic and minor elevation of amylase is however common and the clinical relevance remains uncertain. Nevertheless, screening for amylase may therefore be useful before starting valproate and whenever patients experience severe abdominal pain, nausea and vomiting pancreatitis, although uncommon, needs to be excluded. Valproate should be discontinued immediately in patients suspected to have valproate-induced pancreatitis.

Hypersensitivity reactions

Hypersensitivity reactions to valproate are uncommon. A recent study found no serious cutaneous diagnosis in 1504 new valproate users [57]. In a controlled study, 1.7% of patients developed an exanthematous rash while receiving valproate [58]. Furthermore, it has to be kept in mind that rashes may also be seen in about 3% of patients during exposure to placebo [26]. In exceptional cases toxic epidermal necrolysis, Steven Johnson Syndrome and Erythema multiforme pruritus and photosensitivity have been noted [26].

Hematological adverse effects

Valproate may cause thrombocytopenia and inhibit the secondary phase of platelet aggregation leading to altered bleeding time, petechiae, bruising, hematoma formation and overt hemorrhage in individual patients [59]. Platelet count was inversely correlated to VPA dose and both free and total serum concentration of valproate. In adult patients with refractory partial seizures 27% of those randomized to valproate through concentration of 80–150 mg/l developed thrombocytopenia. All patients had normal platelet counts prior to the divalproex trial. Platelets normalized in all patients whether the drug was withdrawn or not [60]. Platelets may drop as low as 8000 per µl with no symptoms [16]. Relative lymphocytosis, macrocytosis, hypofibrinogenemia, anemia, macrocytic with or without folate deficiency, leucopenia, eosinophilia, bone marrow depression, transient myeloid dysplasia, and acute porphyria have been noted [26]. Acute promyelocytic leukemia was diagnosed in an infant receiving valproate, all hematopoetic abnormalities resolved after valproate was discontinued [61]. These adverse effects were dose-dependent and seen primarily during treatment with high doses. Clinical problems are uncommon, except rarely in patients undergoing surgery. A recent report pointed out however that bleeding complications were not found in 111 patients undergoing cortical sur-gery while receiving valproate treatment at the time of surgery despite abnormalities in laboratory parameters such as lower platelet counts, lower baseline fibrinogen, lower red blood cell counts, hematocrit and hemoglobin levels [62]. There was no significant difference between patients receiving valproate and those receiving other antiepileptic drugs in estimated blood loss during surgery or qualitative wound discharge postsurgery. In the view of the authors there is no need to discontinue valproate therapy prior to elective surgery. Thromboctopenic purpura and anemia has been reported in a breast-fed infant whose mother was treated with valproate [63]. As the mother stopped breastfeeding, the infant recovered. In addition, asymptomatic folate deficiency may develop during exposure to valproate. Folate substitution is therefore recommended in women of childbearing age (see pregnancy).

CNS side-effects including acute and subacute encephalopathy

Sedation has been noted in patients receiving valproate but usually abates after a few days of treatment and rarely requires discontinuation. Sedation is seen more often during rapid titration and during co-administration of phenobarbital [8]. Patients should be informed of the risk of somnolence or sedation especially during the early weeks of treatment with valproate and in those receiving benzodiazepines and barbiturates. Asterixis has been noted with valproate, and tremor is a common transient, dose-related effect reported since the earliest use of the drug [7]. In a controlled study 45% of patients developed tremor [16]. After 12 months of treatment mild occasional tremor was present in 32% of those taking valproate. A tremor readily evident to the patient and the physician was reported or found in 6% of those on valproate and marked dysfunctional tremor was seen in another 3% [16]. Usually it starts within the first weeks of exposure and its occurrence is dose dependent [64]. Most cases have serum concentrations of about 100 mg/l.

Other uncommon and mostly mild CNS effects include ataxia, nystagmus, dysarthria, diplopia, asterixis, and incoordination which often disappear despite continued treatment. Rarely, lethargy, coma, confusion, delirium, nausea, ataxia, asterixis, dysarthria, worsening of seizures, drowsiness, incontinence, lethargy, poor concentration, and not invariably hyperammonemia have been described as symptoms and signs of an acute or subacute encephalopathy associated with the initial treatment with valproate [65, 66]. In 75% of 39 cases reviewed by Bauer and Elger [67] encephalopathy developed within the first week of valproate therapy. In some cases, there was an increase of seizures while on therapy and these problems disappeared in 85% within 5 days when valproate was withdrawn or the dose of valproate was reduced [67–72]. In one recent case fatal encephalopathy with massive cerebral edema and central herniation has been described [72]. Just prior to death, plasma levels of free and acyl carnitines were decreased. The relationship of the encephalopathy to the acute administration of valproate in this case is unclear. Reexposition to valproate may result in a relapse of the encephalopathy [68, 74, 75]. The EEG usually shows background slowing and generalized high voltage theta-delta activity. Although reported to occur more frequently during rapid titration of valproate or during co-administration of phenobarbital, it should be noted that stupor may also be seen, albeit in rare cases where valproate is given alone and during recommended titration schedules and valproate serum concentrations may not be elevated. Hepatic laboratory tests are normal. Measuring ammonia blood concentration in patients with stupor cannot be generally recommended since elevated ammonia levels may also be seen in asymptomatic patients and are not invariably present in patients developing stupor. The encephalopathic effects of valproate have been explained by the urea cycle inhibition resulting in hyperammonemia and a CNS depressant effect of valproate which is synergistic with ammonia [11]. Rapid titration of valproate or comedication with phenobarbital was reported in some, but not all patients. In one case of nonconvulsive status epilepticus, valproate was implicated which was given in a fast dose escalation of up to

1600 mg/d over 5 days. This patient improved after receiving the benzodiazepine antagonist flumazenil and deteriorated after administration of midazolam [76]. The authors suggested enhanced benzodiazepine receptor activity as a possible cause of valproate-induced acute encephalopathy. The mechanism of acute encephalopathy is still an enigma, and may involve hitherto unrecognized (mitochondrial?) metabolic idiosyncrasy of individual patients. It should be kept in mind however that stupor, lethargy and worsening of seizures may also be a symptom of nonconvulsive status epilepticus and may be, although extremely uncommon, one of the early signs of acute liver failure. Finally, it should be noted that exacerbation of seizures may be an important heralding sign of impending acute liver failure as discussed earlier. In addition, worsening of seizures in patients on valproate should alert the physician to suspect an inborn error of mitochondrial metabolism. Mitochondrial myopathy, encephalopathy, lactic acidosis and stroke-like episodes (MELAS) were diagnosed in a patient with worsening of seizures during exposure to valproate [77]. The underlying mitochondrial DNA defects should be sought for family screening and genetic counseling. In addition, the authors suggest that valproate should not be given to patients with suspected mitochondrial diseases. Although not directly related to the development of encephalopathy, a recent study on seven children treated with valproate found valproate-induced ultrastructural abnormalities of mitochondria in skeletal muscle [78] suggesting perhaps a link between valproate treatment and mitochondrial dysfunction. A rapid reduction of the dose of valproate or withdrawal, if necessary, and EEG-monitoring to exclude status epilepticus are recommended in any patient suspected to have valproate-induced encephalopathy.

Late adverse effects

Although most adverse reactions to valproate seen early in the treatment are generally mild to modest in intensity, long-term risks, although altogether uncommon, have raised concerns in recent years. These include reversible cognitive impairment associated with pseudoatrophy of the brain and insidious onset of Parkinsonism, polycystic ovaries and other insulin-related complications, while the development of neural tube defects and other minor or major congenital malformations in the offspring of women exposed to valproate during pregnancy is well known (Tab. 1).

Reversible dementia with brain atrophy and Parkinsonism

Reversible dementia associated with reversible cerebral pseudoatrophy has been reported in children and adults taking valproate [65, 79, 80]. Insidious onset of mental deterioration coincided in one further child with serum valproate concentrations near 100 mg/l. Following discontinuation of VPA repeat MRI showed disappearance of pseudoatrophic changes and disappearance of clinical symptoms with an improvement of IQ testing [81]. Schöndienst et al. [82] reported four cases with reversible pseudo-atrophy documented by repeat computer tomography who had cerebellar dysfunction, dystonic signs, and cognitive impairment after chronic valproate treatment resulting in serum concentrations ranging from 60–150 mg/l. In three of the four patients the symptoms and signs including the pseudo-atrophy were reversible when valproate was withdrawn. A reversible Parkinsonian syndrome was reported from Australia [83] and Japan [84]. Extrapyramidal symptoms have been noted in an elderly manic patient who received valproate in addition to lithium [85]. Armon et al. [86] have presented adult patients with reversible Parkinsonism, subtle cognitive deficits and, at least in one of their patients reversible pseudo-atrophy following treatment with valproate for at least 1 year. The authors warn that the association may be overlooked due to the insidious onset [86]. In their view, mitochondrial respiratory chain dysfunction induced by valproate may be a candidate mechanism for reversible valproate-induced Parkinsonism and cognitive impairment. Discontinuation of valproate leads to subjective and objective improvement at least 3 months later. Although placebo effects may possibly have contributed to the improvement after discontinuation of valproate, the findings are serious enough to merit a controlled investigation. Clinical monitoring for subtle signs of Parkinsonism in patients receiving valproate is advised. The pros and cons of replacing valproate by alternative antiepileptic drugs should be carefully considered.

Psychiatric and cognitive side-effects

Emotional upset, depression, psychotic reactions, aggression, hyperactivity and behavioral deterioration have been noted. Psychotic reactions may occur during treatment with most antiepileptic drugs including valproate [87]. Behavioral changes have been reported in children receiving valproate. In one study, 63% of all children developed behavioral alterations [88].The most common adverse effects were irritability (19%), insomnia (16%), hypersomnia (19%), hyperactivity (11%), and lassitude (8%).

In a careful review of cognitive side-effects of six recent comparative trials of valproate versus other antiepileptic drugs, Vermeulen and Aldenkamp [89] noted that all trials had significant methodological problems rendering meaningful interpretation very difficult. Vining et al. [90] compared phenobarbital and valproate in 21 school children each, and, in the words of the reviewer reported scattered significant findings favoring valproate possibly due to insufficient protection against type I error. Meador et al. [91] compared cognitive adverse effects in adults with long-standing epilepsy and found no difference in a small sample of 15 patients each treated with valproate, phenobarbital and valproate. In newly diagnosed epilepsy in school-age children receiving valproate or phenytoin or carbamazepine in groups of 14 patients each, valproate and phenytoin were favored over carbamazepine. A possible bias was noted against carbamazepine and phenytoin which are not effective against absence seizures in this series [92]. In a trial comparing valproate with phenobarbital in school-age children on chronic treatment, Callandre et al. [93] concluded that phenobarbital, but not valproate induce cognitive impairment. Helmstaedter et al. [94] found no difference in cognitive adverse effects between valproate (5 patients) and carbamazepine (11 patients) in a small sample of previously untreated patients with fairly long-standing epilepsy. Finally, Craig and Tallis [95] looked at elderly patients of 60 years or older with newly diagnosed seizures and found no difference between the phenytoin and the valproate group, again in a small sample. When school-age children who were seizure free were withdrawn from their medication including valproate (n=17), phenytoin (n=10), and carbamazepine (n=56), cognitive monitoring revealed no differences [96]. It has been critically noted however that withdrawal may be insensitive to permanent or slowly reversible effects, and furthermore, that the no difference finding was based on small sample size. In addition, the serum concentration of phenytoin was low and cognitive function was assessed by few measures [89]. The problem with a small sample being of course that a no difference outcome is difficult to interpret due to poor statistical power to detect differences in cognitive adverse effects of clinical significance between the drugs, if they existed. In summary, it is clear that despite the methodological criticism, valproate does not seem to have large, clinically significant differences in cognitive adverse effects compared to other antiepileptic drugs, and may be less prone to induce cognitive effects than phenobarbital in chronic treatment. In a small sample, Brouwer et al. [97] found no difference in cognitive performance when conventional valproate was switched to a slow release formulation of valproate. Performances on attention and vigilance were assessed in children. However, mean diurnal trough and peak serum concentrations were quite similar. In addition, comparison of performance of the epilepsy group during treatment with conventional and controlled release formulation and matched controls showed no differences in a posttest-only comparison.

Hearing loss

Hearing loss, either reversible upon discontinuation of valproate or permanent has been noted in patients while receiving valproate [55]. In one series of 35 patients with valproate-induced Parkinsonism 63% had progressive hearing difficulties and this was documented by audiometry in some patients. Twenty-three patients were studied by audiometry before and after discontinuation of valproate. Ten patients improved, one worsened, ten had no change, and two were unclassifiable [86]. A possible mechanism could be the production of mitochondrial respiratory chain dysfunction by valproate that has been shown to affect Complex 1 activity *in vitro* [98]. It should be noted that progressive hearing impairment is common in patients with mitochondrial encephalomyopathies [99]. Development and progression of hearing impairment may precede some of the cognitive and motor changes, and should prompt evaluation for their presence. If discontinuation is considered, the adverse effects of valproate should be balanced against its effectiveness in seizure control, taking into account the availability of other antiepileptic drugs.

Hair loss

Valproate-associated transient hair loss or change of hair texture have been reported in 2 to 12% of patients, particularly in those given high doses during long-term therapy [16, 32, 100]. In many cases transient hair changes are noted after several months of exposure to valproate. In some patients alopecia requires discontinuation of valproate.

Weight changes

An increase in body weight is seen in as many as 8%–59% of patients receiving valproate and may be severe enough to require discontinuation [8, 14, 101]. In fact, weight increase is probably, next to tremor, the most frequent adverse effect of valproate in clinical practice. In a controlled trial, 20% of the patients on valproate had a weight gain by more than 5.5 kg [16]. In a report from Canada, 70% of the patients on valproate recorded a weight gain of more than 4 kg. Patients with no history of weight gain and those with below or within normal range body mass prior to starting valproate experienced the most severe percentage weight gain [102]. The exact mechanism by which valproate induces weight gain is not known. However, valproate-induced impairment in fatty acid beta oxidation may be of relevance [14, 15]. Others have hypothesized that obesity may result from increased availability of long-chain fatty acids due to competitive valproate binding [103]. In most cases the weight change is noted after 3–5 months of exposure and may be as high as 10% of the body weight. In a recent study on children with epilepsy, aged 4–15 years, there was no difference in weight gain between those receiving valproate or carbamazepine, suggesting perhaps that prepubertal patients are less prone to develop excessive valproate-associated weight gain [104]. The weight gain seen in approximately 50% of women with epilepsy during treatment with valproate may be progressive. Isojärvi and co-workers have shown that in women with epilepsy valproate-related obesity is associated with hyperinsulinemia, polycystic ovaries, and hyperandrogenism, which were found in 10 of 11 women with an indisputable weight gain (mean, 22 kg; range, 8–49 kg) during valproate medication [4]. Low serum concentrations of insulin-like growth factor-binding protein 1 were found which, according to Isojärvi et al. [4], may lead to hyperandrogenism and polycystic ovaries. Since weight gain seems to be of the same magnitude in both genders during valproate treatment [14], studies on possible valproate-induced hyperinsulinemia and related risks are pertinent in men with epilepsy. No data are currently available.

Endocrine and reproductive dysfunction

Irregular menses, secondary amenorrhea, breast enlargement, galactorrhea, and parotid swelling have been reported. Thyroid function tests may be abnormal although thyroid disease has not been described in patients receiving valproate. Reproductive endocrine disorders such as the polycystic ovarian syndrome (POCS) are common in women with epilepsy, especially those with left-sided medial temporal lobe epilepsy [105]. Valproate has been implicated to contribute to the development of POCS, hyperandrogenism and menstrual dysfunction. Isojärvi et al. [106] reported in 1993 that women with epilepsy treated with valproate had a higher incidence of POCS (43%), menstrual dysfunction (45%), and hyperandrogenemia (17%). In a follow-up study including both Finnish and Norwegian patients, POCS were found by the same group in 62% of women taking valproate as compared with 22% taking carbamazepine [107]. Furthermore, more frequent ovulatory dysfunction and POCS occurrence in patients taking valproate was reported in an Italian study [4a]. Experimentally significant changes in sex-steroid hormone levels and ovarian structure were found in female rats treated with valproate [108]. Substituting lamotrigine for valproate in women with valproate-related polycystic ovaries or hyperandrogenism decreased the number of polycystic ovaries from 20 during valproate administration to 11 1 year after replacing valproate with lamotrigine. In addition, the body-mass index and fasting serum insulin and testosterone concentrations decreased during the first year of replacing valproate by lamotrigine [109]. Furthermore the HDL-cholesterol/total cholesterol ratio increased from 0.17 to 0.26. The authors concluded that valproate induces a metabolic syndrome with centripetal obesity, hyperinsulinemia, lipid abnormalities and polycystic ovaries/hyperandrogenism. These valproate-related risks can be partially reduced by substituting lamotrigine for valproate. Seven women had menstrual disturbances during valproate medication, but the menstrual cycle became normal in all but two of them during the first year after valproate was changed to lamotrigine. Most women were obese, perhaps due to valproate, and obesity is implicated in the development of POCS as discussed in the section on weight gain. It is unclear whether the valproate-induced weight gain is the only mechanism responsible for the development of POCS or whether other direct effects of valproate may play a role in the development of POCS. Although it would be premature to restrict the use of valproate in female patients, physicians should be aware of this valproate-related risk, and valproate should be avoided in women with obesity, symptoms and signs of POCS, menstrual dysfunction or signs of hyperandrogenism such as hirsutism. In women receiving valproate showing obesity, signs of polycystic ovaries and hyperandrogenism, valproate should be replaced by lamotrigine. Although infertility is common in women with POCS, pregnancies do occur and they are fraught with a higher rate of obstetric complications. It is pertinent to study if valproate-induced POCS is a cause of complications in pregnancy of women with epilepsy. In addition to infertility due to polycystic ovaries and hyperandrogenism, other features of insulin-resistance such as androgenic obesity and lipid abnormalities with elevated serum triglycerids and a low HDL-cholesterol levels and HDL-cholesterol/total cholesterol ratios are well known risk factors for cardiovascular disease [110]. Clearly, the evidence that valproate may contribute or even induce an insulin-mediated metabolic syndrome in some women is of great clinical relevance. Independent confirmation is urgently warranted to assess both the scope of the problem and the relative contribution of valproate.

Finally, infertility was reported in a male patient receiving valproate who was fertile before when taking carbamazepine and became fertile again after switching to carbamazepine. In this rare case a direct gonadal or sperm membrane effect of valproate was suggested [111]. In a report from India, the total testosterone level may be reduced after 1 year of treatment with valproate [112]. Chronic toxicity studies in juvenile and adult rats and dogs have demonstrated reduced spermatogenesis and testicular atrophy at doses greater than 200 mg/kg day in rats and greater than 90 mg/kg per day in dogs [26]. The effect of valproate on testicular development and on sperm production and fertility in humans is unknown.

Systemic lupus erythematosus

In three cases mild systemic lupus erythematosus (SLE) has been reported in association with valproate [113]. The drug-induced SLE is difficult to distinguish from spontaneous occurring cases, although renal and neurological symptoms and signs appear to be less common compared to spontaneous cases [10]. Valproate should be immediately discontinued, if possible, especially now that new drugs which have not been implicated such as gabapentin are available.

Metabolic changes including hyperammonemia

Hyponatremia [114] and inappropriate ADH secretion have been noted as well as hyperglycinemia which was associated with fatal outcome in a patient with preexistent nonketotic hyperglycinemia. A renal Fanconi Syndrome has been described in two children on valproate [115]. Discontinuation of valproate was associated with a complete recovery of renal tubular function in one patient. The authors believe that children who suffered from perinatal anoxia may be at risk and recommend screening children receiving valproate for this potentially reversible renal tubulopathy. Decreased carnitin concentrations have been noted but their clinical relevance remains unclear. Hyperammonemia has frequently been reported in patients receiving valproate [116]. As most cases are asymptomatic and are not associated with any evidence for liver disease, the clinical relevance of this laboratory finding is very limited, except for rare patients with metabolic disorders such as carbamyl-phosphate-synthetase deficiency [117]. In one child with this deficiency syndrome valproate led to a hyperammonemia (226 µmol/l) with vomiting and lethargy. The mechanism is not well known. It may however involve inhibition of the urea cycle by valproate [11]. Carnitin supplementation may lower serum ammonia concentration in asymptomatic children [118]. However, carnitin substitution should be reserved for those patients with clinical evidence of carnitin deficiency [119]. In patients with stupor and associated hyperammonemia, discontinuation of valproate is recommended.

Miscellaneous adverse effects

Edema of extremities, angioedema, gingival hyperplasia [120], enuresis, and fever have been independently associated with the use of valproate in individual cases.

Overdose

Clinical signs of acute massive overdosage usually include coma with muscular hypotonia, miosis, and respiratory dysfunction including apnea. Myoclonic seizures have been reported in patients after acute massive overdose of valproate [120] possibly as part of a drug-induced encephalopathy as discussed above. Hospital management of overdose should include oral-activated charcoal [122] and gastric lavage, if the patient is seen within 12 h after ingestion, osmotic diuresis, and cardiac and respiratory monitoring. Dialysis and exchange infusion have been successful in very severe cases [123]. Direct hemoperfusion was successful in a patient with coma during an acute valproate intoxication. The authors suggest that the coma may have been related to the inhibition of beta-oxidation in the mitochondria, which was reversible by elimination of plasma valproate by direct hemoperfusion [124]. Naloxone has been successfully employed in one case [29]. The use of flumazenil has not been evaluated. Although rare, fatal outcome of overdose has been reported [26].

Interactions of valproate affecting other drugs

Effects of valproate on other drugs could involve one or more of the following mechanisms. Valproate is a weak inhibitor of different p-450 enzymes. At concentrations of 3000 µM, which is approximately four to five times the effective clinical serum concentration, valproate mildly inhibits (8.1%–17.3%) CYP1A2, CYP3A4, and profoundly inhibits (57%) CYP2C9, while it does not inhibit CYP2D6 or CYP2E1 [13]. In addition to being a weak inhibitor of the epoxide hydrolase and glucuronyl transferase, valproate displaces other drugs from their protein binding and, at the same time may exert pharmacodynamic effects on other drugs. As a consequence of enzymatic inhibition of P-450 enzymes, high doses of valproate may lead to an moderate (10–25%) increase of the serum concentration of a number of drugs including carbamazepine-10,11-epoxide, lamotrigine, lorazepam, nimodipine, phenobarbital, and zidovudine. The list of interactions is not exhaustive, since new interactions are continuously being reported. Among the numerous interactions, many are of little clinical relevance, the exception being the interaction involving valproate plus phenobarbital or lamotrigine and, to a lesser degree, the interaction between carbamazepine and valproate. Finally it is useful to keep in mind that interactions may be not only relevant when adding another drug but also, and this is less well appreciated, interactions may cause side-effects when a drug is withdrawn.

Valproate and phenobarbital

There is evidence that valproate may cause an increase in the serum concentration of phenobarbital, mainly by inhibition of the P-450 isoenzyme CYP2C9 [13]. Coadministration of valproate, 250 mg bid, for 14 days with 250 mg phenobarbital results in a 50% in the half-life, and a 30% decrease in the phenobarbital plasma clearance [13]. Valproate has been shown to inhibit both the direct N-glucuronidation of phenobarbital and the O-glucuronidation of p-hydroxyphenobarbital [136]. Phenobarbital stupor or coma may however also develop without elevated phenobarbital serum concentration. As a consequence, patients, especially children and adolescents, should be clinically monitored for sedation for several weeks following the addition of valproate, and the dosage of phenobarbital should be reduced, if appropriate. Primidone, which is metabolized to phenobarbital, may be involved in a similar interaction.

Valproate and lamotrigine

Valproate acutely inhibits lamotrigine metabolism [137]. When valproate is added, the clearance of lamotrigine is reduced by approximately 21%

within the first hour. Hepatic competition between lamotrigine and valproate for glucuronidation has been implicated as a possible mechanism [137]. The serum concentration of lamotrigine is increased and the risk of lamotrigine-induced rash and, less common and less well documented, tremor is increased in the presence of valproate. The rash rate in patients receiving valproate plus lamotrigine was 19.5% compared to 14.5% when lamotrigine was given alone [138]. In a steady state study of volunteers, the elimination half-life of lamotrigine increased from 26–70 h with valproate co-administration [139]. Lamotrigine should be titrated in 25 mg increments every 2 weeks until week 4, followed by 25–50 mg increments every 1–2 weeks in patients receiving lamotrigine plus valproate without any other (enzyme-inducing) antiepileptic drugs [140].

Valproate and phenytoin

The displacement of proteins from highly bound drugs, e.g. phenytoin, aspirin, carbamazepine and dicumarol is another mechanism of interaction of valproate. The addition of valproate may increase the free concentration of phenytoin in serum. At the same time, the total serum concentration may however increase because valproate inhibits the metabolism of phenytoin [141]. The two conflicting mechanisms may result in unpredictable effects on the serum concentration of phenytoin, and a decrease or no change of the concentration may result in individual patients. Although protein displacement is common and predictable, it alone appears to be rarely responsible for adverse effects. Breakthrough seizures or unexpected phenytoin toxicity may occur when valproate is added to phenytoin. The dosage of phenytoin has to be adapted based on the individual response, if necessary.

Valproate and carbamazepine

Clinical toxicity has been reported when valproate is added to carbamazepine, possibly due an indisputable increase in the serum concentration of carbamazepine-epoxide [142]. Valproate appears to inhibit the glucuronidation of carbamazepine-10,11-trans-diol, and probably also inhibits the conversion of carbamazepine-10,11-epoxide to his trans-diol derivative, rather than simply inhibiting the latter reaction only [143]. Others have suggested that valproate is probably not a selective inhibitor of epoxide hydrolase, and affects nonspecifically all steps of the epoxide-diol pathway of carbamazepine metabolism [144]. In addition, protein-binding displacement results in higher free fractions of carbamazepine and its 10,11-epoxide [145] of carbamazepine, a decrease of carbamazepine serum concentration may however also occur in individual patients. In the maintenance phase of a recent controlled trial no significant changes were seen in the serum concentrations of phenytoin, carbamazepine, or 10,11-carbamazepine epoxide between the divalproex group or the placebo group [22]. The clinical relevance of the interaction may be small in most patients. Clinical monitoring and dosage adaptation, if appropriate, are advised.

Miscellaneous interactions of relevance

Adding valproate to clonazepam has exacerbated absence status in individual cases [26]. Valproate inhibits the glucuronide conjugation of lorazepam resulting in higher serum concentrations of lorazepam [146, 147]. Following the addition of valproate, peak serum concentrations of nortryptilin have increased by 19% [148]. The serum concentrations of neuroleptics and antidepressants may increase when valproate is added. In most cases this is of little concern, although in individual cases it may have clinical relevance and require dose adaptation or discontinuation of valproate. Valproate may increase the serum concentration of clomipramine [149]. The patient went into status epilepticus 12 days after initiation of 75 mg clomipramine to treat depression. Valproate inhibits glucuronidation of zidovudine and increases the concentration of zidovudine in serum and cerebrospinal fluid and may result in zidovudine toxicity [150, 151]. In a child receiving valproate and phenytoin, a 6-day acyclovir treatment reduced valproate and phenytoin serum concentrations to subtherapeutic levels, perhaps due to a reduction in the bioavailability of valproate and phenytoin [152]. Valproate may also potentiate the action of CNS depressants, e.g. alcohol, benzodiazepines and antidepressants. Clinical monitoring is therefore advised. Caution is recommended when valproate is added to drugs affecting coagulation, e.g. aspirin or warfarin. *In vitro* valproate increased the unbound fraction of warfarin, but the clinical relevance is unknown; displacing warfarin from binding sites may cause transient potentiation of anticoagulant effect. In addition, valproate may lead to thrombocytopenia and inhibit the secondary phase of platelet aggregation possibly leading to altered bleeding time, petechiae, bruising, hematoma formation and overt hemorrhage as discussed above. Coagulation tests are recommended if valproate therapy is instituted for a patient on anticoagulation. Clinical monitoring is advisable especially during the first weeks of adding valproate, and the dosage of these compounds should be appropriately adjusted.

Drugs that are not involved in relevant interactions with valproate

None or no likely clinically relevant effects of valproate have been noted for the following drugs, although individual variation may occur: acetaminophen, amytriptyline/nortriptyline, caffeine, clozapine, gabapentin, lithium, lorazepam, oxcarbazepine, paroxetin, piracetam, remacemide, tiagabine, topiramate, vigabatrin, and zonisamide. Last but not least, valproate has not been reported to reduce the efficacy of oral contraceptives because clinically relevant doses of valproate do not have enzyme-inducing properties. Given the teratogenicity of valproate, the lack of interaction between valproate and oral contraceptives is reassuring.

Interactions of co-administered drugs on valproate

Interactions of drugs with valproate can be conveniently classified as those often resulting in reduced steady-state concentrations of valproate, those which will lead to increased serum concentrations of valproate, and finally those that usually do not require any adaptation of the dose (Tab. 8).

Table 8. Effect of other drugs on valproate [26, 125, 135]

Drug	Effect on valproate serum concentration
Enzyme inducing antiepileptic drugs (including carbamazepine, phenobarbital, primidone and phenytoin)	Lower valproate serum concentration, higher dosage of valproate and lowering the dose of the other antiepileptic drug may be necessary
Felbamate	Higher valproate serum concentration, lowering the dosage of valproate is recommended
Mefloquine	Higher valproate serum concentration, mefloquine may exacerbate seizure in addition
Aspirin or other highly protein-bound medications including phenytoin and tiagabine	Free valproate concentrations in serum may be higher, total concentration may be lower
Vitamin K dependent factor anticoagulant	Close monitoring of prothrombin rate is recommended
Cimetidine	Higher valproate serum concentration due to enzyme inhibition, controversial, see text
Erythromycin	Higher valproate serum concentration due to enzyme inhibition
Mefenamic acid	Free valproate concentrations in serum may be higher, total concentration may be lower
Doxorubicin	Lower valproate concentration
Nimodipine	Higher valproate concentration
Folic acid	Lower valproate concentration
Cholecystiramine	Lower valproate concentration

Drugs that will lower the serum concentration of valproate

Drugs that affect the level of expression of hepatic enzymes, particularly those that elevate levels of glucuronyl transferases may increase the clearance of valproate and reduce its serum concentration at steady-state by 30–40% leading to increased dose requirements [153, 154]. Examples are carbamazepine, phenobarbital, primidone or phenytoin that decrease valproate serum concentrations. Conversely, when enzyme-inducing drugs are discontinued, the serum concentration of valproate may increase dramatically in some patients resulting in intoxications with valproate [155]. Carbamazepine and phenytoin induce a two-fold increase in the formation of delta-4-en- valproate and stimulate omega-oxidation and omega1-oxidation. Coadministration of up to 24 mg/d tiagabine lowers the steady-state serum concentrations of valproate by 10–12% [156]. Lamotrigine causes a small decrease of valproate serum concentration by 25% which is probably without clinical relevance in most patients [157]. Ethosuximide has been described to lower the serum concentration of valproate by approximately 30% [158].The mechanism of this interaction is unclear. The antimicrobial agent panipenem/betamipron may decrease the serum concentration of valproate [159]. In general, dosages should be adjusted according to serum concentrations in case of polytherapy if the patient does respond upon the addition of valproate. Failure to reach effective serum concentrations of approximately 100 mg/l may explain the poor efficacy in some patients, it should be noted however, that adding valproate can be expected to result in complete seizure control in approximately 8% of patients with refractory partial seizures [22].

Drugs that will increase the serum concentration of valproate

First of all, it should be noted that the serum concentration of valproate may increase by up to 42% when comedication with enzyme-inducing drugs is discontinued [160]. Unfortunately this common cause of valproate toxicity is often underestimated. Drugs that are inhibitors of P-450 enzymes, e.g. chlorpromazine, cimetidine, erythromycin, fluoxetine, isoniazid and stiripentol may be expected to have only minor effects of valproate clearance in the range of 10–20% because cytochrome P450 microsomal mediated oxidation is a relatively minor secondary metabolic pathway compared to glucorunidation and beta-oxidation (see chapter on metabolism of valproate in this volume). Stiripentol inhibits the formation clearance of the Delta 4-en-valproate by 30%. Isoniazid may inhibit the metabolism of valproate resulting in valproate intoxication [161]. The following list of drugs cannot be exhaustive since new interactions are continuously reported. Aspirin, phenylbutazone, and naproxene coadministration may decrease the protein binding of valproate and inhibit the clearance of valproate [162]. In some patients acetylsalicylic acid has been noted to precipitate valproate toxicity, possibly through additional metabolic inhibition [162]. When felbamate is added to a regimen including valproate, the serum concentration of valproate will increase dose-dependently to up to 25% within days of adding felbamate. Inhibition of β-oxidation of valproate by felbamate was proposed as one mechanism in healthy volunteers [163, 164]. A dose reduction of valproate by 25% has been suggested to compensate for the increase of the serum concentration of valproate [165]. Tiagabine may exceptionally lower the serum concentration of valproate [166]. Mefloquine increases the clearance of valproate, and may, in addition, exacerbate seizures in patients with epilepsy. It should not be used for malaria prophylaxis in patients with epilepsy. Rifampicin may increase the clearance of valproate (26) requiring a higher dose of valproate in patients receiving rifampicin. The effect of cimetidine on valproate serum concentration is controversial; an increase in valproate serum concentrations has been reported but not confirmed by others. A number of additional interactions are listed in Table 8.

Drugs that usually cause no relevant change of the serum concentration of valproate

No or no clinically relevant effects on valproate have been reported during the coadministration of antacids, chlorpromazine, haloperidol and ranitidine.

Pregnancy, lactation and teratogenic effects

Valproate may produce teratogenic effects in the offspring of human females receiving the drug during pregnancy (e.g. [3, 167]). Additional risks include the rare development of hepatic failure and clotting problems in newborns of mothers exposed to valproate during pregnancy [26]. Before discussing the teratogenic risk associated with valproate, it may be useful to very briefly review the risks associated with epilepsy and antiepileptic drugs in general [168, 169]. About 90% of women with epilepsy will deliver healthy children free of birth defects. Nevertheless, the rate of malformations in offspring born to mothers with epilepsy receiving any antiepileptic drugs is approximately 2–3 times higher than in the general population. Malformations most frequently encountered are labial clefts and cardiovascular malformations. The higher incidence of congenital anomalies in antiepileptic drug treated women with epilepsy cannot be regarded as a cause and effect relationship however [168]. There are intrinsic methodological problems including an ascertainment bias, especially epilepsy itself or genetic factors may be as important or even more important than the antiepileptic drug treatment in contributing to congenital anomalies.

The molecular basis of a genetic predisposition is unclear and no tests are yet available for identifying parents or fetuses at particular high risk. Developmental delay or neurological abnormalities have also been reported, the role of antiepileptic drugs including valproate, if any, remains uncertain. The teratogenic risks associated with sodium valproate have been demonstrated experimentally in several species including the mouse, rat and rabbit (see chapter by Nau in this volume). The incidence of neural tube defects in the fetus may be higher in mothers receiving valproate during the first trimester of pregnancy. The risk of valproate-reexposed women having children with spina bifida is approximately 1–2 % [3, 170]. Other congenital anomalies associated primarily with valproate include hypospadias, and bilateral radial aplasia is a rare but specific effect of valproate and craniofacial defects, cardiovascular and urogenital malformations, and anomalies involving various body parts have been reported [168]. Sufficient data to determine the incidence of these anomalies are not available at present, however. The global risk of malformations in women receiving valproate during the first trimester of pregnancy may not be higher than in women receiving other antiepileptic drugs. In fact, a recent metaanalysis of five prospective European studies determined that the increased risk of major congenital malformations for children exposed to valproate was 4.9 (95% Confidence intervals: 1.6–15.0). The risk for carbamazepine was also 4.9 (95% CI: 1.3–18.0). Offspring of mothers using 1000 mg or more valproate per day were at a significantly increased risk of major congenital malformations, especially neural tube defects, compared to offspring exposed to 600 mg/d or less [171]. No difference was found in those exposed to 600–1000 mg/d and those receiving less than 600 mg/d. Cases of facial dysmorphia and limb malformations have been reported following exposition to valproate during early pregnancy. The minor anomalies and dysmorphic features include deformities of the midface and digital and minor skeletal abnormalities. The incidence of these effects is not well established. There are suggestions that the maternal dose of valproate and the maternal serum concentration during early pregnancy is higher in epileptic women with neural tube defects in their offspring compared to epileptic mothers with normal offspring [172]. Spina bifida was associated with a significantly higher average daily dose of 1640 mg/d valproate as compared to normal outcome (941 mg/d). Furthermore, experimental data, reviewed by Nau in this volume, indicate a correlation of peak maternal serum concentrations and the incidence of anencephaly in mice. These data suggest that lowering peak serum concentrations may possibly reduce the incidence of neural tube defects in humans. The Medical Research Council study on the prevention of neural tube defects has shown that folate substitution may reduce the risk of having another child with spina bifida for mothers who had earlier given birth to a child with spina bifida [173]. Experimentally, the addition of vitamin B_6 and B_{12} with or without folinic acid attenuated the rate of anterior neural tube defects [174]. The protection was not complete, however, suggesting the involvement of other factors. Others found no support at all for the role of folinic acid deficiency in valproate-induced open neural tubes in rat embryos [175, 176]. In addition it should be noted that women with epilepsy were excluded from the MRC trial. Nevertheless, implications may exist for the management of women receiving valproate which will be discussed in the next paragraph.

Guidelines for the use of valproate in women with epilepsy, migraine and bipolar disorders

1. In women with childbearing potential valproate should only be given after careful consideration of the teratogenic risk involved. It should be noted that most data, however incomplete, have been accrued in women with epilepsy. It would appear however that valproate may at least in part also be teratogenic in women receiving the drug for migraine or bipolar disorders. Currently there are not enough data on the use of valproate in migraine or bipolar disorders to establish if the rate of malformations differs compared to that in women with epilepsy taking valproate. It would appear therefore prudent to apply the same degree of caution when valproate is used for indications other than epilepsy. Prenatal diagnosis should be discussed before pregnancy. Folate supplementation for reducing the incidence of neural tube defects is recommended for all women even though the evidence is equivocal as discussed in the preceding paragraph. However, in contrast to earlier views folate supplementation is safe and does not exacerbate seizures in women with epilepsy. Although the individual dose of folate may vary, the usually recommended daily dose for women with epilepsy and those on enzyme-inducing agents and valproate is 400 μg/d. It has been pointed out that patients with a 677C > T 5,10-methylene-tetrahydrofolate reductase mutation may need higher doses. About 11% of the Dutch population share this mutation [177]. In addition, multivitamin B_6 and B_{12} supplementation has been suggested to reduce the incidence of other congenital anomalies including microcephaly, and should therefore be given to all women of childbearing potential receiving valproate. Intake of alcohol should generally be avoided during pregnancy. Adequate contraceptive measures need to be taken in view of the teratogenicity of valproate. In that respect it is important to keep in mind that valproate does not reduce the efficacy of oral contraceptives.

2. Necessary changes in antiepileptic medication should be carried out before pregnancy. If left until a woman is pregnant, the most susceptible period for the development of major malformations i.e. the period of organogenesis from 3–8 weeks of gestation is likely to have been passed by the time the woman is pregnant. If pregnancy is planned, an extra effort should be made to reevaluate the need for antiepileptic drug treatment. If valproate

is needed in fact, antiepileptic comedication should be discontinued, if possible. Monotherapy with the minimum effective dose is divided in several doses over the day. Slow release formulations may be expected to be preferable because they avoid the high peak concentrations seen with conventional formulations.

3. During pregnancy, valproate should not be discontinued if it has been effective. As a general rule, the dose of valproate should only be changed based on generally accepted clinical grounds such as an increase in seizures or clinical adverse effects. Drug changes that might worsen seizure control should be avoided. Abrupt discontinuation or dose reduction in women with epilepsy may be harmful to mother and the fetus through seizure exacerbation and, possibly, status epilepticus. Prenatal diagnosis should be offered. It may consist of fetal ultrasound examination during weeks 18–20, and of alpha-fetoprotein analysis in serum and, if necessary, in amniotic fluid during week 16. Prenatal diagnosis should be discussed before pregnancy.

4. During puerperium, the serum concentrations of valproate may increase and adverse effects, mostly tremor, may develop. A moderate dose reduction may be useful. The excretion of valproate in breast milk is minimal, with concentrations of 1–10% of the maternal serum concentration. Adverse effects have not been noted in breast fed infants.

5. Adverse effects in the neonate. Hypofibrinogenemia has been noted in neonates of mothers exposed to valproate during pregnancy. Afibrinogenemia was seen in a neonate of a mother who had low fibrinogen. The child died due to hemorrhages [29]. If valproate is used during pregnancy, clotting parameters should be followed in the neonate. Acute hepatic failure, resulting in the death of one neonate and one child were documented following the use of valproate during pregnancy [26].

Special warnings and precautions for use

Valproate should not be given to patients with acute or chronic hepatitis, a personal or family history of severe hepatic disorders, especially when drug-related, hypersensitivity to valproate or porphyria. In women with polycystic ovaries or hyperandrogenism, valproate should only be given after careful consideration of the risk-benefit involved. In patients suspected to have diseases associated with mitochondrial or metabolic dysfunction or Alpers disease valproate should be avoided if possible. Prior to treatment with valproate, liver function tests and haematological tests including blood cell count, platelet count, bleeding time and coagulation tests are recommended. When an urea cycle enzymatic deficiency is suspected, special metabolic investigations should be performed because valproate may cause symptomatic hyperammonemia in patients with urea cycle abnormalities. In women of childbearing age adequate contraceptive measures need to be taken, and folate and multivitamin B should be given in those who may become pregnant.

Conclusions

Valproate is a very effective first-line drug and is tolerated well in most patients. Most adverse effects of valproate are mild to moderate in intensity and require discontinuation in about 10 percent of patients. Sedation and hypersensitivity reactions are rare. Main areas of concern are the teratogenicity of valproate, acute liver failure, especially in those younger than 3 years, and the newly emerging problems with a reproductive metabolic syndrome with insulin-resistance, polycystic ovaries, hyperandrogenism and lipid abnormalities and late-onset Parkinsonism with or without pseudobrain atrophy in some patients. Nevertheless, valproate remains a very valuable and safe first-line antiepileptic drug if the drug is used with precaution. Careful clinical monitoring allows for early recognition and, at least in part, for prevention of both common and rare adverse effects.

References

1 Wilder BJ et al (1983) Gastrointestinal tolerance of divalproex sodium. *Neurology* 33: 808–811
2 Bryant A E, Dreifuss FE (1996) Valproic acid hepatic fatalities. III. U.S. experience since 1986. *Neurology* 46: 465–469
3 Lindhout D, Schmidt D (1986) In-utero exposure of valproate and neural tube defects. *Lancet I*: 1392–1393 (letter)
4 Isojärvi JIT, Laatikainen TJ, Knip M, Pakarinen AJ, Juntunen KT, Myllylae VV (1996) Obesity and endocrine disorders in women taking valproate for epilepsy. *Ann Neurol* 39: 579–584
4a Murialdo G, Galimberti CA, Magri G, et al (1997) Menstrual cycle and ovary alterations in women with epilepsy on antiepileptic therapy. *Endocrinol Invest* 20(9): 519–526
5 Schmidt D (1982) Adverse Effects of Antiepileptic Drugs. Raven Press, New York
6 Schmidt D (1992) Anticonvulsants. In: MNG Dukes (ed): *Meyler's side-effects of drugs*, 12th Ed. Elsevier Science Publishers BV, New York, 122–143
7 Schmidt D (1984) Adverse effects of valproate. *Epilepsia 25* (Suppl. 1): S 44–S 49
8 Davis R, Peters DH, McTavish D (1994) Valproic Acid. A reappraisal of its pharmacological properties and clinical efficacy in epilepsy. *Drugs* 47: 332–372
9 Dreifuss FE (1995) Valproic acid. Toxicity. In: RH Levy, RH Mattson, BS Meldrum (eds): *Antiepileptic drugs*. Raven Press, New York, 641–648
10 Loiseau P (1996) Tolerability of newer and older antiepileptic drugs: a comparative review. *CNS Drugs* 6: 148–166
11 Stephens JR, Levy RH (1994) Effects of valproate and citrulline on ammonium-induced encephalopathy. *Epilepsia* 35: 164–171
12 Armon C, Shin C, Miller P, Carwile S, Brown E, Edinger JD, Paul RG (1996) Reversible Parkinsonism and cognitive impairment with chronic valproate use. *Neurology* 47: 626–635
13 Hurst SI, Hargreaves JA, Howald WN, Racha JK, Mather GG, Labroo R, Carlson SP, Levy RH (1997) Enzymatic mechanism for the phenobarbital-valproate interaction. *Epilepsia* 38; Suppl 8: 111–112

14 Dinesen H, Gram L, Andersen T, Dam M (1984) Weight gain during treatment with valproate. *Acta Neurol Scand* 70: 65–69
15 Gidal BE, Anderson GD, Spencer WD, et al (1994) Valproic acid (VPA) associated weight gain in monotherapy patients with epilepsy. *Epilepsia* 35: S142
16 Mattson RH, Cramer JA, Collins JF, The Department of Veterans Affairs Epilepsy Cooperative Study No. 264 Group (1992) A comparison of valproate with carbamazepine for the treatment of complex partial seizures and secondarily generalized tonic-clonic seizures in adults. *N Engl J Med* 327: 765–771
17 Verity CM, Hosking G, Easter DJ, The Peadiatric EPITEG Collaborative Group (1995) A multicentre comparative trial of sodium valproate and carbamazepine in pediatric epilepsy. *Dev Med Child Neurol* 37: 97–108
18 Davidson DLW (1989) The adult EPITEG trial: a comparative multicentre clinical trial of sodium valproate and carbamzepine in adult onset epilepsy. Part 2: Adverse effects. In: D Chadwick (ed) Proceedings of the Fourth International Symposium on Sodium Valproate and Epilepsy, International Congress and Symposium Series No. 152, *Royal Society of Medicine Services*, London, 114–121
19 Weber M, Loiseau P, Boon P, Cunha L, Herranz JL, Siebenga E, Stodieck S, Broglin D, Labadie F, Robin JL, Djouadi (1993) Comparative multicentre study of the efficacy and tolerability of controlled release form of sodium valproate (Depakine Chrono R) and carbamazepine (Tegretol LP R) in partial Epilepsy. 20th International Epilepsy Congress, Oslo (Norway) *Sanofi Recherche*
20 Christe W, Krämer G, Vigonius U, Pohlmann H, Steinhoff B, Brodie M, Moore A (1997) A double-blind controlled clinical trial: oxcarbazepine versus sodium valproate in adults with newly diagnosed epilepsy. *Epilepsy Res* 26: 451–460
21 COSTART – coding symbols for thesaurus of adverse reaction terms. 3rd ed. Rockville MD (1989) Food and Drug Administration, *Center for Drug Evaluation and Research*
22 Willmore LJ, Shu V, Wallin B, the M88-194 Study Group (1996) Efficacy and safety of add-on divalproex sodium in the treatment of complex partial seizures. *Neurology* 46: 49–53
23 Giroud M, Gras D, Escousse A, Dumas R, Venaud G (1993) Use of intractable valproic acid in status epilepticus. A pilot study. *Drug Invest* 5: 154–159
24 Devinsky O, Leppik I, Willmore LJ, Pellock JM, Dean C, Gates J, Ramsay RE, Intravenous Valproate Study Team (1995) Safety of intravenous valproate. *Ann Neurol* 38: 670–674
25 Ramsay RE, Uthman B, Leppik IE, Pellock JM, Wilder BJ, Morris D, Cloyd JC (1997) The tolerability and safety of valproate sodium injection given as an intravenous infusion. *J Epilepsy* 10: 187–193
26 Physicians' Desk Reference (1997) 51th Edition, Medical Economics Company, Montvale, NJ
27 Callaghan N, Majeed T, O'Connell A, Oliveira DBG (1994) A comparative study of serum F protein and other liver function tests as an index of hepatocellular damage in epileptic patients. *Acta Neurol Scand* 89: 237–241
28 Schmidt D, Shorvon S (1996) The Epilepsies. In: T Brandt, LR Caplan, J Dichgans, H-CH Diener, C Kennard (eds): *Neurological disorders. Course and Treatment.* Academic Press, San Diego, 159–182
29 König StA, Elger CE, Vassella F, Schmidt D, Bergmann A, Boenigk HE, Despland PA, Genton P, Krämer G, Löscher W, Mayer T, Nau H, Schneble H, Siemes H, Stefan H, Wolf P (1998) Empfehlungen zu Blutuntersuchungen und klinischer Überwachung zur Früherkennung des Valproat-assoziierten Leberversagens. *Schweizerische Ärztezeitung* 79: 580–585
30 Dreifuss FE, Santilli N, Langer DH, Sweeny KP, Moline KA, Meander KB (1987) Valproic acid hepatic fatalities: a retrospective review. *Neurology* 37: 397–400
31 Dreifuss FE, Langer DH, Moline KA, Maxwell JE (1989) Valproic acid hepatic fatalities. III. US experience since 1984. *Neurology* 39: 201– 207
32 Jeavons PM, Clark JE (1974) Sodium valproate in treatment of epilepsy. *Brit Med J* 2: 584–586
33 Loyning Y, Johannessen SI, Ritland LS, Strandjord RE, Koster R (1982) Cases of serious/fatal hepatotoxicity due to valproate.Recommended control scheme and preliminiary results. In: J Oxley, D Janz, H Meinardi (eds): *Chronic Toxicity of Antiepileptic Drugs.* Raven Press, New York, 47–70

34 Stenzel E, Albani M, Doose H, Penin H, Scheffner D, Schmidt D (1987) Valproat-Therapie und Lebertoxizität. *Pädiatr Prax* 35: 159–163
35 Pellock JM, Willmore LJ (1991) A rational guide to routine blood monitoring in patients receiving antiepileptic drugs. *Neurology* 41: 961–964
36 Camfield C, Camfield P, Smith E, Tibbles JA (1996) Asymptomatic children with epilepsy: little benefit from screening anticonvulsant-induced liver, blood, or renal damage. *Neurology* 36: 838–841
37 Fenichel GM, Greene HL (1985) Valproate hepatotoxicity: two new cases, a summary of others, and recommendations. *Pediatr Neurol* 1: 109–113
38 Erasmus C, Hjelm M,Wilson J (1983) The value of routine liver function monitoring during sodium valproate therapy. *Br J Clin Pract* 727 Suppl: 77–78
39 Silberstein SD,Willmore LJ (1996) Divalproex Sodium: Migraine Treatment and Monitoring. *Headache* 239–242
40 Ware S, Milliward-Sadler GH (1980) *Acute liver disease associated with sodium valproate Lancet II*: 1110–1113
41 Appleton RE, Farell K, Applegarth DA, Dimmik JE, Wong LTK, Davidson AGF (1990) The high incidence of valproate hepatotoxicity may relate to familial metabolic defects. *Can J Neurol Sci* 17: 145–148
42 Harding BN, Egger J, Portmann B, Erdohazi M (1986) Progressive neuronal degeneration of childhood with liver disease. *Brain* 109: 181–206
43 Zäh W, Rengeling M, Rühl G, Hackenberg K (1985) Akute Lebernekrose durch Valproinat. *Dtsch med Wschr* 110: 956–959
44 Williams R (1982) Hepatic Disorders: Discussion. In: J Oxley, D Janz, H Meinardi (eds): *Chronic Toxicity of Antiepileptic Drugs*. Raven Press, New York, 79–84
45 Tredger JM, Sherwood RA (1997) The liver. New functional, prognostic and diagnostic tests. *Ann Clin Biochem* 34: 121–141
46 Schmidt D, Krämer G (1994) The New Antiepileptic Drugs. Implications for Avoidance of Adverse Effects. *Drug Safety* 11: 422–431
47 Tiagabin-Product Monograph, Novo-Nordisk, Mainz, Germany
48 Powell-Jackson PR,Tredger JM,Williams R (1984) Hepatotoxicity to sodium valproate: a review. *Gut* 25: 673–681
49 Straßburg HM, Sauer M, Ketelsen UP, Böhm N, Schwab M, Volk B (1990) Letale Valproat-Unverträglichkeit bei progressiver zerebraler Poliodystrophie Alpers. In: J Lütschg (ed): Aktuelle Neuropädiatrie 1990. Springer-Verlag, Berlin, 258–263
50 Dodson WE (1997) Discussion. In: D Schmidt, E Perucca (eds): *The place of felbamate in the treatment of Lennox-Gastaut Syndrome: an update.* Proceedings of a meeting; 1997 April 22–23: Düsseldorf (Germany). s'Hertogenbosch,The Netherlands: *Vanzuiden Comm*, 8
51 Schwabe MJ, Dobyns WB, Burke B, Armstrong DL (1997) Valproate-induced liver failure in one of two siblings with Alpers disease. *Ped Neurol* 16: 337–343
52 Njolstad PR, Skjeldal OH, Agsteribbe E, Huckriede A, Wannag E, Sovik O, Waaler PE (1997) Medium chain acyl-CoA dehydrogenase deficiency and fatal valproate toxicity. *Pediatr Neurol* 16: 160–162
53 O'Neil MG, Perdun CS, Wilson MB, McGown ST, Patel S (1996) Felbamate-associated fatal acute hepatic necrosis. *Neurology* 46:1457–1459
54 Morselli PL, Lloyd KG, Palminteri R (1995) Progabide. In: RH Levy, RH Mattson, BS Meldrum (eds): *Antiepileptic Drugs,* Fourth Edition. Raven Press, New York, 997–1009
55 Asconape JJ, Penry JK, Dreifuss FE, Riela A, Mirza W (1993) Valproate-associated pancreatitis. *Epilepsia* 34: 177–183
56 Zenker M, Metzker M, Wegener E, Heidemann PH (1995) Valproat-induzierte Pankreatitis. *Monatsschr Kinderheilkd* 143: 843–846
57 Gidal B, Spencer N, Maly M, Pitterle M, Williams E, Collins M, Jones J (1994) Valproate-mediated disturbances of hemostasis: Relationship to dose and plasma concentration. *Neurology* 44: 1418–1422
58 Tennis P, Stern RS (1997) Risk of serious cutaneous disorders after initiation of use of phenytoin, carbamazepine, or sodium valproate: a record linkage study. *Neurology* 49: 542–546
59 Richens A, Davidson DLW, Cartlidge NEF, Easter DJ, The Adult EPITEG Collaborative Group (1994) A multicentre comparative trial of sodium valproate and carbamazepine in adult onset epilepsy. *J Neurol Neurosurg Psychiatry* 57: 682–687

60 Carney P, Nasereddine W, Drury I, Varma N, Payne T, Shu V, Beydoun A (1997) Relation between thrombocytopenia and valproate dose. *Epilepsia* 38; Suppl 8: 102
61 Bottom KS, Adams DM, Mann KP, Ware RE (1997) Trilineage hematopoietic toxicity associated with valproic acid therapy. *J Ped Hematol Oncol* 19: 73–76
62 Anderson GD, Lin Y-X, Berge C, Ojemann GA (1997) Absence of bleeding complications in patients undergoing cortical surgery while receiving valproate treatment. *J Neurosurg* 87: 252–256
63 Stahl MM, Neiderud J, Vinge E (1997) Thrombocytopenic purpura and anemia in a breast-fed infant whose mother was treated with valproic acid. *J Pediatrics* 130: 1001–1003
64 Sasso E, Delsoldato S, Negrotti A, Mancia D (1994) Reversible valproate-induced extrapyramidal disorders. *Epilepsia* 35: 391–393
65 Zaret BS, Cohen RA (1986) Reversible valproic acid-induced dementia: a case report. *Epilepsia* 27: 234–240
66 Mantelet S, Gremion J, Feline A, Masnou P (1997) Valproate-associated encephalopathy: A case report. Review of the literature. *Ann Med Psycholog* 155: 510–512
67 Bauer J, Elger CE (1993) Die akute Valproinsäure-Enzephalopathie. *Akt Neurol* 20: 16–21
68 Bauer J (1996) Seizure-inducing effects of antiepileptic drugs: a review. *Acta Neurol Scand* 94: 367–377
69 Chadwick DW, Cumming WJK, Livingstone I, Cartlidge NEF (1979) Acute intoxication with sodium valproate. *Ann Neurol* 6: 552–553
70 Horiuchi M, Imamura Y, Nakamura N, Maruyama I, Saheki T (1993) Carbamoylphosphate synthetase deficiency in an adult: deterioration due to administration of valproic acid. *J Inheri Metab Dis* 16: 39–45
71 Jones GL, Matsuo F, Baringer JR, Reichert WH (1990) Valproic acid-associated encephalopathy. *West J Med* 153: 199–202
72 Zaccara G, Paganini M, Campostrini R, Arnetoli G, Zappoli R, Moroni F (1984) Hyperammonemia and valproate-induced alterations of the state of consciousness. *Eur Neurol* 23: 104–112
73 Triggs WJ, Gilmore RL, Millington DS, Cibula J, Bunch TS, Harman E (1997) Valproate-associated carnitine deficiency and malignant cerebral edema in the absence of hepatic failure. *Int J Clin Pharmacol Therap* 35: 353–356
74 Marescaux C, Warter JM, Micheletti G, et al (1982) Stuporous episodes during treatment with sodium valproate: report of seven cases. *Epilepsia* 23: 197–305
75 Marescaux C, Warter JM, Brandt C, et al (1985) Adaption of hepatic ammonia metabolism after chronic valproate administration in epileptics treated with phenytoin. *Eur Neurol* 24: 191–195
76 Steinhoff BJ, Stodieck SR (1993) Temporary abolition of seizure activity by flumazenil in a case of valproate-induced non-convulsive status epilepticus. *Seizure* 2: 261–265
77 Lam CW, Lau CH, Williams JC, Chan YW, Wong LJC (1997) Mitochondrial myopathy, encephalopathy, lactic acidosis and stroke-like episodes (MELAS) triggered by valproate therapy. *Eur J Pediatr* 156: 562–564
78 Melegh B, Trombitas K (1997) Valproate treatment induces lipid globule accumulation with ultrastructural abnormalities of mitochondria in skeletal muscle. *Neuropediatr* 28: 257–261
79 Papzian O, Canizales E, Alfonso I, Archila R, Duchowny M, Aicardi J (1995) Reversible dementia and apparent brain atrophy during valproate therapy. *Ann Neurol* 38: 687–691
80 Walstra GJ (1997) Reversible dementia due to valproic acid therapy. *Ned Tijd Geneesk* 141: 391–393
81 Guerrini R, Belmonte A, Canapicchi R, Casalini C, Perucca E (1998) Reversible pseudoatrophy of the brain and mental deterioration associated with valproate treatment. *Epilepsia* 39: 27–32
82 Schöndienst M (1988) Hirnatrophien durch Valproat. *Vier Fallberichte Epilepsie-Blätter* 1: 23 (Abstract)
83 Froomes PR, Stewart MR (1994) A reversible parkinsonian syndrome and hepatotoxicity following addition of carbamazepine to sodium valproate. *Aus New Zeal J Med* 24: 413–414
84 Park-Matsumoto YC, Tazawa T (1998) Valproate induced Parkinsonism. *Brain Nerve* 50: 81–84

85 Wils V, Goluke-Willemse G (1997) Extrapyramidal syndrome due to valproate administration as an adjunct to lithium in an elderly manic patient. *Int J Geriatric Psychiat* 12: 272
86 Armon C, Shin C, Miller P, Carwile S, Brown E, Edinger JD, Paul RG (1996) Reversible Parkinsonism and cognitive impairment with chronic valproate use. *Neurology* 47: 626–635
87 Trimble M (1991) *The psychoses of epilepsy*. Wrightson, London
88 Herranz JL, Arteaga R, Armijo JA (1982) Side-effects of sodium valproate in monotherapy controlled by plasma levels; a study in 88 pediatric patients. *Epilepsia* 23: 203–214
89 Vermeulen J, Aldenkamp AP (1995) Cognitive side-effects of chronic antiepileptic drug treatment. A review of 25 years of research. *Epilepsy Res* 22: 65–95
90 Vining EP, Mellitis ED, Dorsen MM, Freeman JM, et al (1987) Psychologic and behavioral effects of antiepileptic drugs in children: a double-blind comparison between phenobarbital and valproic acid. *Pediatrics* 80: 165–174
91 Meador KJM, Loring DW, Huh K, et al. (1990) Comparative cognitive effects of anticonvulsants. *Neurology* 40: 391–394
92 Forsythe I, Butler R, Berg I, et al (1991) Cognitive impairment in new case of epilepsy randomly assigned to carbamazepine, phenytoin and sodium valproate. *Dev Med Child Neurol* 33: 524–534
93 Calandre EP, Dominguez-Granados R, Gomez-Rubio M, et al (1990) Cognitive effects of long-term treatment with phenobarbital and valproic acid in school children. *Acta Neurol Scand* 81: 504–506
94 Helmstaedter C, Wagner G, Elger CE (1993) Differential effects of first antiepileptic drug application on cognition in lesional and non-lesional patients with epilepsy. *Seizure* 2: 125–130
95 Craig I, Tallis R (1994) Impact of valproate and phenytoin on cognitive function in elderly patients: results of a single-blind randomized comparative study. *Epilepsia* 35: 381–390
96 Tonnby B, Nilsson H, Aldenkamp AP, et al (1994) Withdrawal of antiepileptic medication in children – correlation of cognitive function and plasma concentration – The multicentre "Holmfrid" study. *Epilepsy Res* 19: 141–152
97 Brouwer OF, Pieters MSM, Bakker AM, et al (1992) Conventional and controlled release valproate in children with epilepsy: a cross-over study comparing plasma levels and cognitive performance. *Epilepsy Res* 13: 245–253
98 Nicklas WJ, Vyas I, Heikkila RE (1985) Inhibition of NADH-linked oxidation in brain mitochondria by 1-methyl-4-phenylpyridine, a metabolite of neurotoxic 1-methyl-4-phenyl-1,2,3,6-tetrahydropyridine. *Life Sci* 36: 2503–2508
99 Di Mauro S, Bonilla E, Zeviani M, Nakagawa M, De Vivo DC (1985) Mitochondrial myopathies. *Ann Neurol* 17: 521–538
100 Hassan MN, Laljee HCK, Parsonage MJ (1976) Sodium valproate in the treatment of resistant epilepsy. *Acta Neurol Scand* 54: 209–218
101 Egger J, Brett EM (1981) Effects of sodium valproate in 100 children with special reference to weight. *Br Med J* 283: 577–581
102 Corman CL, Leung NM, Guberman AH (1997) Weight gain in epileptic patients during treatment with valproic acid: a retrospective study. *Can J Neurol Sci* 24: 240–244
103 Brodersen R, Jorgensen N, Vorum H, Krukow N (1990) Valproate and palmitate binding to human serum albumin: An hypothesis on obesity. *Mol Pharmacol* 37: 704–709
104 Easter D, O'Bryan-Tear CG, Verity C (1997) Weight gain with valproate or carbamazepine – A reappraisal. *Seizure* 6: 121–125
105 Herzog AG (1993) A relationship between particular reproductive endocrine disorders and the laterality of epileptiform discharges in women with epilepsy. *Neurology* 43: 1907–1910
106 Isojärvi JIT, Laatikainen TJ, Pakarinen AJ, Juntunen KT, Myllylä VV (1993) Polycystic ovaries and hyperandrogenism in woman taking valproate for epilepsy. *N Engl J Med* 329: 1383–1388
107 Isojärvi JIT, Tauboll E, Dale PO, Rättyä J, Harbo HF, Gjerstad L, Myllylä VV, Tapanainen J (1997) Polycystic ovaries in woman taking valproate monotherapy for epilepsy: a two-center study. *Epilepsia* 38; Suppl 8: 102
108 Tauboll E, Isojärvi JIT, Harbo HF, Pakarinen AJ, Gjerstad L (1997) Changes in sex-steroid hormone levels and ovarian structure after valproate treatment in female rats. *Epilepsia* 38; Suppl 8: 47

109 Isojärvi JIT, Rättyä J, Myllylä VV, Knip A, Koivunen R, Pakarinen AJ, Tekay A, Tapanainen (1998) Valproate, lamotrigine, and insulin mediated risk in women with epilepsy. *Ann Neurol* 43: 446–451
110 Moller DE, Flier JS (1991) Insulin resistance – mechanisms, syndromes, and implications. *N Engl J Med* 325: 938–948
111 Curtis VL, Oelberg DG, Willmore LJ (1994) Infertility secondary to valproate. *J Epilepsy* 7: 259–261
112 Nag D, Garg RK, Banerjee A (1997) Sodium valproate monotherapy and sex hormones in men. *Neurology India* 45: 240–243
113 Asconapé JJ, Manning KR, Lancman ME (1994) Systemic lupus erythematosus associated with use of valproate. *Epilepsia* 35: 162–163
114 Branten AJ, Wetzels JF, Weber AM, Koene RA (1998) Hyponatremia due to sodium valproate. *Ann Neurol* 43: 256–257
115 Ryan SJ, Bishof NA, Baumann RJ (1996) Occurrence of renal Fanconi syndrome in children on valproic acid therapy. *J Epilepsy* 9: 35–38
116 Gidal BE, Inglese CM, Meyer JF, Pitterle ME, Antonopolous J, Rust RS (1997) Diet- and valproate-induced transient hyperammonemia: effect of L-carnitine. *Ped Neurol* 16: 301–305
117 Batshaw ML, Brusilow SW (1982) Valproate-induced hyperammonaemia. *Ann Neurol* 11: 319–321
118 Bohles H, Sewell AC, Wenzel D (1996) The effect of carnitine supplementation in valproate-induced hyperammonaemia. *ACTA Paediatr Int J Paediatr* 85: 446–449
119 Warner MH, Anderson GD, McCarty JP, Farwell JR (1997) Effect of carnitine on measures of energy levels, mood, cognition, and sleep in adolescents with epilepsy treated with valproate. *J Epilepsy* 10: 126–130
120 Anderson HH, Rapley JW, Williams DR (1997) Gingival overgrowth with valproic acid: a case report [Review] [20 ref]. *ASDC J Dent Child* 64: 294–297
121 Lortie A, Chiron C, Mumford J, Dulac O (1993) The potential for increasing seizure frequency, relapse, and appearance of new seizure types with vigabatrin. *Neurology* 43: S 24–S 27
122 Farrar HC, Herold DA, Reed MD (1993) Acute valproic acid intoxication: Enhanced drug clearence with oral-activated charcoal. *Crit Care Med* 21: 299–301
123 Pertoldi F, D'Orlando F, Galliazzo S, Mercante WP (1997) Haemodialysis in the management of severe valproic acid overdose. *Clin Intensive Care* 8: 244–246
124 Matsumoto J, Ogawa H, Maeyama R, Okudaira, K Shinka T, Kuhara T, Matsumoto I (1997) Successful treatment by direct hemoperfusion of coma possibly resulting from mitochondrial dysfunction in acute valproate intoxication. *Epilepsia* 38: 950–953
125 Krämer G (1998) Epilepsien im höheren Lebensalter. Klinik und Besonderheiten der Pharmakotherapie. Thieme, Stuttgart
126 Sanofi-Winthrop, Paris, Corporate Data Sheet (1997) Sodium valproate
127 Hansten PD, Horn JR (1997) Hanston and Horn's drug interactions analysis and management. *Applied Therapeutics*, Vancouver
128 Krämer G (1992) Medikamentöse Interaktionen von Antiepileptika. In: HC Hopf, P Poeck, H Schliak (eds): *Neurologie in Klinik und Praxis*, Band 1, 2. Auflage. Thieme Verlag, Stuttgart, 3: 94–97
129 Krämer G (1994) Interaktionen. In: Fröscher W, Vassella F (eds): *Die Epilepsien. Grundlagen, Klinik, Behandlung.* de Gruyter, Berlin, 545–551
130 Krämer G (1997) Interaktionsmatrix Antiepileptika. Wechselwirkungen komprimiert. *Mediamed*, Berkheim
131 Krämer G (1998) Pharmacokinetic interactions of the new antiepileptic drugs. In: H Stefan, G Krämer, B Mamoli (eds): *Challenge epilepsy. New Antiepileptic drugs.* Blackwell Wissenschafts-Verlag, Berlin, 87–103
132 Krämer G, Besser R, Theisohn M (1987) Interaktionen von Carbamazepin mit anderen Medikamenten. In: G Krämer, HC Hopf (eds): *Carbamazepin in der Neurologie.* Thieme Verlag, Stuttgart, 70–90
133 Ouslander JG (1981) Drug therapy in elderly. *Ann Intern Med* 95: 711–722
134 Pitlick WH (ed) (1989) Antiepileptic Drug interactions. Demos, New York
135 Rambeck B, Krämer G (1992) Medikamentöse Interaktionen von und mit Valproinsäure. In: G Krämer, M Laub (eds): *Valproinsäure. Pharmakologie, klinischer Einsatz, Nebenwirkungen, Therapierichtlinien.* Springer Verlag, Berlin 106–112

136 Bernus I, Dickinson RG, Hooper WD, Eadie MJ (1994) Inhibition of phenobarbitone N-glucosidation by valproate. *Br J Clin Pharmacol* 38: 411–416
137 Yuen AW, Land G, Weatherley BC, Peck AW (1992) Sodium valproate acutely inhibits lamotrigine metabolism. *Br J Clin Pharm* 33: 511–513
138 Messenheimer J, Mullens EL, Giorgi L, Young F (1998) Safety review of adult clinical trial experience with lamotrigine. *Drug Safety* 18: 281–296
139 Fitton A, Goa KL (1995) Lamotrigine: an update of its pharmacology and therapeutic use in epilepsy. *Drugs* 50: 691–713
140 GlaxoWellcome (1998) Lamictal Monograph. London,1997
141 Lai M-L, Huang JD (1993) Dual effect of valproic acid on the pharmacokinetics of phenytoin. *Biopharm Drug Dis* 14: 365–370
142 Ramsay RE, McManus DQ, Guterman A, Briggle TV, Vazquez D, Perchalski R, Yost RA, Wong P (1990) Carbamazepine metabolism in humans: Effect of concurrent anticonvulsant therapy. *Ther Drug Monit* 12: 235–241
143 Bernus I, Dickinson G, Hooper WE, Eadie MJ (1997) The mechanism of the carbamazepine-valproate interaction in humans. *Br J Clin Pharmacol* 44: 21–27
144 Svinarov DA, Pippenger CE (1995) Valproic acid-carbamazepine interaction: Is valproic acid a selective inhibitor of epoxide hydrolase? *Ther Drug Monit* 17: 217–220
145 Liu H, Delgado MR, Browne RH (1995) Interactions of valproic acid with carbamazepine and ist metabolites' concentrations, concentration ratios and level/dose ratios in epileptic children. *Clin Neuropharm* 18: 1–12
146 Anderson GD, Gidal BE, Kantor ED, Wilensky AJ (1994) Lorazepam-valproate interaction: Studies in normal subjects and isolated perfused rat liver. *Epilepsia* 35: 221–225
147 Samara EE, Granneman RG, Witt GF, Cavanaugh JH (1997) Effect of valproate on the pharmacokinetics and pharmacodynamics of lorazepam. *J Clin Pharmacol* 37: 442–450
148 Wong SL, Cavanaugh JH, Shi H, Awni WM, Granneman GR (1996) Effects of divalproex sodium on amitriptyline and nortriptyline pharmacokinetics. *Clin Pharmacol Ther* 60: 48–53
149 De Toledo JC, Haddad H, Ramsay RE (1997) Status epilepticus associated with the combination of valproic acid and clomipramine. *Ther Drug Monitor* 19: 71–73
150 Akula SK, Rege AB, Dreisbach AW, Dejace PMJT, Lertora JJL (1997) Valproic acid increases cerebrospinal fluid zidovudine levels in a patient with AIDS. *Am J Med Sci* 313: 244–246
151 Lertora JJL, Rege AB, Greespan DL, Akula S, George WJ, Hyslop NE, Agrawal KC (1994) Pharmacokinetic interaction betweeen zidovudine and valproic acid in patients infected with human immunodeficiency virus. *Clin Pharmacol Ther* 56: 272–278
152 Parmeggiani A, Riva R, Posar A, Rossi PG (1995) Possible interaction between acyclovir and antiepileptic treatment. *Ther Drug Monitor* 17: 312–315
153 Pisani F (1992) Influence of co-medication on the metabolism of valproate. *Pharm Weekbl Sci Ed* 14: 108–113
154 Spina E, Pisani F, Perucca E (1996) Clinically significant pharmacokinetic drug interactions with carbamazepine. An update. *Clin Pharmacokinetics* 31: 198–214
155 Jann MW, Fidone GS, Israel MK, Bonadero P (1988) Increased valproate serum concentrations upon carbamazepine cessation. *Epilepsia* 29: 578–581
156 Gustavson LE, Cato A, Guenther HJ, et al (1995) Lack of clinically important drug interactions between tiagabine and carbamazepine, phenytoin or valproate. *Epilepsia* 36 (Suppl 3): S 159–S 160
157 Anderson GD, Yau MK, Gidal BE, Harris SJ, Levy RH, Lai AA, Wolf KB, Wargin WA, Dren AT (1996) Bidirectional interaction of valproate and lamotrigine in healthy subjects. *Clin Pharm Ther* 60: 145–156
158 Salke-Kellermann RA, May T, Boenigk HE (1997) Influence of ethosuximide on valproic acid serum concentrations. *Epilepsia Res* 26: 345–349
159 Nagai K, Shimizu T, Togo A, Takeya M, Yokomizo Y, Sakata Y, Matsuishi T, Kato H (1997) Decrease in serum levels of valproic acid during treatment with a new carbapenem, panipenem/betamipron [8]. *J Antimicro Chemother* 39: 295–296
160 Duncan JS, Patsalos PN, Shorvon SD (1991) Effects of discontinuation of phenytoin, carbamazepine, and valproate on concomitant antiepileptic medication. *Epilepsia* 32: 101–115

161 Jonville AP, Gauchez AS, Autret E, Billard C, Barbier P, Nsabiyumva F, Breteau M (1991) Interaction between isoniazid and valproate. A case of valproate overdosage [1]. *Eur J Clin Pharm* 40: 197–198
162 Perucca E (1996) Establised antiepileptic drugs. In: MJ Brodie, DM Treiman (eds) *Modern Management of epilepsy*. Baillière's Clinical Neurology International practice and research. Baillière Tindall, London, 693–722
163 Hooper WD, Franklin Me, Glue P, Banfield CR, Radwanski E, Mc Laughlin DB, McIntyre ME, Dickinson RG, Eadie MJ (1996) Effect of felbamate on valproic acid disposition in healthy volunteers: inhibition of beta-oxidation. *Epilepsia* 37: 91–97
164 Wagner ML, Graves NM, Leppik IE, Remmel RP, Shumaker RC, Ward DL, Perhach JL (1994) The effect of felbamate on valproic acid disposition. *Clin Pharmacol Ther* 56: 494–502
165 Felbamate Core Product Information (1998) Schering-Plough
166 Gustavson LE, Cato A, Guenther HJ, et al (1995) Lack of clinically important drug interactions between tiagabine and carbamazepine, phenytoin or valproate. *Epilepsia* 36 (Suppl 3): S159–S160
167 Robert E, Guibaud P (1982) Maternal valproic acid and congenital neural defects. *Lancet* 2: 1142
168 Lindhout D, Omtzigt JGC (1992) Pregnancy and the risk of teratogenicity. *Epilepsia* 33 (Suppl 4) S41–S48
169 Janz D, Dam M, Richens A, Bossi L, Helge H, Schmidt D (eds) (1982) Epilepsy, pregnancy and the child. Raven Press, New York
170 Centers for Disease Control (1983) Valproate: a new cause of birth defects - report from Italy and follow up from France. *Morbidity and Motality Weekly Report* 32: 438–439
171 Samren EB, van Duijn CM, Koch S, Hiilesmaa VK, Klepel H, Bardy AH, Beck-Mannagetta G, Deichl AW, Gaily E, Granstrom ML, Meinardi H, Grobbee DE, Hofman A, Janz D, Lindhout D (1997) Maternal use antiepileptic drugs and the risk of major congenital malformations: A joint European prospective study of human teratogenesis associated with maternal epilepsy. *Epilepsia* 38: 981–990
172 Omtzigt JGC, Los FJ, Grobbee DE, Pijpers L, Jahoda MGJ, Brandenburg H, Stewart PA, Gaillard HLJ, Sachs ES, Wladimiroff JW, Lindhout D (1992) The risk of spina bifida aperta after first-trimester exposure to valproate in a prenatal cohort. *Neurology* 42: 119–125
173 MRC Vitamin Study Research Group (1991) Prevention of neural tube defects: results of the MRC Vitamin Study. *Lancet* 338: 132–37
174 Elmazar MM, Thiel R, Nau H (1992) Effect of supplementation with folinic acid, vitamin B_6, and vitamin B_{12} on valproic acid-induced teratogenesis in mice. *Fund App Toxicol* 18: 389–394
175 Hansen DK, Grafton TF (1991) Lack of attenuation of valproic acid-induced effects by folinic acid in rat embryos in vitro. *Teratology* 43: 575–582
176 Hansen DK, Grafton TF, Dial SL, Gehring TA, Siitonen PH (1995) Effect of supplemental folic acid on valproic acid-induced embryotoxicity and tissue zinc levels *in vivo*. *Teratology* 52: 277–285
177 Wilcken DE (1997) MTHFR 677 $\gg$ T mutation, folate intake, neural-tube defect, and risk of cardiovascular disease. *Lancet* 350: 603–604

Milestones in Drug Therapy
Valproate
ed. by W. Löscher

The future of valproate

Nathan B. Fountain and Fritz E. Dreifuss*

Comprehensive Epilepsy Program, Department of Neurology, University of Virginia School of Medicine, Charlottesville, VA 22908, USA

Introduction

Valproate (VPA) currently has established efficacy and its use is expanding for the treatment of epilepsy, migraine headache and acute mania, as addressed elsewhere in this volume, but the future of VPA is unknown because it depends on whether safer and more effective medications become available for these conditions. Nevertheless, in this chapter we speculate on the future of VPA based on the current knowledge of its properties and use.

If VPA has an assured role in the pharmacopeia of the future, it is because of novel aspects of the mechanism of action and pharmacokinetics which we review here. We review recent insights into the pathophysiology of VPA induced hepatitis which may lead to a VPA derivative devoid of this risk, which is a rare serious idiosyncratic side-effect representing the most significant impediment to the common use of VPA. We discuss the unique niche VPA will continue to occupy for commonly accepted indications. Based on case reports or pharmacologic principles, we further speculate on whether there may be expanded indications in the future for the treatment of other similar conditions including specific epilepsies, pain disorders, movement disorders and psychiatric diseases, for which further systematic investigation of efficacy may be warranted. Overall, VPA's future position in the armamentarium of drugs for the treatment of neurological diseases seems strong.

Unique pharmacologic characteristics of VPA

The mechanism of action of VPA is relatively unique [1]. Any theory of its mechanism must consider VPA's demonstrated efficacy and accepted role in such seemingly diverse conditions as seizures, migraine and mania. All of these conditions are characterized by paroxysmal symptoms, therefore,

---

* Deceased October 18, 1997.

their pathophysiology may be based on "hyperexcitability" of specific neurons, although this has been studied extensively only for seizures. The mechanism is undoubtedly complex and probably works at several sites [2]. The most prominent effect is on inhibition and consists of enhancement of gamma-aminobutyric acid (GABA) mediated inhibition, since GABA levels are increased in human CSF and animal brain after administration of VPA, and the effect of VPA could be enhanced regionally [3, 4]. Unlike benzodiazepines and barbiturates, VPA does not appear to act directly on GABA receptors, but rather affects multiple sites in the inhibitory pathway. It may block GABA degradation by inhibiting GABA-transaminase (GABA-T) and other enzymes responsible for degradation of GABA, and it enhances production of GABA by inducing glutamic acid decarboxylase (GAD), the enzyme responsible for GABA production. It may also block sustained repetitive firing of action potentials which prevents propagation to other neurons in a manner similar to phenytoin and carbamazepine, which is attributed to blocking voltage gated sodium channels and is not related to inhibition [5]. Acting to alter the balance of inhibition and excitation through multiple mechanisms is clearly an advantage for VPA and probably contributes to its advantage as a "broad spectrum" antiepileptic drug (AED) and may be the basis for efficacy in multiple neurologic conditions. Furthermore, it makes VPA a reasonable consideration for treatment of other neurologic conditions in which hyperexcitability is presumed to be the primary pathology.

The unique structure of VPA may contribute to its ability to activate multiple mechanisms of action and result in its diverse utility [6]. VPA is a fatty acid which allows it to affect pathways that are different from those affected by other AEDs, which are generally small organic molecules with a more focused mechanism of action. VPA not only acts directly on the GABA system but also affects basic cellular functions, including impairing the transport of monocarboxylic acids such as pyruvate and hydroxybutyrate, and it affects brain lipid synthesis, and impairs mitochondrial function [7]. The fatty acid structure also leads to extensive metabolism. While these mechanisms and metabolites may cause some of the toxicity of VPA, other metabolites may be anticonvulsant or active in other ways or even synergistic by affecting different pathways. This has led to the investigation of many VPA metabolites and derivatives as potential AEDs.

The pharmacokinetics of VPA are otherwise not unique [8]. However, one of the reasons VPA is so useful is because it is well absorbed and available in more formulations than other AEDs [9]. It is available in various countries as valproic acid, sodium/acid combinations, and magnesium salts, which are formulated as syrup, sprinkles, capsules, tablets, and rectal suppositories, and an intravenous formulation has recently been approved in the United States. The development of a slow-release formulation allows once or twice per day dosing, which furthers the ease of use [10]. This is currently unique to VPA and will contribute to its success in the immediate future, but other AEDs will eventually be available in many formulations, despite the possibility of new CNS drug delivery systems for VPA.

Impediments to future use of VPA

The greatest impediment to the future of VPA is the rare occurrence of serious idiosyncratic side-effects, especially hepatitis [11]. Potentially life threatening hepatitis was initially reported in approximately 1:10000 patients taking VPA and in as many as 1:500 children less than 2 years old on polytherapy [12]. However, the rate was initially found to be 1:45000 for monotherapy patients over 2 years old, but has fallen to less than 1:49000 since the identification of high risk patients [13]. It is important to note that this condition is not always fatal. Pancreatitis is much less common, but also potentially fatal. There are also more common, but less serious idiosyncratic side-effects including gastrointestinal distress, weight gain, thrombocytopenia, hair loss, and tremor. However, it is the potentially life threatening side-effects that make VPA less attractive. Were it not for these risks, some would consider VPA the first line treatment for almost all seizure patients, since it is useful for both generalized and partial-onset seizures, while most other AEDs have demonstrated utility for only one or the other.

Many investigators have attempted to elucidate the risk factors and mechanisms underlying VPA induced hepatotoxicity. The highest risk group is composed of young children with mental retardation who receive polytherapy [12, 14]. Some cases of fatal hepatotoxicity have occurred in children with metabolic defects, particularly ornithine transcarbamylase deficiency [15]. Young children may be at risk because of immature hepatic function and it is possible that some of the cases of mental retardation are due to unrecognized metabolic abnormalities. Taken together these factors suggest that hepatic metabolism plays an important role in the pathophysiology of VPA induced hepatitis.

Several factors point towards a VPA metabolite as the offending agent of VPA induced hepatotoxicity and suggest that the 4-en-VPA metabolite may be responsible, although this metabolite is not necessarily found in high concentrations in patients with hepatotoxicity [6, 11, 16]. The most common pathologic finding is microvesicular steatosis without inflammation, which is relatively unique and found only in Reye's syndrome and Jamaican vomiting sickness [14]. The lack of inflammation and the occurrence of similar pathology in non-inflammatory diseases suggests that the injury is not due to hypersensitivity, but rather to direct toxicity. Furthermore, similar hepatic injury can be induced experimentally by 4-pentanoic acid, which is structurally similar to 4-en-VPA. The strongest evidence that this metabolite is important in the development of hepatotoxicity is that it is a hepatotoxin in animal studies [17]. Supportive evidence includes its ability to inhibit fatty acid β-oxidation, accumulate lipid in liver, and destroy hepatic cytochrome P450. It also forms glutathione conjugates which can link with macromolecules to produce independent toxicity or form immune complexes. There is also evidence that the 2,4-en-VPA metabolite among others may contribute to hepatotoxicity. These findings introduced the possibility that a derivative of VPA may be developed which lacks these toxic effects, but maintains anticonvulsant properties. It also raises the possibility that patterns of metabolite elevations may eventually help identify which patients may be at risk of developing hepatotoxicity [18].

There is also good evidence for the hypothesis that primary or secondary carnitine deficiency contributes to hepatotoxicity, possibly via impaired metabolism of short and medium chain fatty acids [19]. Thus, avoidance of carnitine deficiency by administration of carnitine to high risk patients will hopefully reduce the incidence of hepatotoxicity in the future.

VPA may not appear to induce birth defects with any greater frequency than other AEDs, but the neural tube defects induced are more serious than the milder forms of birth defects induced by other AEDs [20]. This currently limits the use of VPA in women of child bearing potential, and will continue to in the future, despite the use of folate to reduce this risk. The teratogenic effects seem to be due to VPA alone, rather than a toxic metabolite, which again emphasizes the potential utility of a VPA derivative [21].

VPA derivatives

The future of VPA will be expanded if derivatives which lack hepatotoxicity and teratogenicity are found. The first logical source of candidate VPA derivatives which may overcome the above noted limitations has been natural metabolites of VPA. Nau, Löscher and colleagues have investigated many natural derivatives in animal models of seizures and others have been investigated *in vitro* especially those produced by oxidation [22]. The most promising compound is 2-en-VPA which is produced by β-oxidation in humans taking VPA [23, 24]. It is present in relatively small concentrations in humans, but it is more potent than VPA in animal seizure models in rats, mice and dogs, including the pentylenetetrazol (PTZ) seizure threshold and maximal electroshock tests and it is less neurotoxic, which gives it a favorable protective index (TD50/ED50) [4]. In addition, it is differentially concentrated in some brain regions, including hippocampus [22]. It is not hepatotoxic in rats and lacks teratogenicity and developmental neurotoxicity in rats and mice, even at higher doses than VPA [25]. It is potent in blocking experimentally induced epileptiform discharges in hippocampal slices and may be a better inhibitor of GABA breakdown than VPA [26]. At least one analog of 2-en-VPA, cyclooctylideneacetic acid, also has anticonvulsant activities and provides another set of rational candidate compounds to pursue [27]. If the animal findings are substantiated in humans, then 2-en-VPA could replace VPA in the future [23].

Several other VPA derivatives have demonstrated good anticonvulsant and neurotoxic profiles *in vitro* and *in vivo* in animal studies. Valpromide is the primary amide of VPA and has been available in Europe for more than 25 years [28]. It is rapidly metabolized to VPA and thus acts as a prodrug which offers no obvious advantage over the parent compound in preventing VPA induced toxicity, although it may have other benefits. Since glycine is an inhibitory neurotransmitter, glycine conjugates of VPA and 2-en-VPA have been investigated and have demonstrated that the 2-en-VPA glycine conjugate has a better anticonvulsant profile [29]. Amino analogs of VPA have some activity, but have not been investigated in detail [30]. Methyl branching of VPA yields some compounds that have anticonvulsant activity, low neurotoxicity and low teratogenicity in mice [31]. Ester and polyester prodrugs of VPA may have intrinsic activity of their own, but they are also converted to VPA [32, 33]. Alpha-fluoro VPA exploits the strategy of substituting a fluoride to prevent β-oxidation to the toxic 4-en-VPA metabolite. It is at least as anticonvulsant as VPA in the PTZ test and has been demonstrated to increase brain GABA levels [34]. N-(2-propylpentanoyl) urea is more potent than VPA in both maximal electroshock and the PTZ seizure threshold tests and is less toxic, resulting in a much better protective index [35]. Hydroxylated VPA (4-OH VPA and 5-OH VPA), formed in humans by omega oxidation, is a reasonable consideration but is less potent than VPA in electroconvulsive seizures and thus unlikely to be useful in the future [36].

VPA will continue to be used in its current form in the future, but it is possible that derivatives or analogues will eventually become available which could replace it. A VPA derivative without significant toxicity could become the drug of choice for almost all epilepsy patients since the risk/benefit ratio would clearly be in its favor.

Speculation on the future use of VPA for epilepsy

VPA was a revolutionary advancement in the development of AEDs at the time of release because of its efficacy for generalized seizures, especially absence, for which it is still often considered the drug of choice, although it is also useful for myoclonic and generalized tonic-clonic seizures, and infantile spasms [9]. Its future role in this situation may be in question because of the development of more effective and less toxic AEDs. However, it also possesses the unique characteristic of efficacy for both generalized and partial seizures, which is a characteristic that has not been demonstrated for most newer AEDs. Therefore, many clinicians continue to consider VPA as the drug of choice for many epilepsies characterized by the occurrence of mixed seizure types, such as atypical absence, generalized tonic-clonic and atonic seizures occurring as part of the Lennox-Gastaut syndrome; myoclonic and generalized tonic-clonic seizures in association with progressive myoclonic epilepsies; and infantile spasms as part of West's syndrome [37]. This gives VPA special significance for the treatment of highly refractory patients. It is also uniquely useful for myoclonic and generalized tonic-clonic seizures as part of juvenile myoclonic epilepsy, as well as for childhood absence epilepsy and photosensitive epilepsies [37, 38]. The only available alternatives are felbamate, which carries the significant risk of aplastic anemia and hepatic necrosis, and lamotrigine, which carries the risk of potentially fatal rash [39, 40].

VPA would appear to be most effective for seizures occurring as part of defined epilepsy syndromes characterized by generalized seizures. This is probably because patients within each syndrome share a similar pathophysiology beyond the common occurrence of generalized seizures, and this may be similar across these syndromes. There are undoubtedly other epilepsy syndromes yet to be defined which share this pathophysiology and may respond preferentially to valproate. For example, if there is a subset of patients with generalized tonic-clonic seizures on awakening which share a common pathophysiology in addition to the characteristic occurrence of seizures, then these patients may be especially good candidates for VPA.

Although VPA has efficacy for partial seizures, even as monotherapy, it is unlikely to become popular for this indication because of the wide availability of other AEDs with similar efficacy and less risk of serious side-effects.

Use of a single AED as "monotherapy" has been the standard of care for many years, but "rational polypharmacy" is gaining popularity and VPA has a role in it. The premise of rational polypharmacy is that the use of two drugs with different mechanisms of action may have synergistic effects and could be used at lower doses, resulting in less toxicity with synergistic efficacy [41]. VPA could be classified as an "inhibitory" AED because it augments GABAergic inhibition, but its many other different mechanisms of action also make it a rational choice for combination with almost any other AED. This is distinctly different from designer drugs under investigation which have only a single mode of action. Brodie and Yuen found the addition of lamotrigine to valproate provided a better responder rate than the addition of carbamazepine or phenytoin [42].

Beyond the current indications for use of VPA, there are a number of epileptic conditions for which it may be useful. Current AEDs are designed to prevent the expression of seizures in patients who are predisposed, rather than prevent the development of seizures in those who are at risk. In this sense they are "anti-ictal" rather than "anti-epileptogenic." The future of drug development in the treatment of epilepsy is clearly in finding drugs which are anti-epileptogenic by virtue of preventing the neuronal changes that induce an enduring tendency to seizures. VPA prevents or retards kindling which suggests that it may have some antiepileptogenic properties [43]. Although clinical practice has not supported this role for VPA, it is possible that VPA derivatives could have a role in this in the future.

Psychiatric disease

VPA was initially thought to be useful for mania due to bipolar affective disorder based on case reports; subsequently, efficacy has been demonstrated in double-blind placebo controlled trials [44]. It appears to be most effective for the treatment of acute mania, especially for the subset of patients with rapid cycling, a greater number of prior manic episodes, co-morbid substance abuse, and a negative family history [45]. The efficacy of VPA for dysphoric mania and in prophylaxis of manic episodes and depression is less well-established and requires examination in more detail [46].

The pathophysiology of mania is unknown, but mania is clearly somewhat paroxysmal and there is some evidence to suggest that there is impaired GABAergic inhibition, analogous to a relative inhibitory deficit in epilepsy which results in neuronal hyperexcitability [47]. While the pathophysiology is undoubtedly different from epilepsy, the same mechanisms that make VPA useful for epilepsy may also make it useful for mania. In fact, VPA seems to be most effective for mania occurring in specific circumstances, which may be analogous to its syndrome-specific efficacy in epilepsy. As more specific syndromes associated with mania are defined, it may be possible to further define the subset of patients who gain differential benefit from VPA.

The currently accepted therapies for acute mania, as well as depression and psychosis, are safer and generally more effective than VPA, but for the subset of patients who do not respond to lithium, VPA will probably continue to be used. The immediate future role of VPA may be in prophylaxis against further manic episodes, which is the ultimate goal of therapy in this condition. The future awaits a head to head comparison of lithium and VPA for the prophylaxis of mania.

VPA is at least partially effective in a few psychiatric disorders that do not have well-established therapies and VPA may provide a reasonable consideration for these conditions. It has been reported to be effective in cyclothymia, schizoaffective disorder, premenstrual mood cycling, episodic dyscontrol, and in anxiety disorders [48, 49]. VPA reduced withdrawal symptoms in an experimental model of phenobarbital and phenytoin withdrawal [50]. This may suggest it could be useful for the treatment of alcohol withdrawal, where it may also have the potential to prevent alcohol withdrawal seizures. These empiric observations may expand the spectrum of psychiatric diseases which are candidates for VPA therapy.

Migraine and other pain syndromes

VPA has demonstrated efficacy for prophylaxis of migraine [51]. VPA reduces the frequency of migraine headaches by approximately 50% in half of the treated patients in blinded prospective trials with efficacy similar to propranolol. It has an important role in migraine prophylaxis because more than 45% of patients are refractory to propranolol, the gold standard of therapy [52]. VPA is also one of the relatively few effective, well-tolerated prophylactic medications available. There has been relatively little success in the recent development of prophylactic therapies for migraine, despite the explosion of new abortive therapies. Therefore, VPA will continue to be used in the immediate future for prophylaxis against migraine. The greatest drawback is that the patients most likely to use it are young women, for whom the attendant risk of teratogenicity is most troublesome. Given recent progress in the understanding of migraine, more prophylactic drugs will be developed which may be more selective and could have fewer side-effects than VPA and would assume a dominant role in therapy. On the other hand, VPA analogs may assume an important role, if they prove to have fewer serious side-effects.

The mechanism of action of VPA in migraine is unknown, but it may work through central GABAergic mechanisms. VPA reduces the expression of c-fos selectively within the trigeminal nucleus caudalis in an experimental model of trigeminal neuralgia in which meningeal pain afferents are activated by intracisternal injection of capsaicin and acutely and chronically blocks dural plasma protein extravasation in similar models [53]. Furthermore, the effects are selectively reversed by $GABA_A$ antagonists, suggesting that VPA acts through GABAergic mechanisms. Central serotonergic mechanisms are a less likely possibility, but many abortive anti-migraine drugs, such as sumatriptan, are serotonin agonists and VPA increases central serotonin levels, although the direct implication of this is not known [54].

Based on the proposed mechanism of action, VPA may be useful in other pain syndromes in the future. The above experimental model directly suggests that VPA may be useful for trigeminal neuralgia and a small case series demonstrated some benefit in humans [55]. By extrapolation, VPA could have efficacy for other types of neuropathic pain, such as post-herpetic neuralgia and post-stroke thalamic pain (Dejerine-Roussy syndrome), which has been suggested in a few case reports. Anecdotal reports and uncontrolled studies suggest that VPA is effective for cluster headache, a particularly vexing problem [56]. Currently established analgesic and specific drug therapies for most patients with most types of chronic pain are sufficient and will not be replaced by VPA. Although not overwhelmingly useful for chronic daily headache, there are some chronic pain syndromes with few available therapies which are reasonable candidates for systematic study of VPA therapy, including chronic back pain and chronic pain associated with malignancy. The empiric observation that new AEDs, especially gabapentin, are effective for chronic pain syndromes also supports the possibility that VPA may be useful in these conditions, independent from the proposed mechanism of action.

Movement disorders

VPA is not commonly used in the treatment of movement disorders, but there are case reports and there is some experimental evidence that it may be effective for specific movement disorders, presumably on the basis of augmentation of GABAergic neurotransmission. Post-anoxic myoclonus represents the borderland between epilepsy and movement disorders because the myoclonus is most often cortical in origin, but also may arise from subcortical structures. It is a difficult problem because it is often refractory and is disturbing to families and patients. There have been no systematic investigations of drug treatments and no therapies have been obviously effective. However, VPA has been reported to be effective in some case reports [57]. Spinal myoclonus may also respond to VPA [58]. Because there are no obviously effective therapies for myoclonus, this probably warrants more systematic investigation and it is possible that VPA may be effective for myoclonus in other situations, for example, when combined with spasticity. However, studies have not demonstrated utility for Parkinson's disease.

There are anecdotal reports and small case series of the effectiveness of VPA for chorea. Doud et al. found it very effective for Sydenham's chorea and it also has been reported effective for kinesiogenic paroxysmal choreoathetosis and familial paroxysmal choreoathetosis [59, 60]. Therefore, VPA may have a role in the treatment of chorea which is not controlled by conventional drugs.

VPA is occasionally used for dystonia. It is effective in a recently developed animal model of paroxysmal dystonia [61]. Systematic investigation of its efficacy for dystonias with paroxysmal character may be useful, including paroxysmal nocturnal dystonia and hemiballismus.

Conclusion

Widespread use of VPA will continue into the future, especially for the treatment of epilepsy, in which it has been termed the "aspirin of epilepsy" because of it diverse utility and expanding use, despite the availability of newer AEDs. VPA will continue to be used for migraine prophylaxis and for acute treatment of mania. There will undoubtedly be an extension of current indications to include prophylaxis of mania and dysphoric mania. There are also some pain syndromes and movement disorders for which

VPA may be effective and may warrant systematic investigation, especially if more benign VPA derivatives become available.

References

1 Fariello RG, Varasi M, Smith MC (1995) Valproic acid mechanisms of action. In: RH Levy, RH Mattson, BS Meldrum (eds): *Antiepileptic Drugs, Fourth Edition*. Raven Press, New York, 581–588
2 Chapman A, Keane PE, Meldrum BS, Simiand J, Vernieres JC (1982) Mechanism of anticonvulsant action of valproate. *Prog Neurobiol* 19: 315–359
3 Löscher W, Nau H, Siemes H (1988) Penetration of valproate and its active metabolites into cerebrospinal fluid of children with epilepsy. *Epilepsia* 29: 311–316
4 Löscher W, Nau H (1983) Distribution of valproic acid and its metabolites in various brain areas of dogs and rats after acute and prolonged treatment. *J Pharmacol Exp Ther* 226(3): 845–854
5 McLean MJ, MacDonald RL (1986) Sodium valproate, but not ethosuximide, produces use and voltage-dependent limitation of high frequency repetitive firing of action potentials of mouse central neurons in cell culture. *J Pharmacol Exp Ther* 237: 1001–1011
6 Baillie TA, Sheffels PR (1995) Valproic acid chemistry and biotransformation. In: RH Levy, RH Mattson, BS Meldrum (eds): *Antiepileptic Drugs, Fourth Edition*. Raven Press, New York, 589–604
7 Bolanos JP, Medina JM (1997) Effect of valproate on the metabolism of the central nervous system. *Life Sciences* 60(22): 1933–1942
8 Levy RH, Shen DD (1995) Valproic Acid absorption, distribution, and excretion. In: RH Levy, RH Mattson, BS Meldrum (eds): *Antiepileptic Drugs, Fourth Edition*. Raven Press, New York 605–619
9 Bourgeois BFD (1995) Valproic acid clinical use. In: RH Levy, RH Mattson, BS Meldrum (eds): *Antiepileptic Drugs, Fourth Edition*. Raven Press, New York 633–639
10 Klotz U (1982) Bioavailability of a slow release preparation of valproic acid under steady state conditions. *Int J Clin Pharmacol Ther Toxicol* 20(1): 24–26
11 Dreifuss FE (1995) Valproic acid toxicity. In: RH Levy, RH Mattson, BS Meldrum (eds): *Antiepileptic Drugs, Fourth Edition*. Raven Press, New York, 641–648
12 Dreifuss FE, Santilli N, Langer DH, Sweeney KP, Moline KA, Menander KB (1987) Valproic acid hepatic fatalities: a retrospective review. *Neurology* 37(3): 379–385
13 Dreifuss FE, Langer DH, Moline KA, Maxwell JE (1989) Valproic acid hepatic fatalities: II. US experience since 1984. *Neurology* 39(2 Pt 1): 201–207
14 Zimmerman HJ, Ishak, KG (1982) Valproate-induced hepatic injury: analyses of 23 fatal cases. *Hepatology* 2(5): 591–597
15 Kay JD, Hilton-Hones D, Hyuman N (1986) Valproate toxicity and ornithine carbamoyltransferase deficiency. *Lancet* 2: 1283–1284
16 Siemes H, Nau H, Schultze K, et al. (1993) Valproate (VPA) metabolites in various clinical conditions of probable VPA-associated hepatotoxicity. *Epilepsia* 34: 332–346
17 Kesterson JW, Granneman GR, Machinist JM (1984) The hepatotoxicity of valproic acid and its metabolites in rats. I. Toxicologic biochemical and histopathologic studies. *Hepatology* 4: 1143–1152
18 Fisher E, Siemes H, Pund R, Wittfoht W, Nau H (1992) Valproate metabolites in serum and urine during antiepileptic therapy in children with infantile spasms: abnormal metabolite pattern associated with reversible hepatotoxicity. *Epilepsia* 33 (1): 165–171
19 Coulter DL (1984) Carnitine deficiency: a possible mechanism for valproate hepatotoxicity. *Lancet* 1: 689
20 Delgado-Escueta AV, Janz D (1992) Pregnancy and teratogenesis in epilepsy. *Neurology* 42 (Suppl 5): 1–160
21 Nau H, Hauck R-S, Ehlers K (1991) Valproic acid-induced neural tube defects in mouse and human: aspects of chirality, alternative drug development, pharmacokinetics, and possible mechanisms. *Pharmacol Toxicol* 69: 310–321

22 Nau H, Löscher W (1984) Valproic acid and metabolites: pharmacological and toxicological studies. *Epilepsia* 25 (Suppl 1): S14–S22
23 Gram L (1992) Studies on 2-n-propyl-2(E)-pentenoate (delta 2 (E)-valproate) in man. *Pharm Weekbl Sci* 14(3 A): 159–160
24 Löscher W (1992) Pharmacological, toxicological and neurochemical effects of delta 2(E)-valproate in animals. *Pharm Weekbl Sci* 14(3 A): 144–145
25 Vorhees CV, Acuff-Smith KD, Weisenburger WP, Minck DR, Berry JS, Setchell KD, Nau H (1991) Lack of teratogenicity of trans-2-ene-valproic acid compared to valproic acid in rats. *Teratology* 43(6): 583–590
26 Sokolova S, Schmitz D, Zhang Cl, Löscher W, Heinemann U (1998) Comparison of effects of valproate and trans-2-en-valproate on different forms of epileptiform activity in rat hippocampal and temporal cortex slices. *Epilepsia* 39(3): 251–258
27 Palaty JH, Abbott FS (1995) Structure-activity relationships of unsaturated analogues of valproic acid. *J Medicin Chem* 38(17): 3398–3406
28 Bialer M (1991) Clinical pharmacology of valpromide. *Clin Pharmacokinet* 20:114–122
29 Bialer M, Kadry B, Abdul-Hai A, Haj-Yehia A, Sterling J, Herzig Y, Shirvan M (1996) Pharmacokinetic and pharmacodynamic analysis of (E)-2-ene valproyl derivatives of glycine and valproyl derivatives of nipecotic acid. *Biopharmac Drug Dispos* 17: 565–575
30 Scott KR, Adesioye S, Ayuk PB, Edafiogho IQ, John D, Kodwin P, Maxwell-Irving T, Moore JA, Nicholson JM (1994) Synthesis and evaluation of amino analogues of valproic acid. *Pharmaceutic Res* 11(4): 571–574
31 Bojic U, Elmazar MM, Hauck RS, Nau H (1996) Further branching of valproate-related carboxylic acids reduces the teratogenic activity, but not the anticonvulsant effect. *Chem Res Toxicol* 9(5): 866–870
32 Hadad S, Vree TB, van der Kleijn E, Bialer M (1993) Pharmacokinetic analysis and anticonvulsant activity of two polyesteric prodrugs of valproic acid. *Biopharmac Drug Dispos* 14(1): 51–59
33 Hadad S, Vree TB, van der Kleijn E, Bialer M (1992) Pharmacokinetic analysis ester prodrugs of valproic acid. *J Pharmcokinetic Sci* 81(10): 1047–1050
34 Tang W, Abbott FS (1997) A Comparative investigation of 2-propyl-4-pentenoic acid (4-ene VPA) and its α-fluorinated analogue – Phase II metabolism and pharmacokinetics. *Drug Metabo Dispos* 25(2): 219–227
35 Tantisira B, Tantisira MH, Patarapanich C, Sooksawte T, Chunngam T (1997) Preliminary evaluation of the anticonvulsant activity of a valproic acid analog: N-(2-propylpentanoyl) urea. *Research Comm Molecular Path Pharmacology* 97(2): 151–164
36 Sobaniec W, Sobaniec-Lotowska M (1994) The effects of sodium valproate and its metabolites (5-OH-VPA and 4-OH-VPA) on electroconvulsions in rats. *Materia Medica Polona* 26:29–32
37 Dreifuss FE, Fountain NB (1998) Classification of epileptic seizures and the epilepsies with drugs of choice. In: MJ Eadie, F Vajda (eds): *Handbook of Experimental Pharmacology, Antiepileptic Drugs II*. Springer-Verlag, Berlin, in press
38 Jeavons PM, Bishop A, Harding GFA (1986) The prognosis of photosensitivity. *Epilepsia* 27: 569–575
39 Mackay FJ, Wilton LV, Pearce GL, Freemantle SN, Mann RD (1997) Safety of long-term lamotrigine in epilepsy. *Epilepsia* 38(8): 881–886
40 Kaufman DW, Kelly JP, Anderson T, Harmon DC, Shapiro S (1997) Evaluation of case reports of aplastic anemia among patients treated with felbamate. *Epilepsia* 38: 1265–1269
41 Leach JP (1997) Polypharmacy with anticonvulsants – focus on synergism. *CNS Drugs* 8(5): 366–375
42 Brodie MJ, Yuen AW (1997) Lamotrigine substitution study: evidence for synergism with sodium valproate? 105 study group. *Epilepsy Res* 26(3): 423–32
43 Silver JM, Shin C, McNamara JO (1991) Antiepileptogenic effects of conventional anticonvulsants in the kindling model of epilepsy. *Ann Neurol* 29: 356–363.
44 McElroy SL, Keck PE, Pope HG, Hudson JI (1993) Valproate in the treatment of bipolar disorder: literature review and clinical guidelines. *J Clin Psychopharmacol* 12: 42S–52S
45 Post RM, Ketter TA, Denicoff K, Pazzaglia PJ, Leverich GS, Marangell LB, Callahan AM, George MS, Frye MA (1996) The place of anticonvulsant therapy in bipolar illness. *Psychopharmacol* 128(2): 115–129

46 Freeman TW, Clothier JL, Pazzaglia P, Lesem MD, Swann AC (1992) A double-blind comparison of valproate and lithium in the treatment of acute mania. *Am J Psychiatry* 149: 108–111
47 Petty F (1995) GABA and mood disorders: a brief review and hypothesis. *J Affect Disord* 34(4): 275–281
48 Jacobsen FM (1993) Low-dose valproate: a new treatment for cyclothymia, mild rapid cycling disorders, and premenstrual syndrome. *J Clin Psychiatry* 54(6): 229–234
49 Stoll AL, Banov M, Kolbrener M, Mayer PV, Tohen M, Strakowski SM, Castillo J, Suppes T, Cohen BM (1994) Neurologic factors predict favorable valproate response in bipolar and schizoaffective disorders. *J Clin Psychopharmacol* 14(5): 311–313
50 Peichev L (1992) Mechanisms of neurotransmission of valproate sodium in suppressing barbiturate and phenytoin withdrawal syndrome. *Folia Medica* 34(2): 14–19
51 Silberstein SD (1996) Divalproex sodium in headache: literature review and clinical guidelines. *Headache* 36(9): 547–555
52 Rosen JA (1983) Observations on the efficacy of propranolol for the prophylaxis of migraine. *Ann Neurol* 13(1): 92–93
53 Cutrer FM, Moskowitz MA (1996) The actions of valproate and neurosteroids in a model of trigeminal pain. *Headache* 36(10): 579–585
54 Maes M, Calabrese J, Jayathilake K, Meltzer HY (1997) Effects of subchronic treatment with valproate on L-5-HTP-induced cortisol responses in mania: evidence for increased central serotonergic neurotransmission. *Psychiatry Res* 71(2): 67–76
55 Peris JB, Perera GL, Devendra SV, Lionel ND (1980) Sodium valproate in trigeminal neuralgia. *Med J Australia* 2(5): 278
56 Hering R, Kuritzky A (1989) Sodium valproate in the treatment of cluster headache: an open clinical trial. *Cephalgia* 9(3): 195–198
57 Fahn S (1979) Posthypoxic action myoclonus: review of the literature and report of two new cases with response to valproate and estrogen. *Advances in Neurology* 26: 49–84
58 Fouillet N, Wiart L, Arne P, Alaoui P, Petit H, Barat M (1995) Propriospinal myoclonus in tetraplegic patients: clinical, electrophysiological and therapeutic aspects. *Paraplegia* 33(11): 678–681
59 Daoud AS, Zaki M, Shakir R, al-Saleh Q (1990) Effectiveness of sodium valproate in the treatment of Sydenham's chorea. *Neurology* 40(7): 1140–1141
60 Przuntek H, Monninger P (1983) Therapeutic aspects of kinesiogenic paroxysmal choreoathetosis and familial paroxysmal choreoathetosis of the Mount and Reback type. *J Neurol* 230(3):163–169
61 Fredow G, Löscher W (1991) Effects of pharmacological manipulations of GABAergic neurotransmission in a new mutant hamster model of paroxysmal dystonia. *Eur J Pharmacol* 192(2): 107–219

Subject index

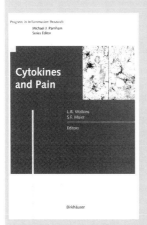